D1660489

Technology, Manufacturing and Grid Connection of Photovoltaic Solar Cells

Technology, Manufacturing and Grid Connection of Photovoltaic Solar Cells

Guangyu Wang
Shanghai Institute of Microsystem and Information Technology,
Chinese Academy of Sciences, China

This edition first published 2018 by John Wiley & Sons Singapore Pte. Ltd under exclusive licence granted by China Electric Power Press for all media and languages (excluding simplified and traditional Chinese) throughout the world (excluding Mainland China), and with non-exclusive license for electronic versions in Mainland China.

Registered Offices
John Wiley & Sons, Inc., 111 River Street, Hoboken, NJ 07030, USA
John Wiley & Sons Singapore Pte. Ltd, 1 Fusionopolis Walk, #07-01 Solaris South Tower, Singapore 138628

Editorial Office
1 Fusionopolis Walk, #07-01 Solaris South Tower, Singapore 138628

For details of our global editorial offices, customer services, and more information about Wiley products visit us at www.wiley.com.

Wiley also publishes its books in a variety of electronic formats and by print-on-demand. Some content that appears in standard print versions of this book may not be available in other formats.

Library of Congress Cataloging-in-Publication Data

Names: Wang, Guangyu, 1943- author.
Title: Technology, manufacturing and grid connection of photovoltaic solar cells / by Guangyu Wang.
Description: Hoboken, NJ : John Wiley & Sons, 2018. | Includes bibliographical references and index. |
Identifiers: LCCN 2017042881 (print) | LCCN 2017055841 (ebook) | ISBN 9781119035190 (pdf) | ISBN 9781119035206 (epub) | ISBN 9781119035176 (cloth)
Subjects: LCSH: Solar cells. | Photovoltaic cells. | Photovoltaic power systems.
Classification: LCC TK2960 (ebook) | LCC TK2960 .W35 2018 (print) | DDC 621.31/244–dc23
LC record available at https://lccn.loc.gov/2017042881

Cover Design: Wiley
Cover Image: © LeeYiuTung/Gettyimages

Set in 10/12pt WarnockPro by SPi Global, Chennai, India

Printed in Singapore by C.O.S. Printers Pte Ltd

10 9 8 7 6 5 4 3 2 1

Contents

About the Author *xv*
Preface *xvii*

1 Basic Physics of Solar Cells *1*
1.1 Development of Solar Cells *1*
1.1.1 Solar Energy Is the Most Promising Renewable Energy Source in the World *1*
1.1.2 Development of Solar Cells *4*
1.2 Solar Radiation and Air Mass *6*
1.2.1 Conversion of Sunlight Into Electricity Using Photoelectric Effect Is an Important Way to Make Use of Solar Energy *6*
1.2.2 Basics of Solar Radiation and Definition of Air Mass *6*
1.2.3 Wavelength of Solar Radiation *7*
1.3 Basics of Semiconductors *8*
1.3.1 Communisation Motion of Electrons in a Crystalline and Formation of Energy Bands *9*
1.3.2 Atomic Structures of Conductors, Insulators and Semiconductors and Energy Bands Image *9*
1.3.3 Energy Band Structure of Dope Semiconductor *10*
1.3.4 Fermi Level *11*
1.3.5 Directional Movement of Electrons and Holes *12*
1.3.6 Generation and Recombination of Carriers *13*
1.4 Light Absorption of Semiconductor Materials *14*
1.4.1 Light Absorption of Semiconductor *14*
1.4.2 Intrinsic and Non-Intrinsic Absorptions of Semiconductor Materials *15*
1.4.3 Light Absorption Coefficient and Semiconductor Materials of Direct/Indirect Transition *16*
1.5 P-N Junctions and PV Effect of Solar Cells *18*
1.5.1 Bending of a P-N Junction Band and Formation of a Built-In Field *18*
1.5.2 Effect of an External Voltage on the P-N Junction Band Structure *20*
1.5.3 Effect of Solar Radiation on a P-N Junction's Band Structure and the PV Effect *20*
1.5.4 Composition of the Photo-Generated Current in the Solar Cell *22*
1.5.5 Key Parameters of the Solar Cell *24*
1.5.5.1 I-V Characteristic Curve of a Solar Cell *24*

1.5.5.2 Relations of the Open-Circuit Voltage and the Height of the P-N Junction Potential Barrier in a Solar Cell *24*
1.5.5.3 Short-Circuit Current of a Solar Cell I_{sc} *24*
1.5.5.4 The Optimum Operation Point of Solar Cells, the Optimum Operation Voltage and Current *25*
1.5.5.5 Filling Factor (FF) *26*
1.5.5.6 Power Conversion Efficiency of a Solar Cell η *26*
1.5.5.7 Temperature Characteristics of a Solar Cell *27*
1.5.6 Application of a Concentration Junction in a PV Cell for Back Surface Field (BSF) *27*
1.5.7 Basic Structure of Homogeneous P-N Junction Crystalline Silicon Solar Cells and Analysis on the Cell's Efficiency *29*
1.6 Solar Cells of Heterojunctions *32*
1.6.1 Composition of Heterojunctions *32*
1.6.2 Construction and Working Principle of the Solar Cell with Heterojunctions *32*

2 Materials of Solar Cells *35*
2.1 Low-Cost Solar-Grade Polycrystalline Silicon *35*
2.1.1 Polycrystalline Silicon—The Most Important Raw Material of the PV Industry *35*
2.1.2 Meaning of Solar-Grade Polycrystalline Silicon *38*
2.1.3 Preparation of Solar-Grade Polycrystalline Silicon (UMG Silicon) by Metallurgical Method *41*
2.1.4 Preparation of Solar-Grade Polycrystalline Silicon by FBR Method *45*
2.1.5 Preparation of Solar-Grade Polycrystalline Silicon by $SiCl_4$ Zinc Reduction Method *45*
2.1.6 Preparation of Solar-Grade Granular Polycrystalline Silicon by VLD Method *47*
2.1.7 Hydrogenation of the Main By-Product $SiCl_4$ Produced in the Production Process of Polycrystalline Silicon by the Siemens Method *48*
2.2 Casting Polycrystalline Silicon *49*
2.2.1 General *49*
2.2.2 Preparation Process of Casting Crystalline Silicon *50*
2.2.3 Impurities and Defects in Casting Crystalline Silicon *54*
2.2.3.1 Non-Metal Impurities in Casting Crystalline Silicon *54*
2.2.3.2 Metal Impurities and Gettering in Casting Crystalline Silicon *56*
2.2.3.3 Crystal Boundaries and Dislocations in Casting Crystalline Silicon *57*
2.2.4 Latest Development of Casting Crystalline Silicon and Wafers *58*
2.2.4.1 Casting of Pseudo-Single Crystal *58*
2.2.4.2 Continuous Output Improvement of Casting Crystalline Silicon Furnaces *59*
2.3 CZ Monocrystalline Silicon *60*
2.3.1 Heat Flow Continuity Equation of Grain Growth Interface and its Application *60*
2.3.2 Heat Conduction in the Melt *62*
2.3.3 Temperature Distribution in the Crystal *63*

2.3.4 Impurity Segregation Between Solid and Liquid *64*
2.4 Nature of a-Si/μC-Si Thin Film *66*
2.4.1 Nature of a-Si Thin Film *66*
2.4.1.1 Basic Nature of a-Si Thin Film and its Application to PV Sector *67*
2.4.1.2 Fermi Level Pinning and Efficiency Degradation Mechanism for a-Si Thin-film Cells *69*
2.4.2 Nature of μC-Si Thin Film *70*
2.5 Preparation Methods of a-Si/μC-Si Film *72*
2.5.1 A Main Raw Material for Silicon Film Preparation—Silane *72*
2.5.2 Introduction to Silicon Thin-film Growth Methods *74*
2.5.3 Preparation of a-Si/μC-Si Thin Film by PECVD Method *74*
2.5.4 Preparation of a-Si/μC-Si Thin Film by HWCVD Method *75*
2.5.5 Growing Silicon Thin Film by Other Methods *76*
2.5.5.1 Direct Preparation of μC-Si Thin Film by LPCVD Technique *76*
2.5.5.2 a-Si Crystallised to Polycrystalline Silicon Thin Film by SPC Technique *77*
2.5.5.3 a-Si Crystallised to Polycrystalline Silicon Thin Film by Metal-Induced Method *77*
2.5.5.4 a-Si Crystallised to Polycrystalline Silicon Thin Film by RTP Technique *77*
2.5.5.5 a-Si Crystallised to Polycrystalline Silicon Thin Film by Linear Laser Technique *78*
2.5.6 Comparisons of Various Silicon Film Growth Methods *78*
2.6 Compound Semiconductor Materials *79*
2.6.1 GaAs and Other Semiconductor Materials *79*
2.6.2 CdTe and CdS Thin Film Materials *80*
2.6.3 $CuInSe_2$ and $CuInS_2$ Thin Film Materials *80*
2.7 Analysis on Impurities in Semiconductor Materials *81*
2.7.1 Glow Discharge Mass Spectrometry (GDMS) Analysis *81*
2.7.2 Secondary Ion Mass Spectrometry (SIMS) Analysis *82*
2.7.3 Infrared Spectroscopy to Detect the Carbon and Oxygen Contents in Silicon Wafer *84*

3 Preparation Methods of Crystalline Silicon Solar Cells *85*
3.1 Preparation Process Flow of CSSCs *85*
3.1.1 Basic Structure of CSSCs *85*
3.1.2 Production Flow of CSSCs *87*
3.2 Performance Detection and Sorting of Raw Silicon Wafer *88*
3.2.1 Measurement of Silicon Wafer Conduction Type *88*
3.2.2 Measurement of Silicon Wafer Resistivity and Thin Layer Square Resistance *89*
3.2.3 Measurement of the Minority Carrier Lifetime *90*
3.2.4 Measurement of Silicon Wafer Thickness *92*
3.2.5 High-Speed Multi-Purpose Silicon Wafer Testers *92*
3.3 Silicon Wafer Surface Cleaning and Texturing *93*
3.3.1 Principles of Chemical Cleaning and Texturing *93*
3.3.2 Production Equipment and Process of Chemical Corrosion Texturing *94*
3.3.3 Laser Texturing and Reactive Ion Etching (RIE) Techniques *95*
3.3.3.1 Laser Texturing *97*

3.3.3.2 RIE *98*
3.4 Junction Preparation by Diffusion *99*
3.4.1 Principles *99*
3.4.2 Process and Equipment *100*
3.4.2.1 Gaseous Diffusion of $POCl_3$ in Tubular Furnace *100*
3.4.2.2 Dilute Phosphoric Acid Doper and Chain-Type Diffusion Furnace *102*
3.4.3 Measurement of Diffusion Layer Thickness (Junction Depth) *103*
3.4.3.1 Measurement of the Longitudinal Distribution of the Phosphorus Concentration on the Diffusion Layer by SIMS Method *103*
3.4.3.2 Measurement of PN Junction Depth by SRP *103*
3.4.4 CSSC Phosphorus Impurity Gettering *104*
3.5 Plasma Corrosion and Laser Edging Isolation *106*
3.5.1 Objectives and Means of Edging Isolation *106*
3.5.2 Principles and Equipment of Plasma Etching *107*
3.5.3 Laser Edging Isolation *109*
3.6 Removal of PSG *110*
3.6.1 Principles and Processes of PSG Removal *110*
3.6.2 Equipment and Production Line for PSG Removal *111*
3.6.2.1 Bath-Type PSG-Removal Integrated Production System *111*
3.6.2.2 Chain-Type PSG-Removal Production System *112*
3.7 Preparation of Anti-Reflection Coating by PECVD and PVD Methods *112*
3.7.1 Objectives and Principles for Anti-Reflection Coating Preparation *112*
3.7.2 Principles of Silicon Nitride Coating Prepared by PECVD *113*
3.7.3 Direct (Tubular) PECVD and Indirect (Plate-Type) PECVD *115*
3.7.4 Typical PECVD Systems *117*
3.7.4.1 Tubular Direct PECVD System *117*
3.7.4.2 Plate-Type Direct PECVD System *117*
3.7.4.3 Plate-Type Indirect PECVD System *118*
3.7.5 Preparation of Silicon Nitride Coating by Physical Vapour Deposition (PVD) *119*
3.7.5.1 Principles of Silicon Nitride Coating Prepared by PVD *119*
3.7.5.2 Comparisons of Silicon Nitride Coating Deposited by PVD and PECVD *121*
3.7.5.3 ATON Sputtering System Produced by Applied Materials, USA *122*
3.7.6 Measurement of the Thickness and Refractive Index of the Anti-Reflection Coating by Ellipsometer *122*
3.8 Preparation of Top/Bottom Electrodes (Surface Metallisation) *123*
3.8.1 Technical Requirements and Production Flow for Top/Bottom Electrode Preparation *123*
3.8.2 Electrode Printing, Drying, Testing and Cell sorting *125*
3.8.3 Fast Sintering Furnace System *126*
3.8.4 Electrode Slurry *128*
3.8.5 Aluminum Impurity Gettering *130*
3.9 Cell Testing and Sorting *130*
3.9.1 Objectives of Solar Cell Testing and Sorting *130*
3.9.2 Cell-sorting Equipment *130*
3.10 Automation of CSSC Production Techniques *133*
3.10.1 Promotion of Cascading/Chain-Type Production Lines *133*

3.10.2 Mounting/Dismounting the Silicon Wafer by Robots Instead of Manual Operation *133*
3.11 Parameter Measurement in CSSC Production Process *136*
3.11.1 Solar Simulator *136*
3.11.2 Measurement of V-I Characteristics and PV Conversion Efficiency for Solar Cells *136*
3.11.3 Measurement of Spectral Response for Solar Cells *139*
3.12 Product Quality Control and Cost Analysis for Solar Cell Production Lines *140*
3.12.1 On-line Inspection of Solar Cell Production *140*
3.12.2 Traditional Process Quality Control on the Solar Cell Production Line *140*
3.12.2.1 Working Environment *140*
3.12.2.2 Quality Control of the Cleaning and Texturing Process *140*
3.12.2.3 Quality Control of Diffusion Process *141*
3.12.2.4 Quality Control in Wafer Edging Isolation and PSG Removal Procedures *141*
3.12.2.5 Quality Control in PECVD Procedure *141*
3.12.2.6 Quality Control in Print and Sintering Procedures *141*
3.12.3 Cost Analysis for CSSCs *142*

4 Preparation Methods of Thin Film Silicon Solar Cells *143*
4.1 Advantages and Prospects of TFSSCs *143*
4.1.1 Advantages of TFSSCs *143*
4.1.2 History and Prospects of TFSSCs *144*
4.2 Structures and Power Generation Principles of TFSSCs *146*
4.2.1 Structures of a-Si:H and μC-Si THSCs *146*
4.2.2 Power Generation Principle of TFSSCs *147*
4.2.3 Light Absorption of a-SiC:H/μC-Si and a-Si:H/a-SiGe:H Stacked Solar Cells *150*
4.3 Preparation Techniques of TFSSCs *151*
4.3.1 TCO Sputtered on Glass Substrate *151*
4.3.2 P-Type (a-SiC:H) Film Deposited by PECVD Method *152*
4.3.3 I (a-Si:H) Intrinsic Zone Deposited by PECVD Method *152*
4.3.4 N-Type (a-Si:H) Layer Thin Film Deposited by PECVD Method *154*
4.3.5 Al(Ag) Back Electrodes Sputtered by PVD Method *154*
4.3.6 Integration of TFSCs and Modules *154*
4.4 Main Production Equipment for TFSSCs *155*
4.4.1 Production System of TFSSCs *155*
4.4.2 Glass Cleaning and Surface Texturing Equipment *158*
4.4.3 TCO Sputtering Equipment and ZAO Target *158*
4.4.4 PECVD System for Thin Film Silicon Deposition *160*
4.4.5 Back Contact Sputtering Equipment *163*
4.4.6 Laser Scriber *164*
4.4.7 Testing Equipment *165*
4.5 Discussion on Some Issues Concerning TFSSC Preparation *166*
4.5.1 Performance, Preparation and Testing of TCO *166*

4.5.2 Influence of PECVD Process Parameters on Deposition and Crystallisation Rates of Silicon Thin Film *167*
4.5.2.1 Hydrogen Dilution *167*
4.5.2.2 Gas Pressure *167*
4.5.2.3 Deposition Temperature *168*
4.5.2.4 Distance Between the Electrode and the Substrate *168*
4.5.2.5 Power Excited by Plasma *169*
4.5.2.6 Frequency Excited by Plasma *169*
4.5.3 VHF-PECVD Method to Deposit Silicon Film *169*
4.5.4 HWCVD Method to Deposit Silicon Film *170*
4.6 Adjustment of TFSSC Energy Band Structure *171*
4.6.1 Methods to Adjust the Band Gap of Thin Film Silicon *171*
4.6.1.1 Significance of Energy Band Structure Adjustment for TFSSCs *171*
4.6.1.2 Gap Adjustment by a-Si Hydrogen Content and Deposition Temperature *172*
4.6.1.3 a-SiC Carbon Material to Widen the Gap *172*
4.6.1.4 a-Si Ge Material to Narrow Down the Gap *172*
4.6.2 a-SiGe TFSCs *172*
4.6.3 Boron, Phosphorous and Hydrogen in a-Si Film *175*
4.7 Physical Principle of PECVD and Deposition of Silicon Thin Film *175*
4.7.1 Glow Discharge and Plasma Generation *175*
4.7.2 Mechanism on a-Si Thin Film Grown by PECVD Method *177*
4.7.2.1 Basic Principles of PECVD *177*
4.7.2.2 a-Si:H (a-Si-Containing Hydrogen) Deposited by PECVD Method *178*
4.7.2.3 Growth Mechanism *179*
4.8 Physical Sputtering Principles and TCO and Back Metal Preparation System *179*
4.8.1 Overview on TCO and Back Metal Deposited by Physical Sputtering *179*
4.8.2 A Simple Parallel Metal DC Diode Sputtering System *182*
4.8.3 RF and MC Sputtering Systems *183*

5 High-Efficiency Silicon Solar Cells and Non-Silicon-Based New Solar Cells *187*
5.1 High-Efficiency Crystalline Silicon Solar Cells (CSSCs) *187*
5.1.1 Selective Emitters and Buried Contact Silicon Solar Cells *188*
5.1.2 Passivation Emitter Silicon Solar Cells *190*
5.1.3 Back Finger Electrodes and Boron Diffusion N-Type Crystalline Silicon (IBC) Solar Cells *192*
5.1.4 Metallisation Wrap-Through (MWT) Silicon Solar Cells *193*
5.2 Production Techniques of HIT High-Efficiency Solar Cells *196*
5.2.1 a-Si/Monocrystalline Silicon Heterojunctions with Intrinsic Layer *196*
5.2.2 Structure and Techniques of Double-Surface HIT Cells *196*
5.2.3 Characteristics of HIT Cells *197*
5.3 Compound Semiconductor Solar Cells *197*
5.3.1 Fabrication Methods of Compound Solar Cells *197*
5.3.1.1 Vacuum Evaporation Technique *197*
5.3.1.2 Liquid Phase Epitaxy (LPE) Technique *198*

5.3.1.3 Metal Organic Chemical Vapour Deposition (MOCVD) Technique *198*
5.3.1.4 Molecular Beam Epitaxy (MBE) *198*
5.3.2 III–V Compound Multi-Junction Crystalline Solar Cells *199*
5.3.3 Cadmium Telluride (CdTe) TFSCs *201*
5.3.4 CIGS TFSCs *204*
5.3.4.1 Vacuum Co-Evaporation and Vacuum-Sputtering Methods *205*
5.3.4.2 Non-Vacuum Method *207*
5.3.4.3 Roll to Roll Method *209*
5.4 Next-Generation Solar Cells *210*
5.4.1 Organic Solar Cells *210*
5.4.2 Dye-Sensitised Solar Cells *213*
5.4.3 Perovskite Solar Cells *215*
5.4.4 Concentrator Solar Cells *217*
5.4.5 Multiple Quantum Well (MQW) Solar Cells *219*

6 Modules and Arrays of Solar Cells *223*
6.1 General *223*
6.1.1 Modules and Arrays of Solar Cells *223*
6.1.2 Packaging Techniques of Several Solar Cell Modules *224*
6.1.3 Packaging Structure of Flat Plate Solar Cell Modules *224*
6.1.4 Solar Cell Modules for Building Integrated PV (BIPV) *225*
6.1.5 Double-Sided Cells and Modules *228*
6.2 Module Packaging Materials *229*
6.2.1 Inspection and Sorting of Cell Wafers *230*
6.2.2 Upper Cover Glass *231*
6.2.3 Adhesives and Modified EVA Film *232*
6.2.4 Back Plate and Localisation *233*
6.2.5 Frameworks and Junction Boxes and Other Materials *235*
6.3 Module Packaging Techniques *237*
6.4 Module Packaging System *239*
6.4.1 Main Equipment in the Production Line of Solar Cell Modules *239*
6.4.2 Laser Scribers *240*
6.4.3 Cell Welders *240*
6.4.4 Solar Cell Module Laminators *242*
6.4.5 Solar Simulators, Turnover Trolleys and Frame Machines *242*
6.5 Reliability of Solar Cell Modules and Inspection After Packaging *242*
6.5.1 Module Packaging and PV System Reliability *242*
6.5.2 Objectives and Descriptions of Solar Cell and Module Tests *244*
6.5.2.1 Indoor Tests of PV Cells *245*
6.5.2.2 Indoor Tests of PV Modules *245*
6.5.2.3 Outdoor Tests of PV Modules *245*
6.5.3 Testing Methods and Verification Standards of Solar Cell Modules *246*
6.5.4 Tests of PV Performance and Macro Defects of Solar Cell Modules *247*
6.5.4.1 Tests of PV Performance of Solar Cell Modules *247*
6.5.4.2 Tests of Macro Defects of Solar Cells and the Modules *247*
6.5.4.3 Testing Principles of Electroluminescence *249*

6.6 Efficiency, Common Specifications and Market Development Trend of Solar Cell Modules *250*
6.6.1 Estimates of Solar Cell Module Power and Efficiency *250*
6.6.2 Common Specifications in the Solar Cell Module Market *252*
6.6.3 Attenuation of Solar Cell Module Power During Usage *253*
6.6.4 Development Trend of Solar Cell Modules in China *253*
6.7 Solar Cell Arrays *254*
6.7.1 Design of Solar Cell Arrays *254*
6.7.2 Array Electrical Connections and Hot Spot Effect *255*
6.7.3 Installation and Measurement of Arrays *256*

7 PV Systems and Grid-Connected Technologies *259*
7.1 Overview on the PV System *259*
7.1.1 Characteristics, Classifications and Compositions of the PV System *259*
7.1.2 Composition and Simple Working Principles of the PV System *263*
7.2 Energy Storage Batteries *265*
7.2.1 Energy Storage Batteries and Their Application to PV System *265*
7.2.2 Lead-Acid Batteries *266*
7.2.3 Lithium Ion Batteries *268*
7.2.4 Liquid Flow Energy Storage Batteries *269*
7.2.4.1 Sodium-Sulphur Batteries *269*
7.2.4.2 Vanadium Redox Batteries *270*
7.2.4.3 Zinc-Bromine Flow Batteries *270*
7.2.5 Super Capacitors *271*
7.2.6 Fuel Cells *272*
7.2.6.1 General *272*
7.2.7 Capacity Design of Battery Packs *274*
7.3 Core of the Inverter—Power-Switching Devices *275*
7.3.1 MOSFET and IGBT and Other Power Electronic Power-Switching Devices *275*
7.3.2 Structure and Working Principles of IGBT *275*
7.3.3 Development History of IGBT *278*
7.3.3.1 Trench Gate Technology *278*
7.3.3.2 Non-Punch-Through (NPT) Technique *278*
7.3.3.3 Filed Stop (FS) Technology *279*
7.4 Inverters *281*
7.4.1 Role of the Inverter in the PV System *281*
7.4.2 Working Principles of the Inverter *282*
7.4.3 Control of the Inverter *285*
7.4.4 Inverter Circuit and Inverter Types *285*
7.4.5 Selection and Requirements of Inverters for PV Applications *286*
7.5 Controllers: Module Power Optimisation and Intelligent Monitoring *287*
7.5.1 Functions of the PV System Controllers *287*
7.5.2 Maximum Power Point Tracking Technology (MPPT) of Solar Cell Controllers *289*
7.5.3 Installation Angle and Position Regulation of Solar Cell Arrays by the Controller *291*

7.5.4 Other Functions of the Controller *293*
7.6 Applications of PV Systems *294*
7.6.1 Classifications of PV Systems *294*
7.6.2 Application Type, Size and Load Types of the PV System *294*
7.6.2.1 Small-Power DC PV Systems *294*
7.6.2.2 DC Power Supply Systems Required of Controllers *295*
7.6.2.3 AC/DC Power Supply Systems Required of Inverters *295*
7.6.2.4 Small-/Medium-Sized Distributed PV Systems with the Grid-Connected Inverter *296*
7.6.2.5 Large-Sized Centralised Grid-Connected PV Stations *296*
7.6.2.6 Hybrid Power System *297*
7.6.3 Energy Storage Device Charging/Discharging by Small-/Medium-Sized PV Systems *298*
7.7 BIPV and Distributed PV Stations *299*
7.7.1 BIPV *299*
7.7.2 Design Principles of BIPV Grid-Connected Power Systems *301*
7.7.3 National Policies and Certification of BIPV in China *301*
7.7.4 Encouragement of Distributed PV Stations by the Chinese Government *301*
7.8 Grid-Connected PV Systems and Intelligent Grids *302*
7.8.1 Grid-Connected PV Systems *302*
7.8.2 Technical Specifications of the Grid on the Grid-Connected PV System *303*
7.8.3 Significance of the Intelligent Grid on PV Power and Other New Energy Utilisation *305*
7.8.4 Development of China's PV Industry in the Past 10 Years and Its Outlook *307*
7.9 Codes and Test Verifications of the PV System *309*
7.9.1 Necessity and Main Contents of PV Product Certification *309*
7.9.2 TUV Certification Oriented to the European Market *311*
7.9.3 UL Certification Oriented to the U.S. and Canadian Market *311*
7.9.4 Certification of PV Products in China *313*

Bibliography *319*
Index *321*

About the Author

Guangyu Wang was born in 1943, Suzhou, China. He was graduated from Suzhou High School in 1960, and Peking University of Science and Technology in 1965, respectively. After graduation, he joined Emei semiconductor material research institute and worked on the process development and production. He received his M.Sc. and Ph.D. degrees in semiconductor material and device from Shanghai Institute of Metallurgy, Chinese Academy of Sciences in 1981 and 1985, respectively. He is now a research professor at Shanghai Institute of Microsystem and Information Technology, Chinese Academy of Sciences (CAS). He is the director of East-China High-Tech Innovation Park in Jiangsu province, China. He is a vice chairman of Shanghai Optoelectronics Trade Association (SOTA), responsible for PV industry. He has won several National Awards and enjoyed special government allowances of the State Council.

The author is grateful to Dr. Bingwen Liang for his assistance with translating and proof-reading. Dr Liang received his Ph.D. degree in Applied Physics from University of California at San Diego (UCSD) in 1993. After graduation, he joined HP's Optoelectronics Division (OED) in San Jose, California and worked on super-high-brightness light-emitting diode (LED) process development and production. In 2008, as a full professor Dr. Liang joined Suzhou Institute of Nano-Tech and Nano- Bionics (SINANO), Chinese Academy of Sciences (CAS) and later was appointed as the Director of Technology Transfer Center. He has more than 60 journal and conference publications and has given a number of conference presentations including invited talks. He co-authored the Chapter of "Semiconductor Lasers" in the Handbook of Optical Components and Engineering, John Wiley, New York, 2003. He has been granted six (6) US patents, more than sixty (60) China patents and about twenty five (25) China pending patents. He has more than twenty three (23) years of experience in III-V compound semiconductor materials and devices. His core expertise includes optoelectronic materials and devices for lighting, displays and green-energy applications.

Preface

In August 2014, through the recommendation of China Electric Power Press, the English version of *Technology, Manufacturing and Grid Connection of Photovoltaic Solar Cells* won the favour of U.S. publisher, John Wiley & Sons, and both parties decided to jointly publish the book. The original Chinese version of the book was completed in September 2010 and since then the photovoltaic (PV) industry has developed very rapidly. In 2010, polycrystalline silicon, the main raw material of PV solar cells, was mainly produced in foreign countries and China's PV power generation accounted for a small proportion of the world PV market. However, in 2014, both China's production of polycrystalline silicon and newly installed capacity of PV power were ranked first in the world. In addition, as China made major progress in the R&D of solar cell research, the core component of PV power generation, efficient silicon solar cell, compound thin-film solar cell, organic solar cell, dye solar cell and perovskite solar cell were developed one after another. PV internet technology was also beginning to take shape and the cost of PV power generation was close to the cost of conventional energy power generation.

This book has a total of seven chapters. Chapter 1, Basic Physics of Solar Cells, describes the PV effect resulted from solar spectral radiation to semiconductor materials and illustrates the basic principle of P-N junction and heterojunction semiconductor solar cell power generation. Chapter 2, Materials of Solar Cells, introduces polycrystalline silicon, monocrystalline silicon, thin-film silicon and compound materials commonly used in the PV industry and impurity analysis methods, especially emphasising low-cost and low-pollution solar-grade materials. Chapter 3, Preparation Methods of Crystalline Silicon Solar Cells, elaborates on the production process and equipment of crystal silicon solar cell which boasts the largest production and widest application in the current PV industry, with a market share of 80%. Chapter 4, Preparation Methods of Thin-film Silicon Solar Cells, discusses the characteristics of thin-film silicon, solar cell optical attenuation mechanism and the device and the principle of thin-film deposition, making theoretical preparations for the introduction of non-silicon thin-film solar cells in Chapter 5, High Efficiency Silicon Solar Cells and Non-Silicon-Based New Solar Cells, which can be divided into two sections. The first section mainly focuses on introducing efficient silicon solar cells such as passivation emitter solar cell, the interdigitated back contact (IBC) solar cell and amorphous silicon/crystalline silicon heterojunction solar cell based on the introduction of the traditional silicon solar cells in Chapters 3 and 4. The second section introduces the non-silicon-based III-V compound semiconductor multi-junction solar cell, II-VI semiconductor thin-film solar cell, organic solar cell, dye-sensitised solar cell, perovskite solar cell, concentrator solar cell and multiple quantum well solar cell, some of which are likely to develop very well in the next 10 years. Chapter 6, Modules and Arrays of Solar Cells, mainly describes module packaging materials, process and equipment. As modules are the major products exchanged by solar cell manufacturers in the market, this chapter focuses on the introduction of the reliability of the modules and the detection following packaging. Chapter 7, PV Systems and Grid-connected Technologies, first introduces multiple PV systems applicable to different environments and then introduces the main components in the PV system, except social cell modules such as energy storage batteries, inverters and controllers. In addition, it discusses the possible impact on the public power grid of the PV and other new energy source grid integration and how to pass product testing and certification and get access to the smart grid. Finally, it provides an overview of China's uneven development course of PV power generation in the past 10 years.

The PV industry has developed at a rapid speed that is rare in the modern industry and is even unmatched by the semiconductor industry. Since the beginning of the twenty-first century, all countries in the world have attached great importance to the imminent exhaustion of the world's fossil energy and the problems, such as environmental pollution and climate change resulted from it, and have fully realised it is urgent and necessary to vigourously develop renewable energy such as solar energy. It's true that China still has a long way to go before we can use new energy, including PV energy, as alternative energy sources, but PV power generation has promising prospects and is developing at an amazing speed. As new energy is one of the heatedly discussed topics in the twenty-first century, I believe PV power generation and grid connection technology will be constantly updated by new research and development.

In the end, I would like to thank Dr. Liang Bingwen for reviewing some chapters of this book and Professor Jiang Xinyuan's valuable advice for this book. It makes me confident that I can finish the book.

1

Basic Physics of Solar Cells

1.1 Development of Solar Cells

1.1.1 Solar Energy Is the Most Promising Renewable Energy Source in the World

The amount of solar radiation striking the earth per second is equivalent to the energy obtained by burning 500 tons of coal. The thermonuclear reactions inside the sun can last 6×10^{10} years and generate 'inexhaustible' energy. Many energy sources on the earth including wind energy, hydro energy, ocean thermal energy, tidal energy and biomass energy derive their energy from the sun. The currently most widely used fossil fuels such as oil, natural gas and coal are also the forms of the stored energy originally obtained from the sun. As shown in Figure 1.1, based on the existing proved reserves and the current consumption rate of the conventional fossil energy, the world's primary energy source from fossil fuels will be exhausted soon, and China in particular will face a severe shortage of fossil energy sources. In addition, the excessive use of fossil fuels has led to environmental problems and climate change, which has attracted growing attention from governments all around the world. With the outbreak of the energy crisis and the impending depletion of fossil energy resources on the earth, people are increasingly aware of the urgent need to develop solar energy and other renewable energy sources. Both the signing of the Kyoto Protocol in 1997 and the holding of the 2009 UN Climate Change Conference in Copenhagen indicate most governments in the world have regarded the renewable energy use as their national energy strategy. Particularly solar energy is of the greatest strategic significance as it is the ideal renewable and sustainable energy source.

According to statistics, in 2006, the world population already exceeded 6.5 billion and the world energy demand converted into the installed capacity was 14.5 terrawatt (TW); by 2050, the world population will reach 90 to 10 billion and the world energy demand converted into the installed capacity will approximate 60 TW. By that time the world will have almost used up the primary energy resources and have to rely on renewable energy sources. It's reported that the world has a potential hydropower capacity of 4.6 TW only 0.9 TW of which can be actually exploited; the world has 2 TW of actually exploitable wind energy and 3 TW of biomass energy. It means that with a potential capacity of 120,000 TW and the actual available capacity of 600 TW, solar energy will remain the only energy source that can meet the future world energy demand. In this sense, photovoltaic (PV) power will be a key part of the future world energy consumption structure.

Technology, Manufacturing and Grid Connection of Photovoltaic Solar Cells, First Edition. Guangyu Wang.

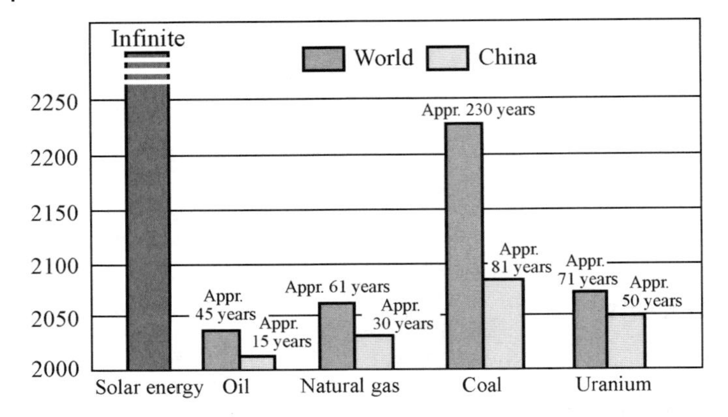

Figure 1.1 Timescale for depletion of conventional fossil energy resources in China and across the world (based on the proved reserves and current consumption rate).

Table 1.1 Future world dnergy demand and renewable resources.

Actual World Energy Consumption in 2004	13 TW
Estimated World Energy Consumption in 2050	30 TW
Estimated World Energy Consumption in 2100	46 TW
Undeveloped Hydropower	<0.5 TW
Ocean energy (tides, waves, currents)	<2 TW
Terrain Energy	12 TW
Available Wind Energy	2~4 TW
Total World Solar Energy	120,000 TW

The U.S. Department of Energy also released a similar report in 2005. Although the long-term world energy consumption forecasting varies from country to country, the prediction of the future world energy consumption trend remains identical (see Table 1.1 for detailed information).

The Joint Research Centre made a similar forecast according to which PV power will become the most important part in the world energy consumption. By 2030, renewable energy will account for more than 30% of the total world energy consumption while over 10% of the world power supply will come from PV power. By 2040, renewable energy will account for more than 50% of the total world energy consumption while over 20% of the world power supply will come from PV power. By the end of this century, renewable energy will account for more than 80% of the total world energy consumption while over 60% of the world power supply will come from PV power. Judging from the development in the past 20 years, PV power development will be further accelerated.

With the gradual depletion of the fossil energy resources, human beings must accelerate developing renewable energy resources to achieve sustainable energy resource development. It's predicted that by 2050 PV power and solar thermal power will account for a higher proportion than fossil energy resources and other renewable energy resources in the world energy consumption structure. Figure 1.2 indicates the forecast of the development trend of world energy consumption structure in the twenty-first century.

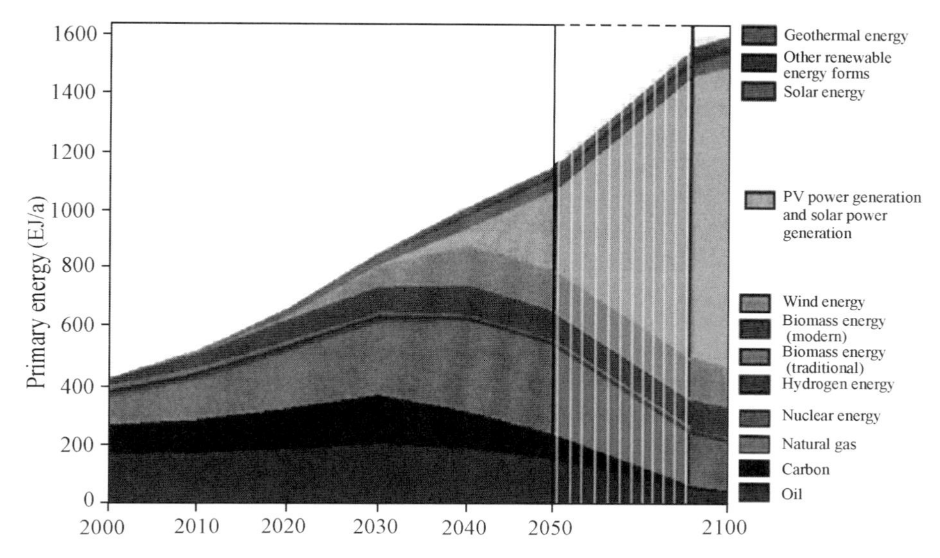

Figure 1.2 Forecast of world energy consumption structure in the twenty-first century.

PV power generation technologies were first applied in space. At present it has been widely used on the earth. Governments all around the world have given support in policy to the development of PV power generation technologies and the application of these technologies in architecture. Grid-connected PV power generation and PV power stations will become the inevitable trend of PV power development. A lot of countries are competing with each other to develop various PV materials and high-efficiency PV power generation technologies to expand the application fields of solar energy.

The rapid development of the solar cell industry is quite unique in the modern industry and unmatched by even the semiconductor industry. In the last 15 years the compound annual growth rate of the world solar cell output has exceeded 30%. At present, Germany and other European countries remains the largest PV market in the world and in these countries, PV power accounts for about 4% of the national electricity consumption. On the other hand, PV markets in Asia, America, Africa and Australia are rapidly rising. Since 2009, China's solar cell and module production has been ranked the first in the whole world. However, grid-connected PV power generation has just started in China. Since 2013 the Chinese government has issued a series of policies to encourage the development of domestic PV applications. In 2013, China's new installed PV capacity was ranked the first in the world. According to the latest data of China's electric power sector, in 2013, total power generation in the Chinese mainland reached 5347.4 terrawatt hours (TWH) while PV power generation only accounted for 0.16% of it, namely, 8.7 TWH. There is a great gap between it and the proportion of PV power in the world energy consumption in the twenty-first century predicted by the intelligence department (see Figure 1.2). As a result, the PV industry has tremendous potential markets and still has a long way to go.

It should be pointed out that due to the currently quite high PV power generation cost, it will be very difficult for PV power generation to compete with conventional

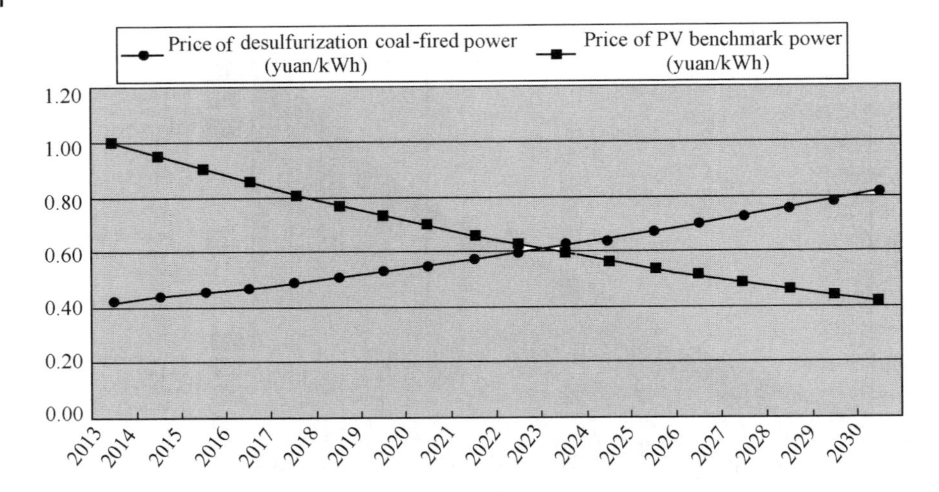

If the coal-fired power price increases by 2% annually and the photovoltaic power price reduces by 5% annually: by 2023 photovoltaic power will reach grid parity on the generation side in some regions rich in energy resources in China.

Figure 1.3 China's PV parity price development roadmap, issued by China's relevant department.

power generation through the business accumulation and technological progress of the PV industry. The governments should give some policy support. Take China as an example. In 2013 China's coal-fired power price was RMB 0.42/kilowatt hours (kWh) while the PV power price was RMB 1.00/kWh. If the coal-fired power price rises by 2% annually and the PV power price reduces by 5% annually, then by 2023, PV power will reach grid parity on the generation side in some regions rich in energy resources in China.

Shown in Figure 1.3 is *China Roadmap of Photovoltaics Development—A Pathway to Grid Parity* released in 2013.

1.1.2 Development of Solar Cells

Nowadays the solar cell industry has become one of the world's fastest-growing high-tech industries. In the long run, the research and development of solar cells will go through three stages: crystalline solar cells, thin-film solar cells and quantum devices. Through in-depth R&D the production of crystalline and thin-film solar cells has been characterised by industrialisation, large scale and commercialisation. Categories of crystalline and thin-film solar cells are shown in Table 1.2.

The 'monocrystalline silicon solar cells' in Table 1.2 have been developed very rapidly in recent years and there have emerged some high-efficiency cells such as the selective emitter cell, the full back electrode silicon solar cell, the passivation film solar cell and the HIT silicon solar cell which are not listed in the table and will be introduced in Chapter 5.

Starting from the basic principles of the solar spectrum and semiconductor, this book will elaborate on solar cell materials, manufacturing processes and equipment of solar cells and modules as well as basics about PV power system and integration.

Table 1.2 Solar cell development stages.

Solar cells	Composition	Conversion efficiency	Commercialisation	Characteristics
	Monocrystalline silicon *	18–20%	Commerciallised	Longest application
Crystalline cells	Polycrystalline silicon	16–18%	Commerciallised	Largest output
	Compound semiconductor (GaAs)	28–35% (GaAs)	Space application	High efficiency, high cost
	Amorphous silicon thin film	8–10%	Industrialisable	Promising but unstable
	Amorphous/microcrystalline silicon thin film	11–13%	Industrialised	Competitive
Thin-film cells	Compound thin film(CdTe)	10–12%	Industrialised	Toxic raw materials
	Compound thin film (CIGS)	10–15%	Industrialising	Shortage of raw materials
	Organic compound thin film	4–6%	R&D	Cheap, capable of large-scale application
	Dye sentisised solar cell ($Ti0_2$)	11% (lab)	R&D	Cheap but unstable
New concept cells	Effiecient cells, quantum well cells, etc.		Being industrialised R&D	High efficiency, advanced technology

1.2 Solar Radiation and Air Mass

1.2.1 Conversion of Sunlight Into Electricity Using Photoelectric Effect Is an Important Way to Make Use of Solar Energy

The sun is the most important star to human beings because it provides human beings with light and heat and constantly radiates a huge amount of energy out to the space. In addition, solar energy is a clean energy source free from pollution.

Solar energy can be made use of in many ways including photochemical reaction, photo-thermal conversion and photoelectric conversion. In the nineteenth century, human beings already realised light that shining onto semiconductor materials can generate current. In the 1950s the first silicon solar cell was produced, which started the photoelectric conversion of solar energy. Through deepening research, more and more new high-efficiency semiconductor materials have been used to produce solar photoelectric materials. At present, polycrystalline silicon, Czochralski monocrystalline silicon, thin-film silicon and compound semiconductor materials have become important solar cell materials.

1.2.2 Basics of Solar Radiation and Definition of Air Mass

To make use of the solar energy, it is necessary to know the basic knowledge on solar radiation. It is known that when the solar rays reach the earth, some of them will be reflected, scattered or absorbed by the amosphere, and only about 70% of them can penetrate the atmospheric layer and reach the ground. In addition, since the geographic positions and landforms are different on the earth surface, the solar rays reaching various regions are not the same, and they also vary with seasons, time and the weather.

Since the actual application of the solar cell has close relations with its position away from the earth, the concept 'air mass (AM)' is defined according to the influence of the air on the solar rays received by the earth surface. It is a dimensionless quantity. The solar rays received by the outer space are defined as zero air mass (AM0), and the total energy flow density of the total sunshine spectrum is 1368 W/m^{-2}, which is applicable to the man-made satellite and the space craft and other similar applications. It is defined as 1 air mass (AM1) that the solar rays directly shine on the earth surface after scattering, reflection and absorption by the atmosphere and the cloud layer, and it is equivalent to the solar rays received by the sea level in the sunny summer. When the incidence angle of the solar rays is θ with the ground, the air mass $AM = 1/\cos\theta$. It is defined that when $\theta = 48.2°$, the air mass is AM1. 5, which is usually the case when the solar rays shine on the ordinary ground in the typical sunny day where the radiation total is 1000 W/m^{-2}. AM1.5 is often used as the criterion for efficiency test of the solar cell and the module. Figure 1.4 shows the air mass diagramme.

Table 1.3 shows the total solar radiation of the regions in China throughout a year. It shows that strong solar radiation is available in Qinghai-Tibet Plateau, Xinjiang and Gansu, and so on. Classes I, II, III and IV in the table stand for the most abundant solar rays, the very abundant solar rays, the abundant solar rays and the ordinary solar rays. Obviously, China is endowed with very abundant solar energy resources, which is good for China to fully develop solar power, completely change the energy portfolio and achieve sustainable energy development from a strategic point of view.

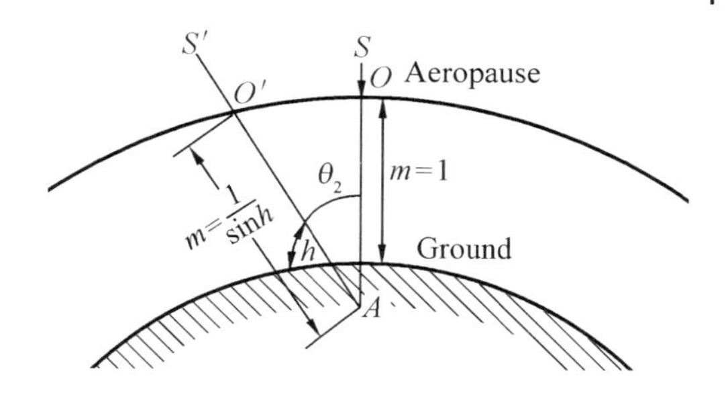

Figure 1.4 Air mass diagramme.

Table 1.3 Solar radiation of the regions in China throughout a year.

Class	Region	Annual sunshine hour, hr
Most abundant, Class I	Qinghai-Tibet Plateau, north Gansu, north Ningxia and south Xinjiang	3200–3300
Very abundant, Class II	Northwest Hebei, north Shanxi, south Inner Mongolia, south Ningxia, central Gansu, east Qinghai and southeast Tibet	3000–3200
Abundant, Class III	Shandong, Henan, southeast Hebei, south Shanxi, north Xinjiang, Jilin, Liaoning, Yunnan, north Shaanxi, southeast Gansu, south Guangdong, south Fujian, north Jiangsu and north Anhui	2200–3000
Ordinary, Class IV	The lower and middle reaches of the Changjiang River, some regions in Fujian, Zhejiang and Guangdong	1400–2200
Poor, Class V	Sichuan, Guizhou	1000–1400

1.2.3 Wavelength of Solar Radiation

'Photovoltaic' refers to the PV conversion effect after the solar light wave is absorbed by the semiconductor material. Since it has close relations with the solar wavelength, we should further discuss the solar wavelength distribution.

It is known that the solar light comes in various wavelength, and 97% of solar spectrum falls in the range of 0.09–3.0 μm. Figure 1.5 shows the solar radiation wavelength distribution of AM0 and AM1.5.

The figure shows, quite a lot of solar rays with some wavelength are scattered and absorbed at AM1.5 due to the action of various gas compositions in the atmosphere where the ozone layer has the strongest absorption of the ultraviolet rays, steam has the largest absorption of the visible light energy and about 20% of the solar energy absorbed by the atmospheric layer is due to the action of steam and the dust can absorb and reflect the solar rays. The ground solar spectrum consists of the ultraviolet light, the visible light and the infrared light, as shown in Figure 1.6.

In addition, the solar radiation spectrum is also different at various moments in a day. Figure 1.7 shows the variation of solar spectral intensity of some region from 8:00 a.m. to 4:00 p.m.

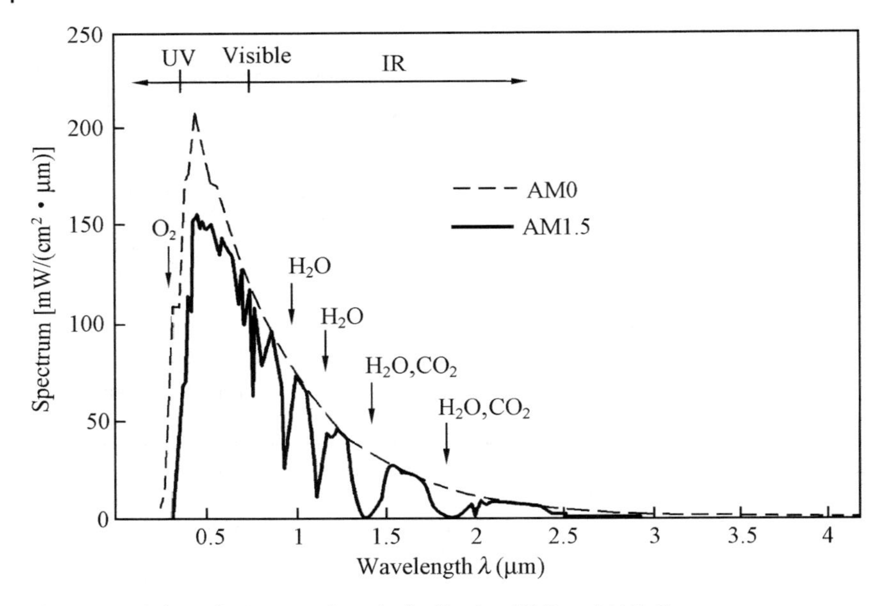

Figure 1.5 Solar radiation wavelength distribution (AM0 and AM1.5).

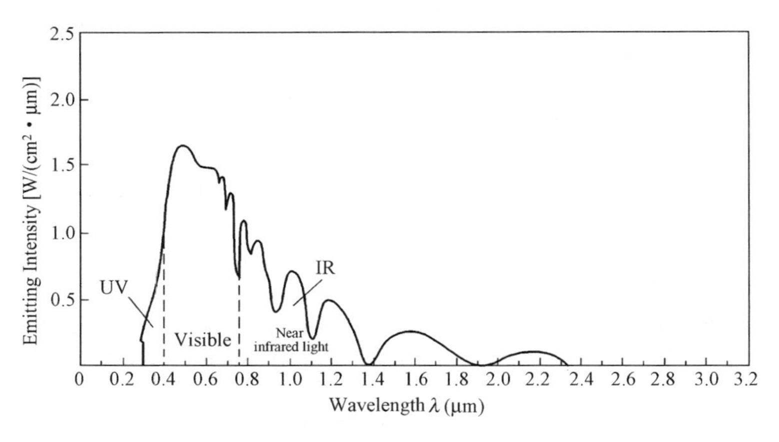

Figure 1.6 Change in the solar radiation intensity onto the earth's surface with the wavelength.

1.3 Basics of Semiconductors

The solar PV energy conversion can be realised after the PV materials are produced to solar cells. In the process, the most important material is semiconductor. As a result, it is a must to discuss the fundamental properties of semiconductors, including the principle of electron and hole generation, the energy band structure, the energy levels produced by impurities and defects, carrier distribution and Fermi level as well as generation, recombiniation, diffusion and drift of minority carriers and so on. These are the physical basics of semiconductor PV effect.

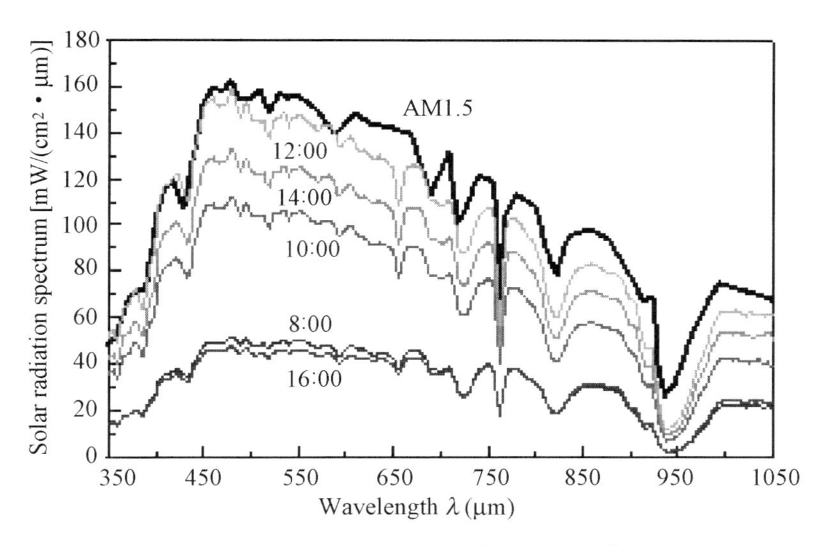

Figure 1.7 Variation of solar spectral intensity of some region from 8:00 a.m. to 4:00 p.m.

1.3.1 Communisation Motion of Electrons in a Crystalline and Formation of Energy Bands

In a crystalline, the atoms are close to each other and arranged in periodic orders. The inner/outer electron shells of different atoms are to some extent overlapping.The overlapping shells are no longer a part of any single atom, and they can transfer to the shell of the adjacent atom with the same number of electrons and thus belong to the atom of the whole crystalline. This is the communisation motion of electrons in a crystalline. The result of electron communisation motion will make the single energy level of the isolated atom split into energy bands. And each energy band consists of several adjacent energy levels. The energy bands that can be taken by the electrons are called as the allowed bands, and no electron is allowed to exist in the gap between two allowed bands, which is called as the forbbiden band (or the band gap). Figure 1.8 shows the energy band splitted from an energy level due to atom communisation motion.

1.3.2 Atomic Structures of Conductors, Insulators and Semiconductors and Energy Bands Image

Figure 1.9 shows the energy band image of conductor, insulator and semiconductor. In the figure, (a) shows insulators where the forbidden band is wide and the valence band is full of electrons while the conduction band is almost empty. Usually, the field cannot make electrons in the valence band transfer to the conduction band. As a result, insulators cannot conduct, (b) shows the energy band overlapping and forbidden band disappearance of metal conductors where electrons in the valence band can freely enter the conduction band under the action of field, that is, extremely small amount of external energy can result in conduction, (c) shows semiconductors where there are a few electrons in the conduction band and a few holes in the valence band and the forbidden band is not so wide as that of an insulator. Some electrons in the valence band can transfer to the conduction band due to thermal motion and solar radiation.

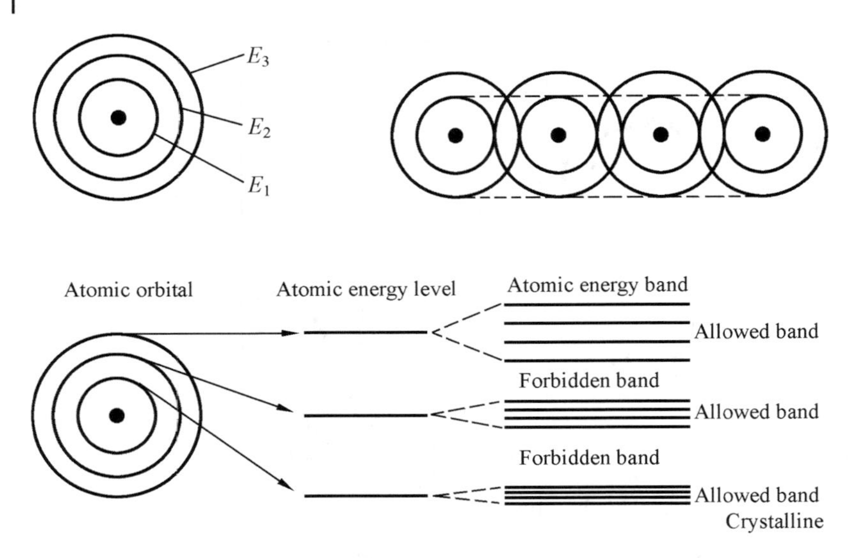

Figure 1.8 Electron communisation of the isolated atom and the atom in a crystalline.

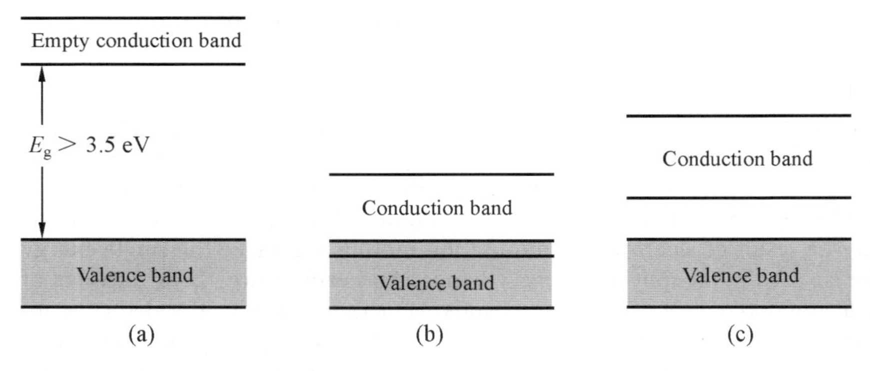

Figure 1.9 Energy band image of insulator, conductor and semiconductor.

Accordingly, semiconductors have some conduction capacity which will be improved with temperature rise and solar radiation intensity growth.

(a) Metal—The energy bands are stacked and even the extremely small external energy can result in conduction;
(b) Insulator—The forbidden band is wide and it cannot conduct;
(c) Semiconductor—There are a few electrons in the conduction band and a few holes in the valence band. It has some conduction capacity.

1.3.3 Energy Band Structure of Dope Semiconductor

The electron/hole concentration of the intrinsic silicon is about $10^{10}/cm^{-3}$ at the ambient temperature. Figure 1.10 shows the atomic structure and energy band image when the semiconductor crystalline silicon contains impurities. Figure 1.10(a) shows the atomic structure of the doped silicon when group V elements (e.g., phosphorus) are doped. Since the energy level position of phosphorus in the silicon is Ec-0.044 eV, which

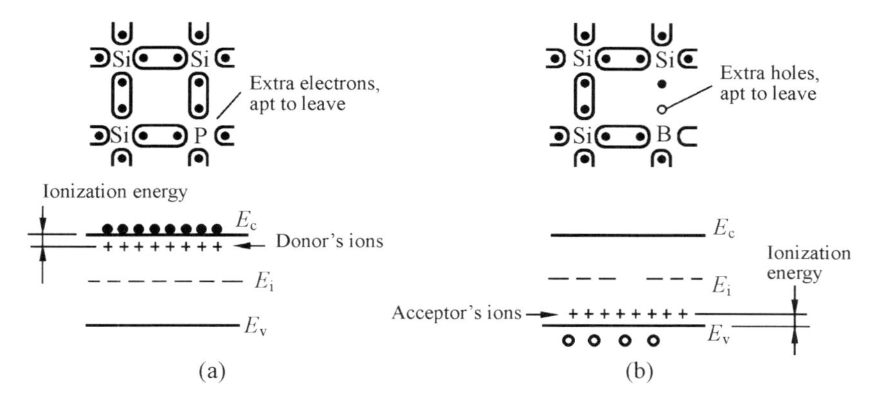

Figure 1.10 Atomic structure and energy band image of doped silicon (a) N-type semiconductor; (b) P-type semiconductor.

is close to the bottom of the conduction band, the small energy can make the electron of the phosphorus jump to the conduction band. The phosphorus atoms can be all ionised to offer the conduction electrons of the identical quantity at the ambient temperature, and the impurity supplying the electrons is called as the donor and the semiconductor as N-type semiconductor. In the figure, E_i is the middle position of the forbidden band.

Figure 1.10(b) shows the atomic structure of the doped silicon when group III elements (e.g., boron) are doped. When the boron atom forms a covalent bond with the neighbouring four silicon atoms, it lacks one valence electron, and thus it is apt to capture one valence electron, and then it is ionised into an anion. The boron atom can be viewed as carrying a hole apt to ionise, and its ionisation energy is only 0.045 eV. It is close to the top of the energy band E_V in the energy band image, and the thermal motion can make the hole jump to the valence band. As a result, the boron atoms can be all ionised to supply the conduction holes of the identical quantity, and the impurity supplying the electrons is called as the acceptor and the semiconductor as the P-type semiconductor. In the figure, Ei is the middle position of the forbidden band.

In the N-type silicon, the concentration of the electrons is far larger than that of the holes, and the current is mainly transferred by electrons. Here the electrons are the majority carriers, and the holes the minority ones. In contrast, in the P-type silicon, the concentration of the holes is far larger than that of the electrons, and the current is mainly transferred by holes. Here the holes are the majority carriers, and the electrons the minority ones. In fact, the silicon material generally contains both the donor and the acceptor impurities. In this case, the conduction type is dependent on the impurity with higher concentration. The concentration of the majority carriers in the N/P-type semiconductors are given in Formulas (1.1) and (1.2):

$$\text{Electron concentration in a N-type semiconductor: } n \approx N_D - N_A(N_D \gg N_A) \quad (1.1)$$

$$\text{Hole concentration in a P-type semiconductor: } p \approx N_A - N_D(N_A \gg N_D) \quad (1.2)$$

1.3.4 Fermi Level

Since energy states are present in both conduction and the valence bands of a semiconductor material, the energy state will also be present in the forbidden band after a donor or an acceptor impurity is doped (Formula 1.3). Let's take a look at the term of 'Fermi level'. Fermi level is defined as the energy of f(E) = l/2, marked as E_F. It indicates the total chemical potential of electrons in a semiconductor material. Based on the minimum energy principle, electrons distribution f(E) in an energy band of a semiconductor can be expressed as

$$f(E) = 1 \Big/ \left[1 + e^{(E-E_F)/kT}\right] \tag{1.3}$$

Where, E is an energy level of electrons in a semiconductor; T is absolute temperature; and K is the Boltzmann constant.

In a N-type semiconductor, electrons are generally distributed close to the conduction band when the electron concentration is high (Formula 1.3). As a result, the Fermi level of a N-type semiconductor is located in the upper half of the forbidden band. In a P-type semiconductor, electrons are distributed close to the valence band, and thus its Fermi level is located in the lower half of the forbidden band. Figure 1.11 shows the relationship of the electron distribution of a semiconductor material f(E) as a function of temperature. At absolute zero temperature, the energy levels below E_F will be all occupied by electrons, and those above E_F will be all empty. When $T > 0$ K, the electrons with energy less than E_F will transfer to the higher energy level larger than E_F after they absorb energy, and the distribution probability of the electron with energy E in the semiconductor energy band is f(E). Consequently, the distribution probability of the holes is [1-f(E)]. The rise of temperature is good for the electrons to transfer towards the higher level and the holes towards the lower level.

1.3.5 Directional Movement of Electrons and Holes

At the ambient temperature, the following two factors can result in directional movement of electrons and holes in semiconductors: One is external field, and the other concentration difference of the carriers. The directional movement caused by the former is called as 'drift', and the directional movement caused by the latter as 'diffusion'.

During drift, the carriers often change their direction after collided by impurities and lattices. Two key physical quantities are used to describe the drift movement: the electron mobility μ_n and the hole mobility μ_p. They are defined as the directional movement speed of electron or hole at the specific field strength in the unit time. Obviously, the mobility is mainly dependent on the total concentration of the ionised impurity at a given temperature, that is, the sum of the donor and acceptor concentration (Formula 1.4).

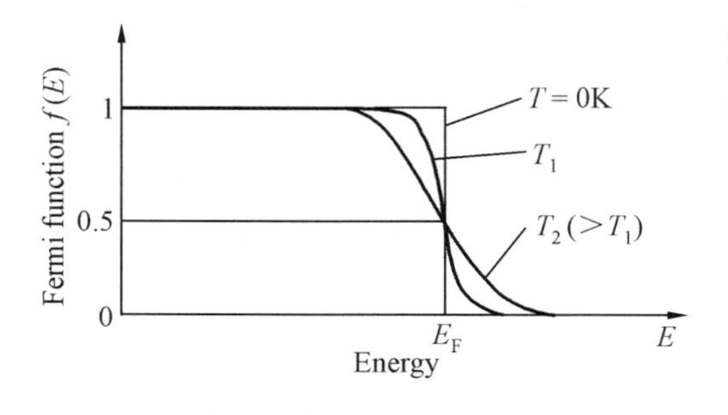

Figure 1.11 Electron distribution function f(E) in the semiconductor material, and its relationship with temperature.

The resistivity of semiconductor material ρ can be easily derived by the mobility:

$$\rho = 1/q(n\mu_n + p\mu_p) \tag{1.4}$$

Where, q is the electron charge.

The relationship between the resistivity and the doped concentration can be obtained from the associated semiconductor manual.

The key physical quantity to describe the diffusion process of the carrier is the diffusion coefficient D which is defined as the directional movement speed of electron and hole at the specific concentration in the unit time. Similar to the carrier drift, it has relations with the thermal movement of electron and hole. Accordingly, the mechanism affecting the mobility (e.g., impurity scattering, lattice scattering) also has an impact on the diffusion coefficient. The diffusion coefficient has inherent relations with the mobility, which can be shown by Formulas (1.5) and (1.6) for the electron and the hole:

$$D_n = kT\mu_n/q \tag{1.5}$$

$$D_p = kT\mu_p/q \tag{1.6}$$

Where D_n, D_p is the diffusion coefficient of the electron and the hole, respectively.

1.3.6 Generation and Recombination of Carriers

The 'generation—transfer—recombination' process of the carrier almost reflects the whole process of most semiconductor devices, including the solar cell.

When the semiconductor is in the radiation of the solar rays (solar ray injection), electrons in the valence band will absorb the photon energy and transfer to the conduction band, generating the holes of identical quantity in the valence band. The photo-generated electrons and holes larger than the concentration of the equilibrium carrier can be also called as 'the nonequilibrium carrier' or 'the excess carrier'. The process where the nonequilibrium carrier is generated in the semiconductor due to change of external conditions is called as injection of carriers.

When the concentration of the carrier is deviated from its equilibrium value, it has a tendency to regain the equilibrium. In case of injection, the equilibrium is recovered by recombination. One important physical quantity to describe the recombination process of the carrier is the carrier lifetime. The survival time of an electron from generation to recombination is called as the lifetime of the electron ζ_n, and the survival time of a hole from generation to recombination as the lifetime of the hole ζ_p. In case of small injection, it generally only takes into account the lifetime of the minority carrier.

The micro-process of recombination is very complicated. The study shows, there are usually three recombination mechanisms: ① direct recombination; ② recombination via the recombination centre; and ③ surface recombination, as shown in Figure 1.12. Three types of recombination often occur simultaneously in the same semiconductor. The measured lifetime of the minority carrier is the comprehensive result of the surface lifetime and the bulk lifetime.

In the figure, Et is the energy level of the recombination centre and Es is the energy level of the surface recombination centre.

In design and manufacture process of solar cells, the diffusion length of carriers $L_n = (D_n \zeta_n)^{1/2}$ and $L_P = (D_P\zeta_p)^{1/2}$) are generally used. It is defined as the average length

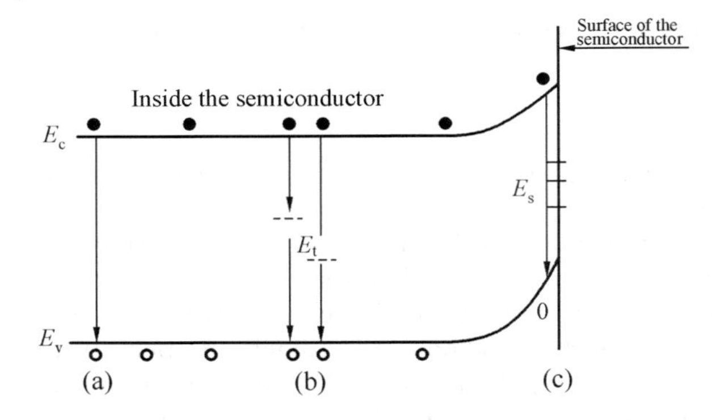

Figure 1.12 Three types of carrier recombination process: (a) direct recombination; (b) recombination via the recombination centre; and (c) surface recombination.

of the carrier diffused in the semiconductor. It refers to the average distance that the carrier can cover in the semiconductor in the process of diffusion and recombination. The diffusion length increases with growth of the minority carrier lifetime.

1.4 Light Absorption of Semiconductor Materials

1.4.1 Light Absorption of Semiconductor

When solar rays radiate on the semiconductor surface, some will be reflected and some absorbed (Formula 1.7). Suppose the intensity of the incident light is I_0 and the reflectivity of the semiconductor surface is R, the intensity of the reflected light is RI_0, and the light intensity entering the semiconductor is I_0 (1−R). Suppose the light absorption coefficient of the semiconductor is α, the light intensity at position x from the said surface in the semiconductor I_x can be determined by the absorption law:

$$I_x = I_0 (1 - R)\ e^{-ax} \tag{1.7}$$

For semiconductor material in the shape of wafer, the incident light shall be first absorbed by the wafer with thickness of d and then reach the back surface of the semiconductor where some light will be reflected and others will penetrate the semiconductor (see Figure 1.13 for the physical image).

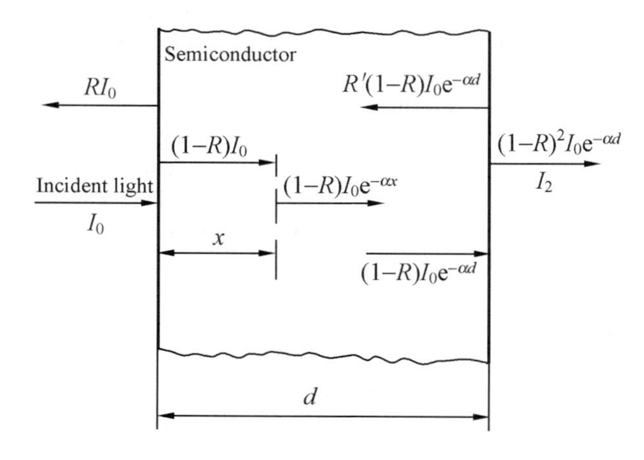

Figure 1.13 Light reflection, transmittance and absorption when the solar rays radiate vertically on the semiconductor.

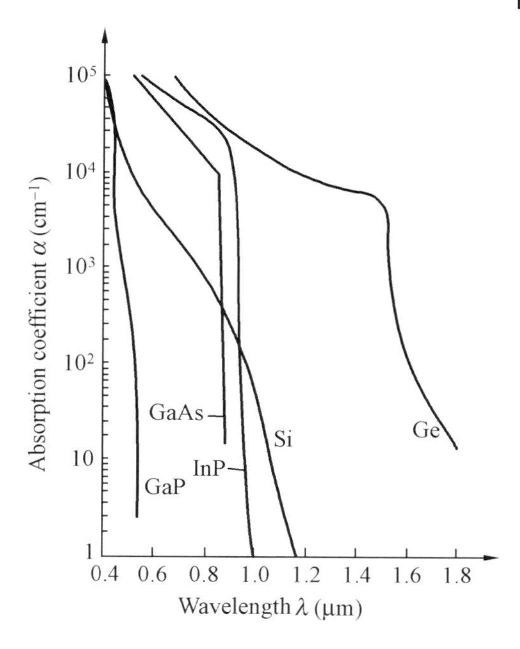

Figure 1.14 Relationship of the absoprtion coefficient of some key semiconductor materials and the wavelength.

The intensity of the transmittance light can be expressed by Formula (1.8):

$$I = I_0(1 - R)^2 e^{-\alpha x} \tag{1.8}$$

This is to say that the incident light will be absorbed by the semiconductor material in the range $1/\alpha$ from the semiconductor surface. It shall be pointed out that the light absorption coefficient is also related to the wavelength of the incident light and the characteristics of the semiconductor material. Figure 1.14 shows the relations between the absorption coefficient and the wavelength for the key solar cell materials, including monocrystalline, germanium, gallium arsenide, gallium phosphide and indium phosphide and so on.

1.4.2 Intrinsic and Non-Intrinsic Absorptions of Semiconductor Materials

The absorption process of light in semiconductor material consists of intrinsic and non-intrinsic absorptions. The so-called intrinsic absorption refers to the process where an electron in the valence band absorbs the photon energy and then overrides the forbidden band to enter the conduction band. Based on the atomic image, the intrinsic absorption can be viewed as that one substrate atom absorbs one photon, which makes one covalent electron change to a free electron, generating a hole in the break of the covalent bond.

Some experiments show, only those photons whose energy $h\upsilon$ is larger than the forbidden band gap E_g can generate the intrinsic absorption, that is, the incident photon must meet the condition $h\upsilon > h\upsilon_0 = Eg$ or $hc/\lambda \geq hc/\lambda_0 = Eg$. Where, υ_0 is the light frequency that can just generate intrinsic absorption (frequency absorption limit), and λ_0 is the light wavelength that can just generate intrinsic absorption (wavelength absorption limit). For a semiconductor material with a forbidden band gap of Eg, there must be a limit frequency υ_0 (or a limit wavelength λ_0).

It shall be pointed out, based on the above model hypothesis of electron-hole pairs generated by light absorption, one photon may generate an electron-hole pair (photon energy > Eg) or not (photon energy < Eg). The energy larger than Eg will be transformed to heat and become relaxation losses. The latest experiments and theories show that for some efficient solar cells such as the multiple quantum well (MQW) and the like, some conditions can be created to make those energy larger than Eg excite another electron-hole pair in the mode of Auger excitation, which doubles the photo generation rate of carriers (so called multiple exciton generation effect) and significantly improves the conversion efficiency of solar cells. This is the so-called third-generation new-concept solar cells.

The non-intrinsic absorption of a semiconductor material includes exciton absorption, impurity absorption and lattice vibration absorption and so on. For a semiconductor material, the most important light absorption, however, is the intrinsic absorption. The intrinsic absorption falls in the range less than the limit wavelength λ_0, and other absorption types all in the range larger than λ_0 and even extending beyond the far infrared region. For the silicon, the intrinsic absorption coefficient is larger than other absorption coefficients by dozens of, or even thousands of, times. Accordingly, it only takes into account the intrinsic absoprtion in most cases.

1.4.3 Light Absorption Coefficient and Semiconductor Materials of Direct/Indirect Transition

Based on the energy band structure of the semiconductor material, there are two types of transitions for electrons from the valence band to the conduction band excited by the photon (intrinsic absorption process): direct and indirect transitions. For example, the gallium arsenide (GaAs) material has an energy band structure of direct transition, and the crystalline silicon material has an energy band structure of indirect transition (Figure 1.15).

Based on the semiconductor energy band theory, the process of photon absorption by electrons shall follow the energy conservation and momentum conservation laws, that is,

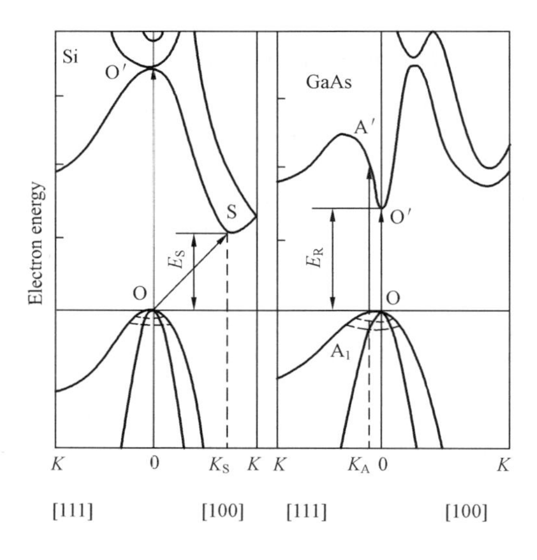

Figure 1.15 Comparisons on energy band structure of crystalline silicon and GaAs.

Energy difference of the electron before/after transition = energy of the photon (hυ)
Momentum difference of the electron before/after transition = momentum of the photon (hυ/C)

That is to say, for the material with energy band structure of direct transition, for example, GaAs and so on, the bottom of the conduction band and the top of the valence band fall at the positions with the same momentum K (K = 0). After the GaAs absorbs the photon h$\upsilon \geq$ Eg, the electron at Point A or 0 will transit to Point A' or 0'. In this case, the momentums before and after transition are identical. For the material with an energy band structure of indirect transition, including silicon and so on, the bottom of the conduction band and the top of the valence band are not at the positions with the same momentum K (K = 0). The momentum at Point S of the bottom of the conduction band $K_s > 0$. After the electron absorbs the photon h$\upsilon \geq$ Eg, the transition from Point 0 to Point S can still occur but it must absorb simultaneously the phonon to make up the momentum difference before and after transition. The direct transition from Point 0 to 0' can also occur but the electron must absorb sufficient photon energy.

Figure 1.16 shows the relationship between the thickness of silicon and GaAs and solar energy absorption at AM0 and AM1.0. The figure shows, for the material of direct transition such as GaAs and the like, a very thin wafer (e.g., several μm) can absorb effectively the solar rays; and for the material of indirect transition such as silicon and the like, the slice shall be very thick (about 100 μm) so as to fully absorb the solar rays.

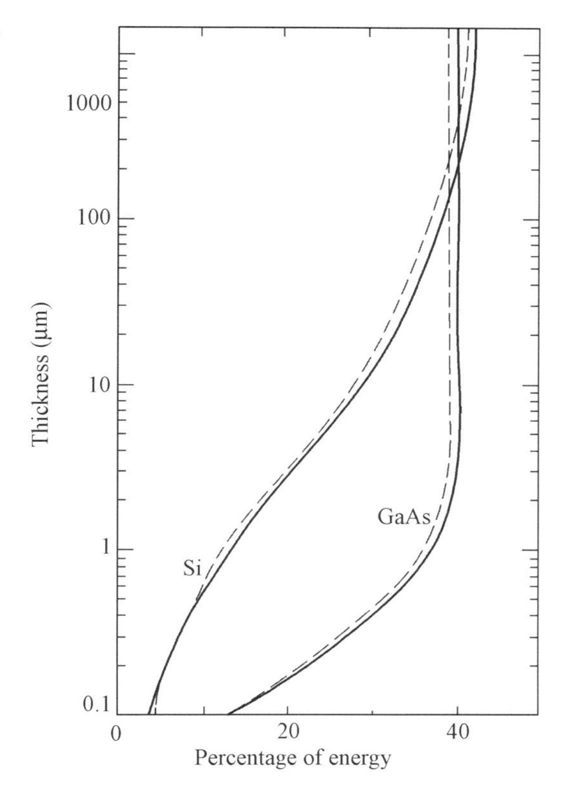

Figure 1.16 Relationship between available solar energy and the thickness of Siand GaAs materials and at AM0 and AM1.0.

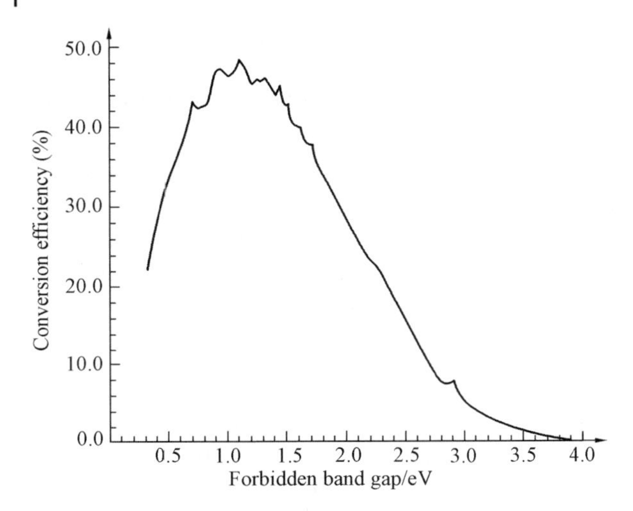

Figure 1.17 Maximum solar cell conversion efficiency of materials in different forbidden band gap.

In Section 1.4.1 (Figure 1.14), we discuss the relationship between the absorption coefficients of semiconductor materials with the wavelength. It shows that GaAs of direct transition see a sharp rise of the absorption coefficient curve, and for Si of indirect transition, the absorption coefficient will rise with the photon energy after the absorption limit λ_0 and it will reach the range of $10^4 \sim 10^5/\text{cm}^{-1}$ at α with photon energy growth, indicating that direct transition has occurred.

Let's further suppose that an electron-hole pair will be generated when the semiconductor absorbs one photon $h\upsilon \geq Eg$ (i.e., the quantum yield is 1), and that each photo-generated carrier pair can make contributions to the photon current (i.e., the photoelectron potential after crossing the forbidden band is equal to the electric energy offered outwards), theoretically, we can work out the maximum solar cell conversion efficiency of the materials with different forbidden band gap at AM1.5 (see Figure 1.17 for the calculation results). Obviously, the maximum conversion efficiency of the solar cell can be achieved when the forbidden band gap of the semiconductor material falls in the range of 1.0–1.5 eV.

1.5 P-N Junctions and PV Effect of Solar Cells

At the radiation of the solar ray, the semiconductor P-N junction can result in the 'PV effect'. Accordingly, these PV components are often called as the 'solar PV cell'. Since the P-N junction is the basic structural unit of the solar cell, it is fundamental to understand the working principle of PV conversion in the solar cell.

In this section, it will introduce the energy band structure of the P-N junction and describe the principle of the PV effect by means of structural variation of the P-N junction energy band in the solar ray.

1.5.1 Bending of a P-N Junction Band and Formation of a Built-In Field

Figure 1.18 shows a semiconductor P-N junction and formation of its band structure. In the N-type zone, the electron concentration is high and the Fermi level E_{Fn} is at

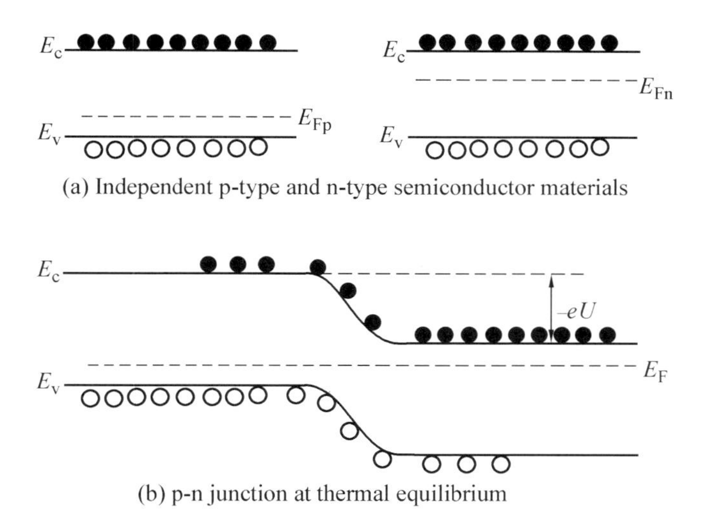

Figure 1.18 Bending of P-N junction bands and formation of built-in field.

a high position; and in the P-type semiconductor, the hole concentration is big, and the Fermi level E_{Fp} is at a low position. When the two are in close contact, electrons will flow from the N-type zone to the P-type semiconductor, and holes will move in the opposite direction. Accordingly, the valence band and the conduction band near the P-N junction will bend to form barrier potential until the movement stops and it reaches equilibrium. For the silicon, it can be viewed as that in the N-type zone near the P-N junction, only a thin layer of ionised phosphorus atom P^+ unable to move is left, and that in the P-type zone near the P-N junction, only a thin layer of ionised boron atom B^- unable to move is left. And an electric double layer is formed in the +/− charge zones on both sides of the P-N junction, which can stop the directional movement of the electrons and the holes. Accordingly, the zone is also called as the barrier layer. Besides, since all the electrons and holes in the electric double layer are flowed away or recombined, it is also knownas the depletion layer or the space charge zone. And a built-in field U_D, which points from the N zone to the P zone, is present. Figure 1.19 shows, when the N/P-type semiconductors form a P-N junction, and E_F is the chemical potential, the chemical potentials of the N/P-type semiconductors shall be equal at thermal equilibrium. As a result, the energy bands of the independent N/P-type semiconductor materials will bend at the P-N junction. And the potential difference at both ends of the space charge zone of the P-N junction V_0 is equal to the difference of the Fermi levels of the original N/P-type semiconductors, forming the 'built-in field' of the P-N junction.

In an equilibrium P-N junction, on both sides of the electric double layer are present with the charges of identical quantity and opposite symbols. The width of the barrier potential is that of the barrier (depletion) layer. Table 1.4 shows the relationship between the base resistivity and the width of the P-N junction depltion layer in the crystalline silicon solar cell.

Generally, the silicon sheet resistivity of the P-type silicon crystalline solar cell is about 1 Ω•cm.

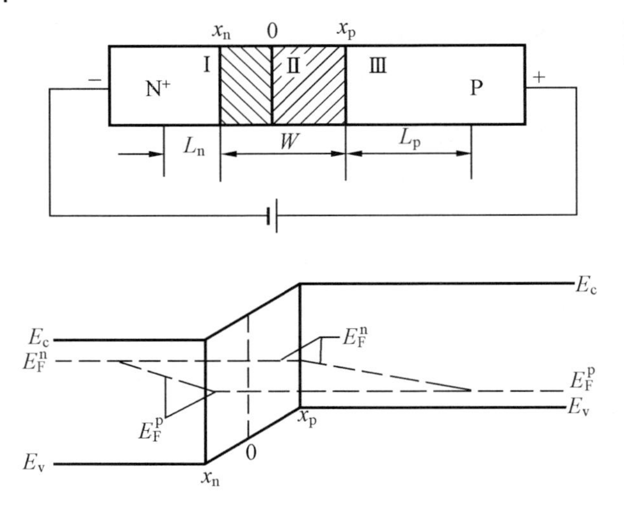

Figure 1.19 P-N junction and its band diagramme at a forward bias.

Table 1.4 Relationship between the base resistivity and the width of the P-N junction depeltion layer in the crystalline silicon solar cell.

Resistivity of the base material (Ω•cm)	10	1	0.1
Width of delpletion layer (μm)	0.75	0.28	0.098

1.5.2 Effect of an External Voltage on the P-N Junction Band Structure

If an external voltage applied U_F is opposite to the built-in field U_D (i.e., the P zone is connected to the anode, and the N zone to the cathode), U_F is called as the forward voltage. In this case, the potential barrier height of the P-N junction will be reduced to $q(U_D-U_F)$, generating a forward bias current. As the forward voltage rises, the P-N junction will have the forward band variation, as shown in Figure 1.19. If the external voltage applied U_F is in the same direction with the built-in field U_D (i.e., the P zone is connected to the negative pole, and the N zone to the positive), U_F is called as a reverse voltage. In this case, the potential barrier height of the P-N junction will rise to $q(U_D+U_F)$, generating a reverse current. Because minority carriers are few, the reverse current is generally small. Figure 1.20 shows a V-I characteristic curve of the P-N junction in the forward/reverse cases. The forward/reverse conduction performances are significantly different from each other for a P-N junction, which is known as a P-N junction's rectification characteristic.

1.5.3 Effect of Solar Radiation on a P-N Junction's Band Structure and the PV Effect

When solar rays radiate on a P-N junction, it will excite photo-generated electron-hole pairs in the N-type, p-type and the depletion regions. After generated, the electron-hole pairs will be immediately separated by the built-in field where the photo-generated electrons will be transmitted to the N-type zone, and the holes will be sent to the P-type zone. If the N-type layer is used as the irradiating surface, a lot of electrons will

Figure 1.20 V-I characteristic curve of the P-N junction.

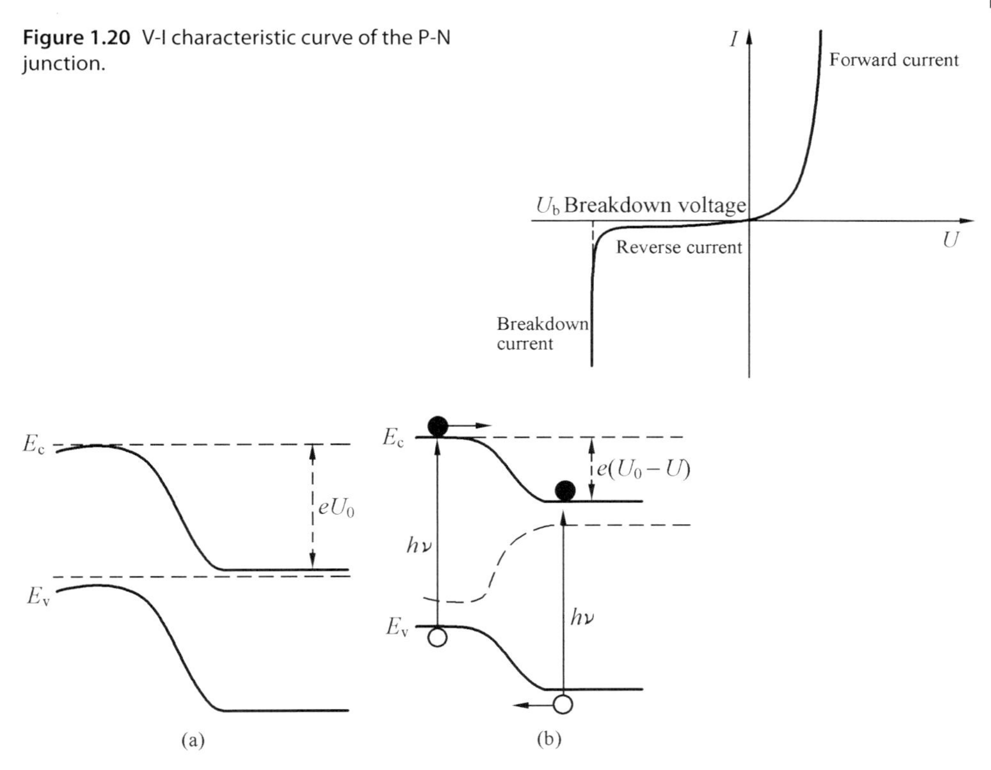

Figure 1.21 Band structure change of a P-N junction with or without solar radiation: (a) band diagramme of P-N junction before solar ray radiation; (b) band diagramme of P-N junction after solar ray radiation.

accumulate there, and a lot of holes will accumulate in the P-type zone. Accordingly, the P-N junction's band structure will change, as shown in Figure 1.21.

The positive/negative charges accumulate on both sides of a P-N junction, which will generate the photovoltage, which is known as the 'PV effect'. The short wavelength light mainly excites electron-hole pairs in the N-type region (the window zone), and the holes move to the P-N junction due to the built-in field. The longer wavelength light mainly excites electron-hole pairs in the P-type region (the base zone), and the electrons also move to the P-N junction due to the built-in field. The carriers make drift movement in the depletion region and diffusion movement outside the depletion zone. If a back surface field (BSF) is built near the back surface, it can force the minority carrier in the P-type zone to be reflected back so as to increase the number of electrons in the P-N junction. Finally, a lot of electrons will accumulate on the irradiating surface of a solar cell, and a lot of holes will accumulate on the back surface. In this case, if metal electrodes are made on both the irradiating and the back surfaces, a photo-generated current will flow from the P layer to the N layer via an extenal load, and the power will be output in the load when an external load is connected. As long as the solar ray is not interrupted, the load will be always present with current. Obviously, the maximum open-circuit voltage of the solar cell is dependent on the potential of the built-in field U_D. This is the fundamental generation principle of the crystalline silicon solar cell (Figure 1.22).

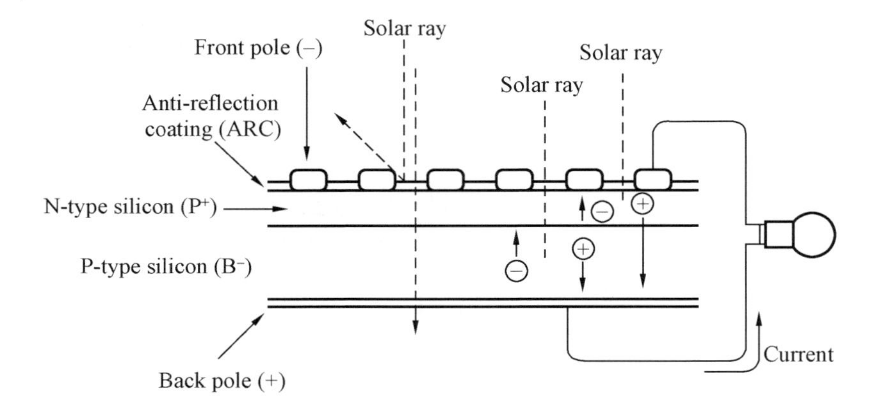

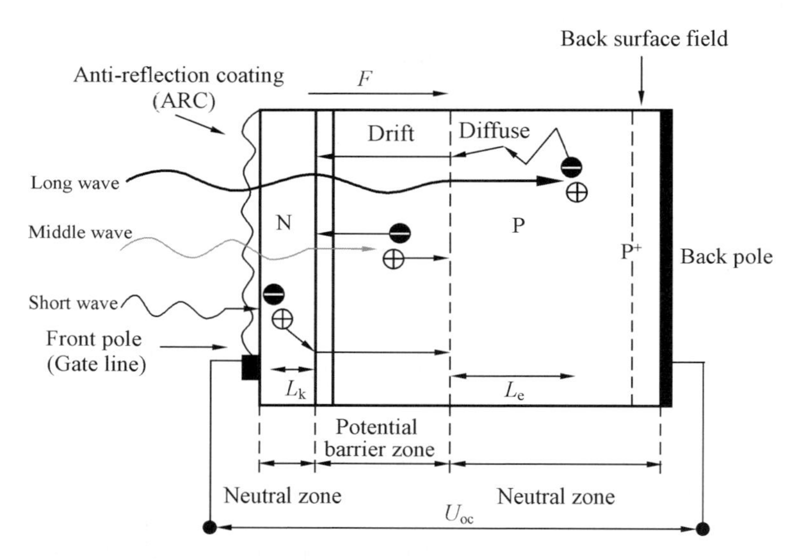

Figure 1.22 Generation principle diagramme of crystalline silicon solar cells.

Figure 1.23 shows an I–V characteristics of solar cells under solar radiation. Both the short-circuit current and the open-circuit voltage rise with the light intensity. It is worth noting that the open-circuit voltage increases exponentially with the solar radiation. The short-circuit current rises with the solar radiation and it is in good linearity in case of strong solar rays. As a result, the crystalline silicon solar cells can be used as an illuminometer after some appropriate modifications to the spectrum.

1.5.4 Composition of the Photo-Generated Current in the Solar Cell

The photo-generated current is a directional movement of the photo-generated carriers. The photo-generated carriers can be generated in the N zone, the depletion zone and the P zone, but they must cross the depletion zone before recombination and then make contributions to the photo-generated current. Obviously, in a solar cell,

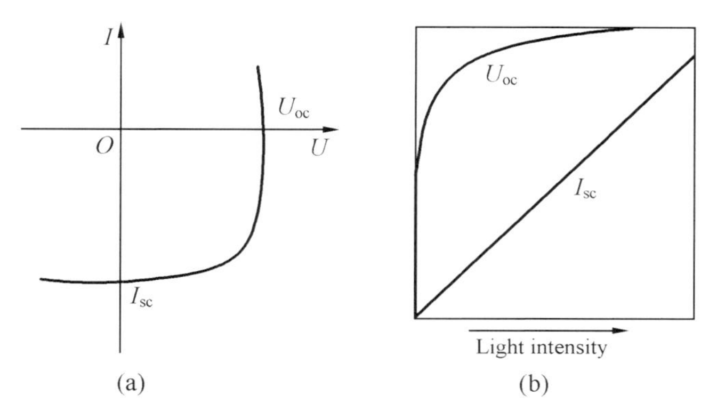

Figure 1.23 Electrical properties of solar cells in solar rays: (a) V-I characteristics of the solar cell in the solar ray; (b) both the short-circuit current and the open-circuit current rise with light intensity growth.

the lifetime and the diffusion length of minority carriers in the P/N zones has a close relation to the PV conversion efficiency.

Since silicon is a material with an indirect band gap, most of photons shall move for a rather long distance in the material before being absorbed when a crystalline silicon solar cell is in operation. As a result, the solar absorption in a silicon solar cell is dependent to a great degree on the base zone (the P-type zone). Photons in a direct band-gap semiconductor has far less movement distance, and the window layer (the N-type zone) in a GaAs solar cell has a bigger contribution to the cell performance.

A theoretical analysis shows, each zone in a solar cell has different contribution to the photo-generated current, which has been proved by experiments. For traditional crystalline silicon solar cells, the photo current generated in the window zone is sensitive to the violet, accounting for about 5% to 12% of the total photo-generated current (varying with the thickness of the zone top), the photo-generated current generated in the depletion zone is sensitive to the visible light, accounting for about 2% to 5% of the total photo-generated current and the photo-generated current generated in the base zone is sensitive to the infrared light, accounting for about 90% of the total photo-generated current, and it is the main composition of the total photo-generated current.

The light absorption efficiency of the N-P junction solar cell has relations to the light wavelength. This is because in the intrinsic semiconductor, the photons whose energy is less than the forbidden band gap cannot excite an electron-hole pair, and only the photon whose energy is larger than or equal to the forbidden band gap can excite an electron-hole pair. That is to say, the photons of different wavelengths have different excitation capacity of electron-hole pairs, which is generally expressed by 'spectral response'. The spectral response of the solar cell has relations to the structure, material performance, junction depth and surface optical characteristics of the solar cell. The average number of carriers generated by one photon (spectral response) is generally less than 1 when the light of a certain wavelength radiates on a P-N junction.

1.5.5 Key Parameters of the Solar Cell

1.5.5.1 I-V Characteristic Curve of a Solar Cell

The I-V characteristic curve is a key to analyse the working characteristics of a solar cell. The load I-V characteristic curve of a solar cell is shown in Figure 1.24. The parameters related to the load I-V characteristic curve include: (1) open-circuit voltage U_{oc}, (2) short-circuit current I_{sc}, (3) I-U characteristic curve, (4) maximum power P_{Max}, (5) voltage at the maximum power point U_m, (6) current at the maximum power point I_m, (7) filling factor FF, (8) PV conversion efficiency and (9) temperature characteristic of a solar cell.

1.5.5.2 Relations of the Open-Circuit Voltage and the Height of the P-N Junction Potential Barrier in a Solar Cell

A photovoltage is generated in a solar cell when it is irradiated by solar rays. Similar to a positive bias applied to a P-N junction, it reduces height of the potential barrier in the depletion zone and make it thinner. The photovoltage of the solar cell under an open-circuit condition is called as the open-circuit voltage U_{oc}, and the point marked as U_{oc} in Figure 1.24. As a key parameter of a solar cell in mV or V, it shows the voltage between its anode and cathode when the load circuit of the solar cell is disconnected. Theoretically, the maximum open-circuit voltage shall be the voltage of the built-in field of the P-N junction. It is mainly dependent on the band gap width of a semiconductor and the Fermi level. Series resistance and some other factors in a cell will reduce its open-circuit voltage.

Generally, the open-circuit voltage of a solar cell U_{oc} rises with the height of its potential barrier of the P-N junction U_D. The previous energy band diagramme of a P-N junction shows, U_{oc} is close to, but smaller than U_D at a strong light irradiation. As a result, a large band gap helps to improve the open-circuit voltage of a cell.

1.5.5.3 Short-Circuit Current of a Solar Cell I_{sc}

When a solar cell in solar rays is short-circuited, its P-N junction is at zero bias. In this case, the current of the solar cell is defined as the short-circuit current I_{sc}. It is also a key parameter for solar cells. All recombination of photo-generated carriers inside the solar cell reduces the short-circuit current. Ideally, the short-circuit current is equal

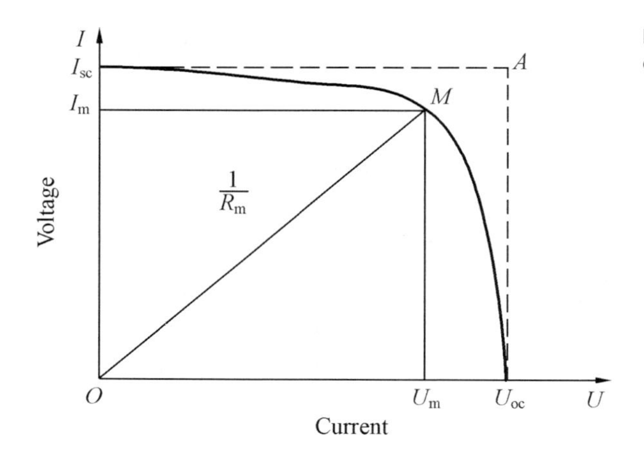

Figure 1.24 Load V-I characteristic curve.

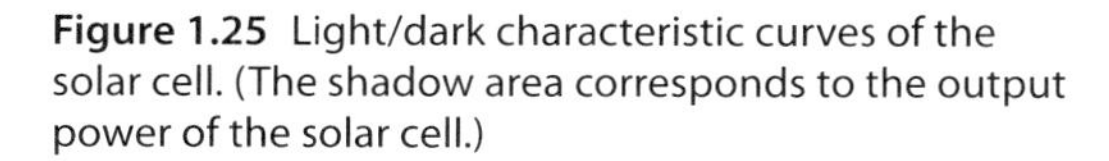
Figure 1.25 Light/dark characteristic curves of the solar cell. (The shadow area corresponds to the output power of the solar cell.)

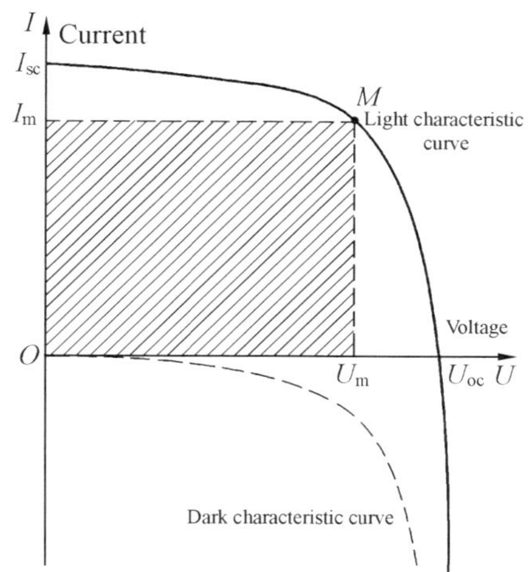

to the photo-generated current, and the point is marked as I_{sc} in Figure 1.25 with the unit in mA or A. The short-circuit current of solar cells per unit cross-sectional area is expressed in the short-circuit current density in mA/cm^{-2}, and so on.

Although the larger band gap width of semiconductors helps to raise the open-circuit voltage, it reduces the proportion of photons that can generate photo-generated carriers in the solar spectrum, and thus reduces the photo-generated current. Increase of the base zone's resistivity and decrease of the ionised impurity scattering can help raise the diffusion length and minority carrier lifetime, which increases the photo-generated current (See Figure 1.17).

1.5.5.4 The Optimum Operation Point of Solar Cells, the Optimum Operation Voltage and Current

Any point on an I-V characteristic curve of a solar cell can be called as an operating point. The line between the working point and the original point can be called as the load line, and the reciprocal of the slope of the load line is equal to R (R_m in Figure 1.24 is the load resistance of the available maximum power of a solar cell). The horisontal/vertical coordinates corresponding to the working point are the working voltage/current.

To better understand the working status of the solar cell, the light/dark V-I characteristic curves of the solar cell are plotted, as shown in Figure 1.25. The dark V-I characteristic curve is actually the V-I characteristic of the ordinary diode. The light characteristic curve will rise when the solar cell is in solar rays. In Figure 1.25, the shadow area corresponds to the output power when the solar cell is working (Formula 1.9).

When the load resistance R_L equals to R_m, we can find an optimum operating point of the solar cell on the curve (or the maximum power point) M, and the corresponding product of the operating current I_m and the voltage U_m is the maximum, that is, the maximum power point P_{max} is:

$$P_{max} = I_m U_m \tag{1.9}$$

The so-called optimum operating voltage V_{op} and current I_{op} are the voltage and the current at the maximum output power point of the solar cell. When the solar ray becomes stronger, the I–V curve will rise sharply and the output power will become larger.

1.5.5.5 Filling Factor (FF)

The ratio of the maximum output power $I_m U_m$ to the product of the open-circuit voltage and the short-circuit current ($I_{sc}U_{oc}$) is called as the filling factor, which is also the ratio of the area of the quadrangle OI_mMU_m to the area of the quadrangle $OI_{sc}AU_{oc}$.

$$FF = P_{max}/I_{sc}U_{oc} = I_m U_m/I_{sc}U_{oc} \tag{1.10}$$

Where, I_m is the optimum working current, U_m is the optimum working voltage, P_{max} is the is the optimum output power, I_{sc} is the short-circuit current and U_{oc} is the open-circuit voltage. As the filling factor rises, the output power will approach to the limit power. FF serves as a key index to evaluate the output performance of a solar cell. Obviously, as FF grows, the shape of the curve will become more 'quadrate' and the output power will become larger. FF is a function of I_{sc}, U_{oc}, irradiation of the incident light and the load. Other parameters such as series resistance also affect the filling factor because the series resistance changes the short-circuit current as shown in Figure 1.26. In addition, the doping concentration in the base zone can also affect the filling factor.

1.5.5.6 Power Conversion Efficiency of a Solar Cell η

The power conversion efficiency refers to the percentage of the incident light that is converted to the electric energy, which can be expressed by Formula (1.11):

$$\eta = (U_m I_m/P_{in} S)^{*}\ 100\% \tag{1.11}$$

Where, P_{in} is the energy density of the incidence light, U_m is the maximum output voltage, I_m is the current at the maximum output power, and S is the area of the solar cell (note: If S is the area where the total area of the solar cell is deducted by the gate lines, the efficiency of the solar cell will be larger).

The power conversion efficiency is a key parameter of a solar cell, which can be affected by many factors, including the structure of the solar cell, the P-N junction characteristics, the material quality, the operating temperature of the cell and the like.

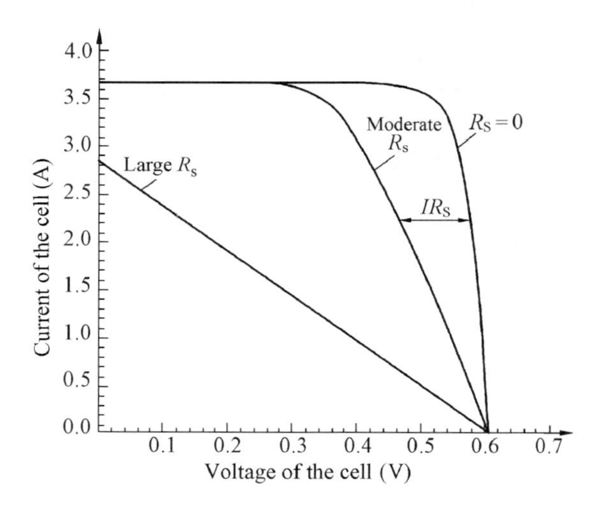

Figure 1.26 Influence of series resistance on the V-I curve of the solar cell.

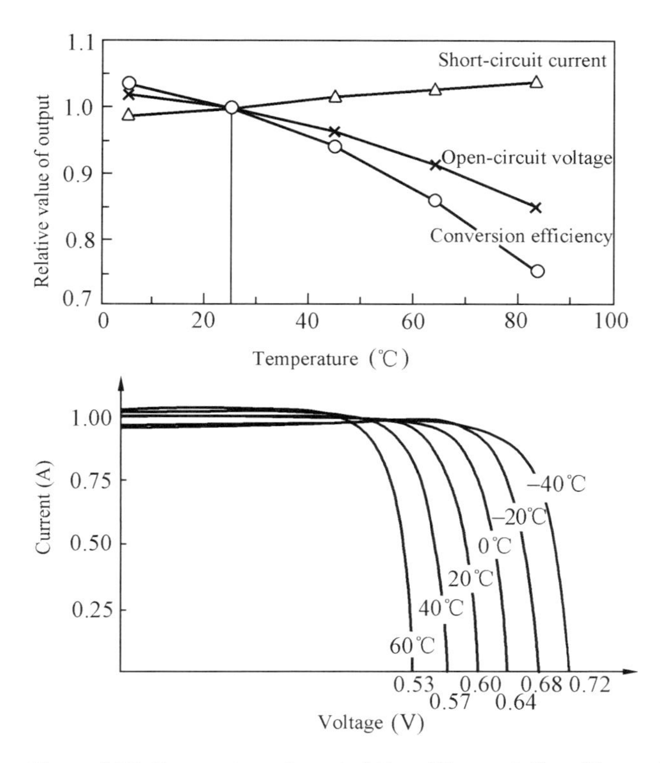

Figure 1.27 Temperature characteristics of the crystalline silicon solar cell.

1.5.5.7 Temperature Characteristics of a Solar Cell

Figure 1.27 shows the temperature characteristics of a single junction crystalline silicon solar cell. Obviously, as the temperature rises, the open-circuit voltage will fall while the short-circuit current will rise slightly but the open-circuit voltage will see a more obvious drop. As a result, the output power of the solar cell (directly affecting the efficiency) will fall as the temperature rises. Generally, the temperature coefficient of the conversion efficiency of the solar cell is negative.

1.5.6 Application of a Concentration Junction in a PV Cell for Back Surface Field (BSF)

It is known that when two pieces of semiconductor with the same conduction type but different doping concentrations (e.g., a P/P^+ junction) are in close contact, will form an electric dipolar layer will be formed due to a presence of an ionised impurity in the higher concentration layer with the majority carriers entering the lower concentration layer, which generates a built-in field and a potential barrier Ug. This potential barrier is called as a concentration junction. In a production process of solar cells, in order to prevent minority carriers from recombination in electrode areas, a P^+ layer is generally added to the back electrode side of a N^+P junction to form N^+PP^+, that is, to stack the P/P^+ concentration junction and the N^+/P to build a back surface field (BSF), which will reflect the minority carriers (electrons) at the bottom of the P-type layer to the P-N junction and increase the number of electrons reaching the P-N junction. The BSF structure can improve the cell's response in the long wavelength range and increase the photo-generated current, which is because the crystalline silicon has low absorption coefficient for the light with long wavelength and the light can reach the bottom of the cell and generate electron-hole pairs there. It shall be pointed out, the P/P^+ layer can increase the photo-generated current and the open-circuit voltage and improve the ohmic contact.

The BSF application has following advantages:

(1) It can reduce recombination of photo-generated minority carriers and the dark current and increase the photo-generated current. It can reflect the photo-generated minority carriers reaching the bottom surface back to be recollected by the PN junction. The BSF is mainly applied to thin cells and the high material resistance (e.g., resistance > 0.5 Ω cm) cases. If the minority carrier (electron) diffusion length is less than the base zone thickness the minority carriers reflected back will be recombined before reaching the P-N junction, the BSF will not work then in this case.
(2) The BSF can be stacked with the P-N junction, which increases the open-circuit voltage.
(3) It can improve the ohmic contact between electrode metals and semiconductor materials, which help reduce the series resistance, and improve the filling factor too.

Figure 1.28 shows the influence of the carrier recombination speed near the cell's back surface on short-circuit current, open-circuit voltage and FF of a solar cell. Obviously, as the recombination speed rises, the short-circuit current and the open-circuit voltage

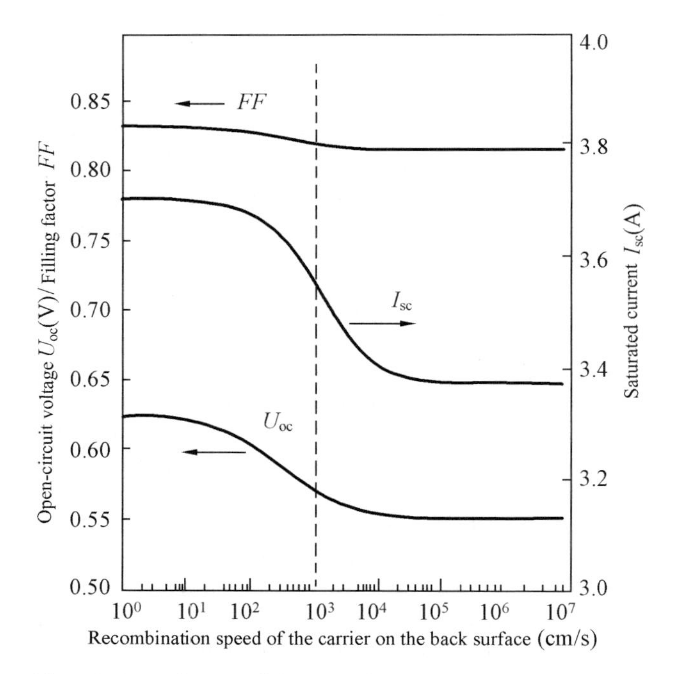

Figure 1.28 Influence of the recombination speed of the carriers of the cell back surface on the solar cell performance.

will be significantly reduced. Therefore, reducing carrier recombination near the back surface is significant to the solar cell's performance.

1.5.7 Basic Structure of Homogeneous P-N Junction Crystalline Silicon Solar Cells and Analysis on the Cell's Efficiency

Figure 1.29 shows an outline and the basic structure of a traditional silicon solar cell. The substrate is P-type monocrystalline silicon, and on top of it is a N^+-type of silicone layer to form a P-N^+ junction. On the top surface grid metal electrodes are formed and the back surface is a metal-based electrode. The top/bottom electrodes are in ohm contact with the N^+ and P layers, respectively, and the whole top surface is covered evenly by anti-reflection coating.

The sunlight has a very broad spectrum, and thus it is impossible for a single cell to convert all the sunlight into the photo-generated current. In the solar spectrum, the wavelength longer than 1.1 μm ($h\nu < Eg$) accounts for about 23%, which cannot be used by crystalline silicon solar cells to generate electron-hole pairs and thus becomes heat eventually. For the short wavelength part, since one photon can only generate one electron-hole pair and the rest can only be converted to heat, the surplus energy is called as the short wave loss, which accounts for more than 30%. For the traditional crystalline silicon solar cells, only a part of the solar radiation energy is actually useful. Figure 1.30 shows the spectral response of a monocrystalline silicon solar cell at AM1.5.

A lot of work has been done to improve the solar cell efficiency. The avoidable losses in PV conversion lie in the optical losses and the electrical losses. The optical losses refer to the reflection on the wafer surface, the shading losses of the front electrode and the spectral response mismatching of the cell materials and the like. The electrical losses refer to the carrier losses (recombination) and the ohmic losses (electrode-crystaline contact), and so on (see Figure 1.31).

The application of single-junction crystalline silicon solar cell over the past years shows that the maximum efficiency at AM1.0 has not been larger than 25%. It is generally recognised in the industry that when the efficiency of the solar cell rises by

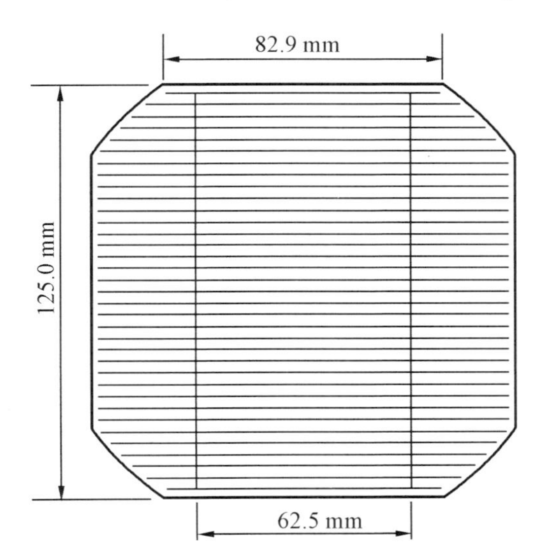

Figure 1.29 Outline and basic structure of silicon solar cells.

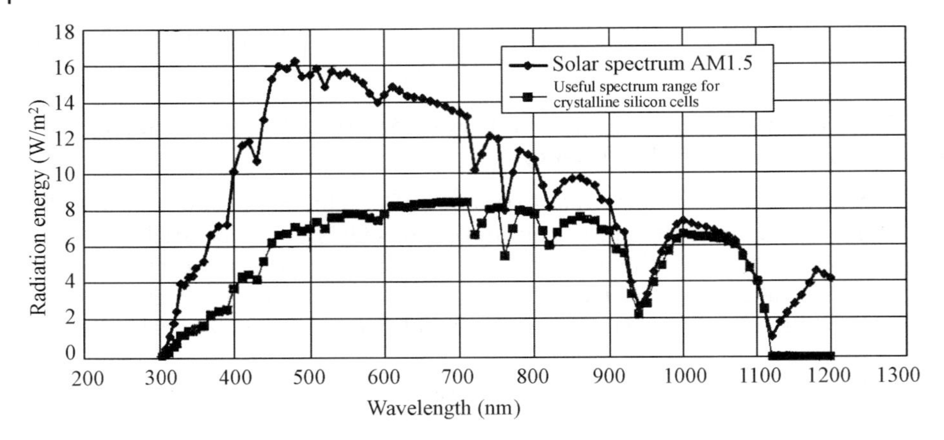

Figure 1.30 Spectral response of a monocrystalline silicon solar cell.

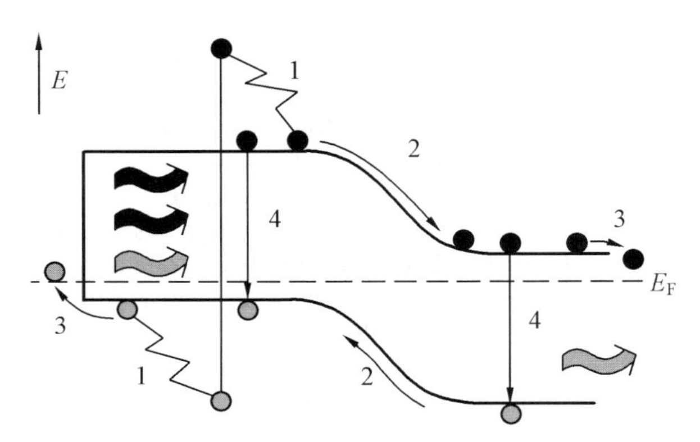

Figure 1.31 Energy losses of traditional solar cells: (1) Grid heat vibration losses; (2) Voltage losses of P-N junctions; (3) Voltage losses of Ohm junction contact; (4) Recombination losses.

1%, the total generation cost can be reduced by 7%. Therefore, researches have been committed to explore approaches to improve solar cell efficiency. For optical utilisation, a solar cell shall absorb to the maximum extent and make a full use of solar photons. Accordingly, the focus shall come to widen the spectral response of the cell, and reduce the optical reflection losses on the irradiating surface and the transmittance losses on the back surface. For PV conversion, the electron-hole pairs generated by the photons entering the cell shall be all collected at both ends of the electrode and then form the current. In consequence, it shall reduce the voltage or current losses due to various recombination of the photo-generated carrier during transmittance. For the crystalline silicon solar cell, the following actions can be taken:

(1) Preparation of texturing surface: The irradiating surface of the solar cell shall be made in a conic structrue like a pyramid so that the light can be reflected many times on the wafer surface. It can reduce the reflection losses and change the forward direction of the light in the silicon to raise the light course (raising the production ratio of the photo-generated carrier). In addition, the uneven texturing surface can increase the area of the junction and further raise the collection rate of the photo-generated carrier, increasing the short-circuit current by 5–10%.

(2) Preparation of anti-reflection coating on the irradiating surface: The anti-reflection coating can be installed on the irradiating surface of the solar cell to increase the photons entering the cell and thus raise the short-circuit current.
(3) Preparation of the light reflection layer on the shading surface: The metal electrode with smooth surface can be used on the shady surface of the cell to reflect the light with long wavelength and raise the long-wavelength optical response and the short-circuit current of the cell.
(4) Annealing and impurity gettering: The appropriate actions such as thermal annealing, hydrogen annealing, laser annealing or impurity gettering and the like can be taken to improve the lifetime of the minority carrier in the cell zones and raise the photo-generated current and photonvoltage.
(5) BSF: The bottom of the solar cell can be built into a concentration high/low junction, that is, introducing the $P\text{-}P^+$ junction to the traditional N^+-Pcell to form a N^+PP^+ cell, which can improve significantly the short-circuit current, the open-circuit voltage and the conversion efficiency of the cell.
(6) The quality semiconductor material of long diffusion length shall be used.
(7) The thickness of the silicon wafer shall be minimised to reduce recombination of the photo-generated carriers before reaching the electrode.
(8) For the solar cell in multi-layer structure, it shall consider the optical absorption characteristics of the intermediate layers in the cell and the reflection and transmittance characteristics at the interfaces. (When the solar cell is assembled to module, it shall also take into account the reflection and transmittance characteristics of the upper cover glass and the coating of the cell assembly.)

Based on the texturing technique, the BSF technique and the optical inner reflection technique as well as the measures to improve the life of the carriers and reduce the recombination rate on the surface, an ideal model of single junction silicon solar cells can be built: It has a thin, slightly doped irradiating surface layer ($<2\ \mu m$), a narrow depletion zone (0.05–0.06 μm) and a thick base zone of long diffusion length (50–100 μm), and the recombination rate on the positive surface is zero. The front is built with etching structure, and the back with optical inner reflection layer. The electrode and wafer of the irradiating and shady surfaces are ingood ohm contact, and the shady area of the grid electrode on the irradiating surface is minimised. To achieve high U_{oc} and U_m, the doping concentration in the P/N zones shall be all less than the limit doping concentration so as to achieve the maximum efficiency.

Based on the generation principle of the solar cell, we can carry out analysis, judgment and inference according to the actual testing data of the solar cell.

For example, if the open-circuit voltage and the short-circuit current are all relatively low, the reasons may be that the series resistance is big or the diffusion length of the carrier in the cell is insufficient. If the measured spectral response of the target cell in the short-wave range is poor, it may be caused by excessively high doping concentration in the window zone, which can result in excessively short diffusion distance of the minority carrier in the short-wave range. If the measured spectral response of the target cell for the infrared light and the near infrared light are poor, it is probably because the diffusion length of the minority carrier in the base zone is insufficient and the carrier generated in the base zone cannot reach the P-N junction. If the filling factor FF is above 0.75, it is accepted for the commercialised solar cell.

1.6 Solar Cells of Heterojunctions

1.6.1 Composition of Heterojunctions

The heterojunctions consist of the abrupt and the graded junctions. If the transitional layer of two materials has the thickness less than 1μm, it is called as an abrupt junction; and if the transitional zone covers the diffusion length of several minority carriers, it is called as a graded junction. Nowadays, the junctions applied to the solar cell are mostly the abrupt ones. Based on the conduction type of the two semiconductor materials forming the heterojunction, it can be divided into two types: the inverted junction and the identical junction. For example, in the thin-film CdTe (cadmium telluride) solar cell, the P-type CdTe and the N-type CdS (cadmium sulphide) will form a inverted heterojunction. In the symbol for heterojunctions, the material with small forbidden band gap is generally written in the front or on the top. In the solar cell with heterojunctions, the material with small forbidden band gap generally serves as the base zone, and the material with big forbidden band gap generally serves as the top zone (the irradiating surface). This is to reduce the light absorption of the top zone and make the base absorb more photons. As a result, the 'top material' is also called as the 'window material'. Figure 1.32 shows the equilibrium energy band diagramme before and after the heterogeneous P and N materials form the abrupt P-N heterojunction at the contact position.

1.6.2 Construction and Working Principle of the Solar Cell with Heterojunctions

The heterogeneous solar cell has similar outline to that of the homogeneous solar cell except that its base and window zones are made of two different materials. When the P-type base zone material is coated with a layer of N-type window zone material, a heterojunction will be formed at the interface. A grid metal electrode can be built on the window zone and the metal electrode can be installed on the back surface of the base zone to form the heterogeneous solar cell. To reduce the reflection losses, the

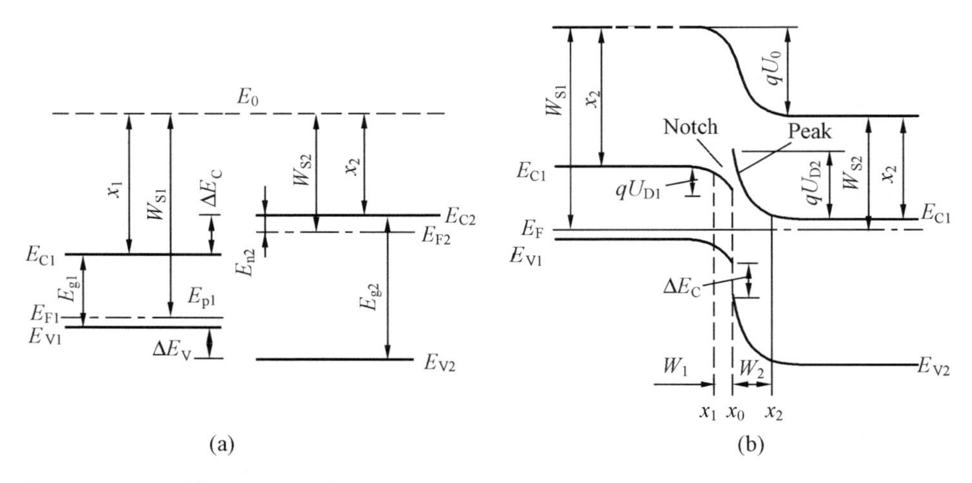

Figure 1.32 Equilibrium energy band diagramme before/after the abrupt P-N heterojunction is formed: (a) before contact (b) after contact.

window zone shall be also coated with a transparent anti-reflection layer, and a BSF shall be added to the back surface of the base zone.

When the heterogeneous solar cell is in the solar rays and the rays transmits the reflection film, the photons in the spectrum ($h\nu \geq E_{g1}$) will first be absorbed by Material 1 to generate the photo-generated electron-hole pair, and the light not absorbed will march forward, and the photon of Material 2 ($h\nu \geq E_{g2}$) will be absorbed to generate the photo-generated electron-hole pair. Because the depletion zone of both Materials 1 and 2 falls at the heterojunction interface X_0, the photo-generated electron-hole pair generated will be immediately separated by the built-in field where the holes will be swept to Zone 2 and the electrons to Zone 1. Accordingly, both sides of the heterojunction will be accumulated with photo-generated charges, generating the photovoltage and the PV effect of the heterojunction. When a load is connected between the top and the bottom electrodes, the photo-generated current flowed from the bottom electrode will generate voltage on the load and output the power. Obviously, the working principle of the heterogeneous solar cell is almost the same with that of the homogeneous solar cell. Figure 1.33 shows the energy band diagramme of the heterojunction solar cell.

Since the short-wave part of the spectral response of the heterojunction is dependent on Material 1 and the long wave part on Material 2, the combinations of materials with various Eg_1, $\alpha_1(\lambda)$, Eg_2, $\alpha_2(\lambda)$ and thickness can be selected to make the spectral response curve better match with the solar ray.

Since Zone 1 (the window material) is generally thin, the main components of the photo-generated current is supplied by Zone 2 (the base zone) with long lifetime minority carriers ($Eg_1 > Eg_2$). In fact, since the window material has big Eg, small concentration of equilibrium minority carriers and short lifetime of minority carriers, the window material can collect few photo-generated carriers and the main spectral response is still dependent on the nature of the base zone.

The photonvoltage of the heterogoneous junction solar cell is dependent on the energy band structure but the maximum open-circuit voltage cannot exceed the built-in electromotive force U_D. Since the lattice matching issue is present in the interface of the heterojunction, the dark current is very complicated and difficult to determine, and

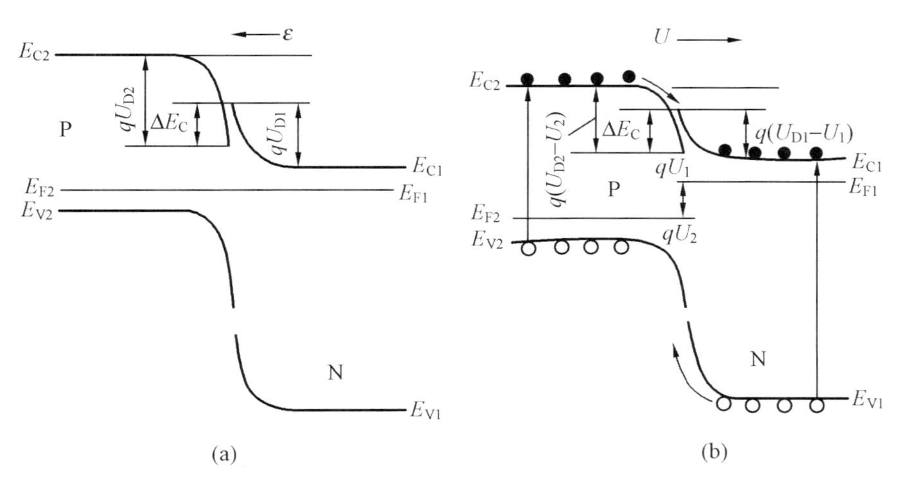

Figure 1.33 Energy band diagramme of the heterojunction solar cell (a) in the solar ray; (b) without solar ray.

the computation formula for the open-circuit voltage of the homogeneous junction is only suited to the ideal heterojunction solar cell with good latticematching and few interfacestate. If the heterojunction is provided with BSF, it will, similar to the homogeneous junction, reduce recombination of minority carriers in the base zone and the dark current, and raise the open-circuit voltage and decrease the contact resistance via additional inner reflection.

Compared with that of homogeneous junctions, the solar cell with heterojunctions has the following significant virtues: 1) The material of wide forbidden band is used as the window layer to make more light transmit the base layer. If the window layer is very thin and the material of the base layer is selected properly to achieve adequate lifetime of the minority carrier and the appropriate energy band gap, the solar rays can be sufficiently absorbed in the base zone; 2) Big energy band gap and high doping can reduce the resistance of the surface thin layer and the series resistance of the cell (in the condition that no high doping effect occurs); 3) The energy band of two and more materials can be used to adjust the spectral response range of the heteorogeneous junction solar cell and it is easy to use the material of variable forbidden band gap for preparation of solar cells.

As for the factors influencing the efficiency of the heterojunction solar cell, some are the same with that of the homogeneous junction solar cell and some are different, which are described below:

1. In addition to the surface reflection, less than 3–4% interface reflection losses may be present due to different refractivity of the two materials on the interface of the heterojunction.
2. For the solar cell of homogeneous junctions, the dark current has generally the following order for the three components: diffusion current (or injection current) > recombination current > tunnel current. For the solar cell of heterojunctions, however, the order is inversal, especially the tunnel current and the recombination current, which have close relations to the material of the cell and the production-process and have the maximum impact on the output characteristics of the cell and thus become the main reason for low efficiency.

2

Materials of Solar Cells

2.1 Low-Cost Solar-Grade Polycrystalline Silicon

2.1.1 Polycrystalline Silicon—The Most Important Raw Material of the PV Industry

Before 2002, the solar-grade polycrystalline silicon was mainly supplied from the scraped and rejected products of the semiconductor electronic-grade polycrystalline and monocrystalline silicon, as shown in Figure 2.1.

Over the past 10 years, the PV industry has developed at an annual growth rate of about 30%, which is far more than that of the semiconductor industry (5–6%). Since the proportion of silicon used by the PV industry has been gradually increased, some quality electronic-grade polycrystalline silicon must be used. Since 2006, the proportion of the solar-grade silicon in the polycrystalline silicon has exceeded that of the electronic-grade silicon (EGS) across the globe, becoming the largest market of the polycrystalline silicon. The silicon demand of solar cells has rapidly grown, which has become the main force for the global silicon market growth. At present, the silicon-based solar cell accounts for about 90% of the global solar cells (monocrystalline silicon solar cell: about 35%; polycrystalline silicon solar cell: about 50%; thin-film silicon solar cell: about 5%). In the silicon applications across the globe, 80–90% is currently used for the solar cells, and the proportion will rise in the future. Obviously, it has completely changed the situation that the polycrystalline silicon for solar cells was mostly from the scraped and rejected products of the semiconductor electronic-grade polycrystalline and monocrystalline silicon a decade ago.

Over the past decade, the output of the polycrystalline silicon across the globe has grown significantly due to development of the PV industry, as shown in Table 2.1. Take mainland China as an example. Its annual output of polycrystalline silicon was only 80 t in 2001, the output rose to 100 t in 2006, 1000 t in 2007 and 10000 t in 2009, listed to the giants of polycrystalline silicon in the world. Table 2.2 shows the polycrystalline silicon output of mainland China in 2001–2013, and Table 2.3 shows the comparisons between the polycrystalline silicon output of mainland China and the world in 2009–2013. The polycrystalline silicon output of mainland China fell in 2012 due to the world economic and trade friction, but it picked back rapidly soon after that. In the end of 2013, the annual output of polycrystalline silicon in China exceeded 80,000 t, which accounted for one-third of the world total output and nearly doubled compared with that in 2010.

Technology, Manufacturing and Grid Connection of Photovoltaic Solar Cells, First Edition. Guangyu Wang.

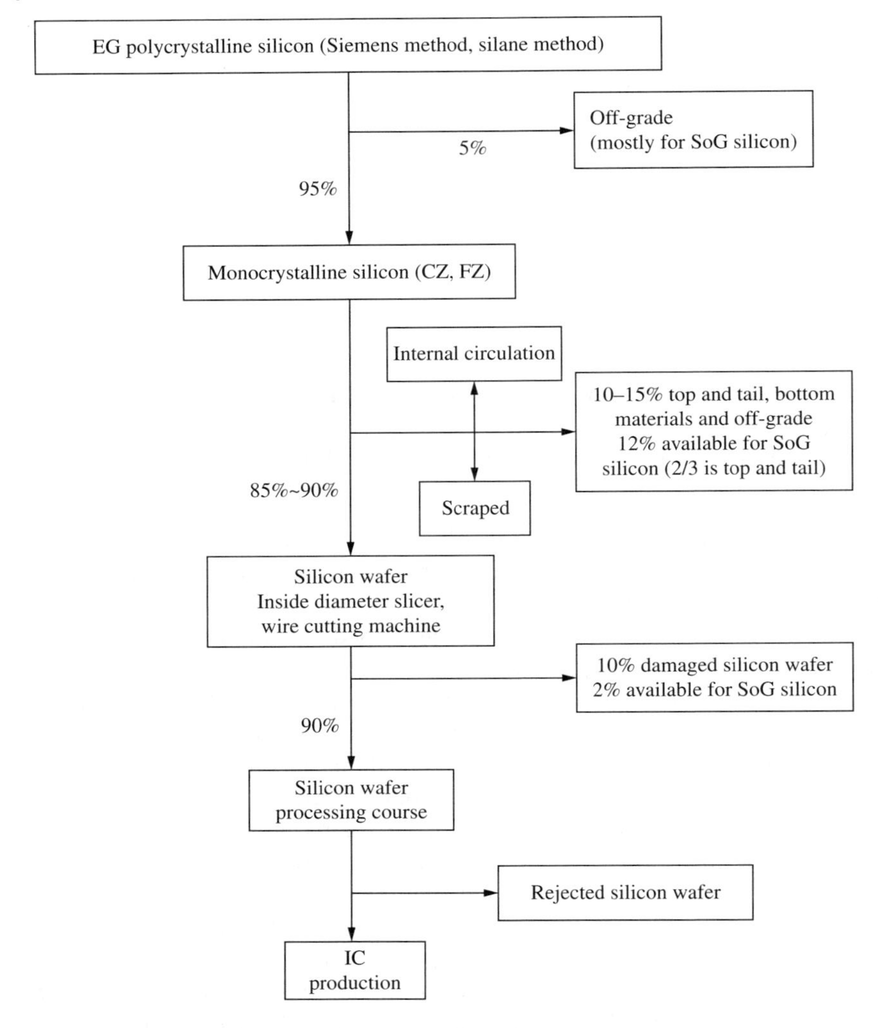

Figure 2.1 Sources of solar-grade polycrystalline silicon in 2002.

The production capacity will exceed 130,000 t in 2014 under the push of the technological support programme of the 'Eleventh-Five-Year Plan' and the 'Twelfth-Five-Year Plan'. This is because the crystalline silicon PV cell is suited for industrialisation. In the upcoming 10–20 years, the crystalline silicon cell will play a predominating role in the market.

The market price of the polycrystalline silicon was up to USD400$/kg in 2008, which reduces to USD20$ nowadays. This has also significantly reduced the cost of the polycrystalline silicon module since 2008. As the new cells, for example, the thin film and the compound cells, come into existence, the demands for various PV cells will coexist. This, together with the PV conversion efficiency, cost, lifetime and flexibility, and so on, will become the key challenges for technical development and industrialisation.

Table 2.1 Output (10,000 t) and applications of polycrystalline silicon across the globe, 2005–2015.

Year	2005	2006	2007	2008	2010	2011	2013	2015(E)
Electronic grade	1.8	1.9	2	2.15	2.4	2.56	2.87	3.23
Solar grade	1.4	2.2	3	4.39	9.6	12.44	22.13	36.77
Total	3.2	4.1	5.0	6.5	12	15	25	40

Table 2.2 Output of polycrystalline silicon (t), mainland China, 2001–2013.

Year	2001	2002	2003	2004	2005	2006	2007	2008	2009	2010	2011	2012	2013
t	80	76.8	71.5	57.8	80	287	1139	4685	20357	46000	82768	63500	80000
Growth %		−4	−7	−19	38	257	297	311	335	126	80	− 23	25

Table 2.3 Comparisons of polycrystalline silicon output of mainland China with the world, 2009–2013.

Year	2009	2010	2011	2012	2013	2014
World output	9.11	18.2	24.1	24.0	25.0	29.0
Output of China	2.03	4.60	8.27	6.35	8.0	13.0

The solar cell has seen rapid development and extended applications. To compete with the fossil-based power, the focus shall be put on generation cost and unit price in addition to achieve governmental policy support by its environmental advantages and world energy crisis. Although the silicon price has been significantly reduced in recent years, the cost of crystalline silicon still accounts for about 40% of the production cost of the silicon solar cell. As a result, the human being has been dedicated to research and industrialisation of the low-cost solar-grade silicon, which mainly includes the following:

(1) Improved Siemens method: The materials are in closed circulation to reduce material consumption and discharge fees; the reflux ratio for rectification and refinement is reduced, the number of silicon core pairs in the reduction furnace is increased and the operation pressure in process is raised to reduce the power consumption and raise the productivity.
(2) The fluidised bed reactor (FBR) polycrystalline silicon deposition technique is used, for example, the trichlorosilane or the silane gas is directly deposited on the silicon powder particles via chemical reaction. Since the silicon deposition surface is raised, it can reduce significantly the production cost and improve the production efficiency.
(3) The polycrystalline silicon can be refined by the metallurgical method. Instead of converting the industrial silicon to trichlorosilane, it can refine the industrial silicon to 6 N by the metallurgical method, which is often called as 'UMG (upgraded metallurgical grade) silicon' and can be directly used to produce the solar cell.
(4) Based on the working principles of the solar cell modules, the polycrystalline silicon, similar to the monocrystalline silicon, can be also used to produce the solar cell.

Although the efficiency of the solar cell is slightly lower than that of the monocrystalline silicon solar cell, it is unnecessary to be drawn to monocrystalline, and the polycrystalline silicon can be cast. Accordingly, the cost of raw materials can be dramatically reduced. In recent years, the cast process has been improved and the mono-like silicon can be grown, whose efficiency is similar to that of the monocrystalline silicon solar cell. Moreover, the oxygen content of the mono-like silicon is less than that of the CZ silicon.

(5) The thin silicon wafer is used. On one hand, the diamond wire of small diameter (<120 μm) is used to cut the monocrystalline silicon column or the polycrystalline silicon casting ingot in order to reduce the losses of silicon wafers. On the other hand, the cell production technique is improved, and the production line automation is increased to raise the finished product rate of the thin silicon wafer. At present, the thickness of the silicon wafer on the production line is generally less than 200 μm, and the crystalline silicon for every watt cell has been reduced from 13 g/watt in 2004 to the current 7 g/watt.

(6) Since the solar-grade silicon has the sources more complicated than the electronic-grade silicon and it contains more impurities, the cell manufacturers have developed the new ‘impurity gettering’ and ‘defect engineering’ in the component technique to tackle them. The silicon of slightly higher impurity content (the so-called ‘dirty silicon’) can be used, which can dramatically reduce the cost of the silicon raw materials and it is hopeful to promote the development of solar silicon cells.

The current supply chain of the solar-grade silicon material can be shown in Figure 2.2. The countries across the world are exploring the production method of low-cost solar-grade polycrystalline silicon (see Table 2.4).

2.1.2 Meaning of Solar-Grade Polycrystalline Silicon

The solar-grade silicon (SOG silicon) refers to the silicon material whose performance is qualified to produce the crystalline silicon solar cell. It contains two meanings: First, since the basic unit of the solar cell is the semiconductor P-N junction capable of generating PV effect, the solar-grade silicon must be able to create a built-in field in the P-N junction with a certain height. For the photo-generated carriers, the electrons and the

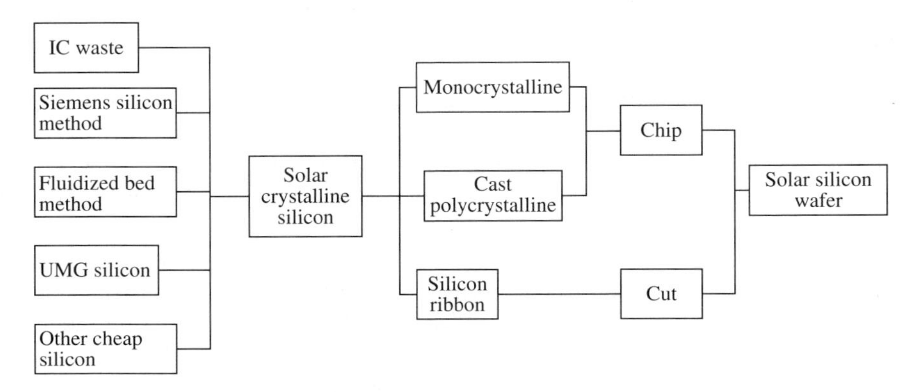

Figure 2.2 Industrial chain of the solar cell silicon.

Table 2.4 Countries and their production methods for low-cost solar-grade polycrystalline silicon in the world.

Company	Country	New techniques
Wacker	Germany	With trichlorosilane as the raw material, FBR is used to produce particle polycrystalline silicon
Elkem Solar	Norway	Metallurgical grade silicon → slag making → crashing → grinding → chemical leaching → directional solidification
Dow Corning	USA	After slag making and smelting, the metal is subjected to water vapour reaction to remove Boron and phosphorus impurities, and HEM furnace is used for directional solidification
Hemlock	USA	FBR
Invensil	France	Metallurgical method: plasma refinement
Steel	Japan	Electron beam, plasma metallurgical techniques, and directional solidification
MEMC	USA	With silane as the raw material, FBR is used to produce particle polycrystalline silicon
NTNU SINTEF	Norway	Molten electrolytic process
REC Silicon	Norway	FBR polycrystalline silicon deposition technique
Tokuyama	Japan	VLD: The gaseous trichlorosilane enters the heated graphite pipe from the top of the reactor, and the hydrogen reduces the liquid silicon drip, which can be continuously produced
Ferro Atlantic (FA)	Spain	Metallurgical grade silicon → oxygen blowing → solid/liquid separation and solidification → slag making → directional solidification
Chisso	Japan	Zinc is reduced to SiCl4

holes will gather in the N/P-type zones, and the band structure of the P-N junction will be bent, generating the photovoltage to be used externally. Since the photovoltage can be formed only after the photo-generated carrier reaches the P-N junction, the solar-grade silicon shall be the silicon material with sufficient purity, and impurities in the silicon, especially those that can affect the diffusion length of the photo-generated carriers, shall be removed to a certain degree. Second, since the solar cell is intrinsically a diode and it is demanding on frequency and voltage withstand performance, the requirements of the solar-grade silicon on impurities are generally less than that of MOS and bi-polar devices. Refinement of silicon materials shall focus on removing the impurities that can affect the diffusion length of the photo-generated carrier and the resistivity.

The PV effect of a semiconductor P-N junction serves as the basis of the working principle of the crystalline silicon solar cell. When the solar rays radiate on both sides of the P-N junction, the non-equilibrium minority carriers (photo-generated electron and hole pairs) will be generated. These electrons and holes, under the action of the built-in field of the P-N junction, will move towards both sides of the P-N junction, forming the photo-generated field. Having deducted the built-in field of the P-N junction, the rest of the photo-generated field will become the PV electromotive force. The heavy metal and the defects become the recombination centre of the minority carrier, which can obviously reduce the minority carrier lifetime and further the PV conversion efficiency. The influence of the metal impurity content on the solar cell conversion efficiency was reported at the international academic conference, as shown in Figure 2.3. And it is known as the Westinghouse curve. They believed it has great influence over the cell efficiency when the concentration of Mo, Ta, Nb, Zr, W, Ti, Fe, V and so on, falls in a range of 10^{13}–10^{14} atom/cm^3; it has obvious influence on the PV conversion efficiency when the Mo content reduces to 10^{12} atom/cm^3; it has influence on the cell efficiency only when the concentration of Ni, Al, Co, Mn, and Cr is above 10^{15} atom/cm^3; it has influence on the cell efficiency only when the concentration of P, Cu is up to 10^{18} atom/cm^3. Although the modern study has made corrections to the influence of some impurities, it is certain that the various impurities in the silicon have different influences on the solar cell conversion efficiency and it is unnecessary and also impossible in the actual production to remove all the impurities in the silicon. To reduce the production cost of the solar cell, the production methods have been developed for low-cost solar-grade polycrystalline silicon.

It is said that the solar power can be viewed as clean energy, safe energy, portable energy, poverty-alleviating energy, desert-based energy, building material energy,

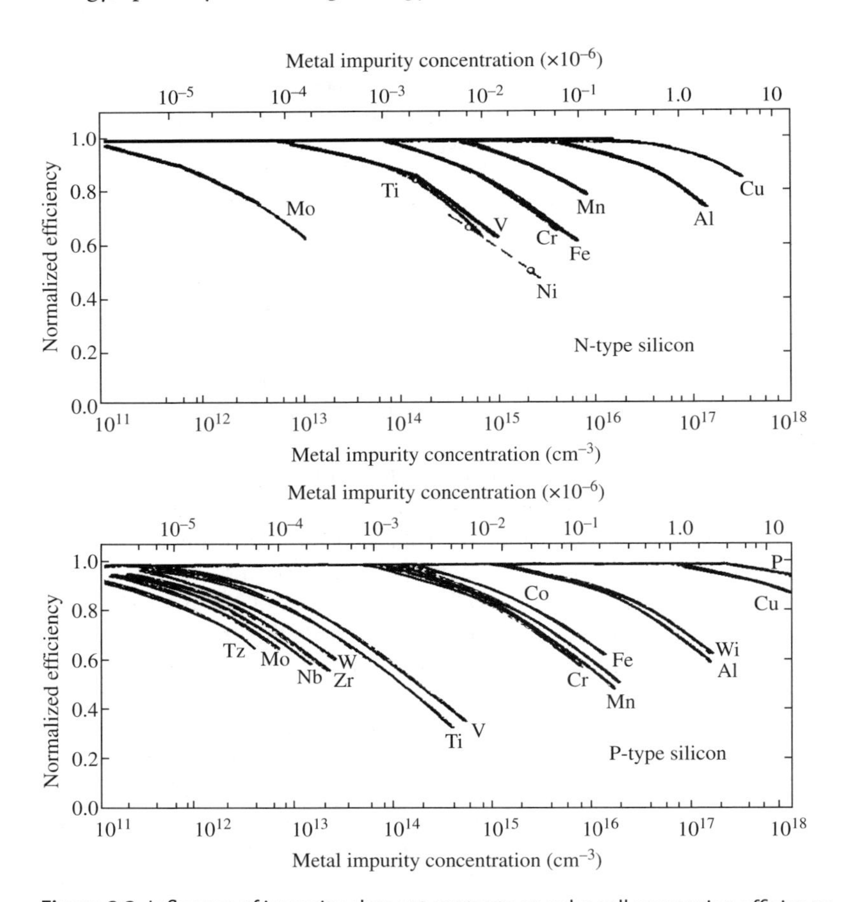

Figure 2.3 Influence of impurity element contents on solar cell conversion efficiency.

emergency energy, peak-clipping energy, military energy and space energy. All in all, the solar power must become the most important strategic energy in the twenty-first century when the fossil energy becomes less and less. The key to tackle the challenge lies in how to achieve the low-cost solar cell materials with low energy consumption and low pollution. Over years, scientists and companies across the world have been exploring with all efforts and many methods have been developed. After the meaning and requirements of the solar-grade silicon are discussed, several promising methods will be introduced.

2.1.3 Preparation of Solar-Grade Polycrystalline Silicon (UMG Silicon) by Metallurgical Method

The polycrystalline silicon of upgraded metallurgical grade silicon (UMG silicon) has been applied to the PV industry for many years, and most of them are doped with the polycrystalline silicon produced by Siemens method. In 2008, the UMG silicon achieved great breakthroughs in applications to the solar cell. In June 2008, CSI Cells Co. Ltd., China-based, first used 100% of the UMG polycrystalline silicon produced by a Canadian company to make the solar cell in the world, which was accepted by the international leading manufacturers. In September, the same year, Ningxia Yinxing Polycrystalline Silicon Company announced, it had produced the cell with the UMG independently developed by itself and built the first 330 kVA HV PV grid-connecting power station in China that is made of UMG polycrystalline silicon. Following it, many companies both at home and abroad have developed the UMG silicon for solar cells (see Table 2.5), winning wide attention for the technology and product. The efficiency of the cell made of UMG polycrystalline silicon has also achieved great improvement in recent years. It is reported that the conversion efficiency of the solar cell was 10–11% in 2007, and it could reach 15–17% and even up to 18% after light attenuation in 2009. This, of course, also owes to the improvement of the solar cell production technology.

The so-called 'metallurgical refinement' to produce polycrystalline silicon shall be viewed as the 'method to remove impurities in the metallic silicon state'. Compared with the Siemens method, the base silicon has no phase change, and the means to remove impurities are limited, that is, the process window is small. As a result, when the ore (SiO_2) is reduced to produce industrial silicon, the SiO_2 ore of low phosphorus/boron and the carbonaceous reducing agent with low ash content should be selected. The reductant charcoal and coke generally used in industrial silicon production contain high

Table 2.5 Main techniques and purity of UMG silicon manufacturers both at home and abroad.

Manufacturer	UMG silicon purity	Main techniques
Timmin Co., Canada	$5N^+$	Oxygen blowing, slag making, directional solidification
JFE, Japan	6 N	Ion beam, electron beam
Elkem, Norway	$6N^-$	Pickle, wet-process metallurgy
Yinxing, China	$6N^-$	vacuum plasma, wet process
Jinxun, China	$6N^+$	slag making, solid/liquid equilibrium

ash content (charcoal: 3–5%; coke: 2%), it is recommended to use the petroleum hydrocarbon resin, the by-product of the petroleum industry (only containing several ppm of phosphorus).

The metallurgical refinement can be based on the following basic principles:

1) Vacuum method (to volatilise the impurity of high vapour pressure by means of the vapour pressure difference between the impurity and the silicon);
2) Acid washing pickling (segregation and dissolution of impurities as the silicon compound at the crystal boundary);
3) Extraction of impurities at the interface of the molten silicon and the slag (segregation of impurities at the interface of the molten silicon and the slag);
4) Segregation of impurities at the solid/liquid interface (Difference of the segregation coefficients of impurities at the interface of the molten and solid silicon);
5) Gas blowing in silicon molten pool (the gas such as oxygen, chlorine, argon, hydrogen or vapour and so on, reacts with the impurities in the silicon liquid).

Based on the above principles, many methods for silicon refinement can be developed. It shall be noted that none of the following methods is effective to all impurities. Accordingly, they shall be jointly used, especially for the impurities with equilibrium distribution coefficient close to 1, which are more difficult to remove.

Method I: Vacuum method: The variation of the vapour pressure of some impurities in the silicon with temperature rise is listed in Table 2.6. Obviously, above the silicon melting temperature (1683 K), impurities such as phosphorus, aluminum, calcium and the like have higher vapour pressure and they are volatilised to the gas phase for removal.

Method II, pickle: Based on the metallurgical principle, when the molten silicon is cooled down and solidified, it is apt to enrich Fe and other impurities at the crystal boundary (known as 'impurity segregation' in the metallurgy). As a result, the metallurgical grade silicon of purity about 2 N can be smashed into small particles by mechanical means (cleaved along the grain surface if available), and then soaked for hours in hydrochloric acid, nitric acid and hydrofluoric acid to dissolve the metal impurities enriched at the 2 N silicon grain interface by acid and improve the purity of the metallurgical grade silicon from 2 N to 4 N. The method, however, may result in environmental pollution due to waste acid emission.

Table 2.6 Variation of vapour pressure of various elements with temperature rise Unit: Pa.

Temperature (K)	1723	1773	1823	1848	1873
Si	0.087	0.189	0.392	0.556	0.781
Al	1222.915	1545.390	2464.762	3594.471	6034.388
Ca	80268.70	----	----	----	----
Fe	0.724	1.508	2.735	3.764	5.136
Cr	2.740	5.759	11.599	16.223	22.484
Ti	0.016	0.038	0.089	0.133	0.197
B	8.56E−05	2.48E−04	6.78E−04	1.10E−03	1.76E−03
P	19413.970	18289.200	34532.150	45657.120	59914.430

Method III, slag making. First, based on the relationship of the free energy of oxide formation with temperature, find out the slag system in the silicon possibly capable of removing some impurities. For example, the boron-removing reaction in the slag system Na_2O-CaO-SiO_2 may be $2[B] + 3(Na_2O) = 6[Na] + (B_2O_3)$; then, based on the ternary phase diagramme of Na_2O-CaO-SiO_2, study the variation rule of the fuse point, the specific weight and the viscosity of the slag system to determine the appropriate kinetics reaction conditions and the metallurgical process for impurity (boron) removal. The experimental result shows, the slag system has good impurity (boron) removal effect (as shown Table 2.7).

The relationship of the free energy of formation of oxide with temperature, including Na_2O, CaO, SiO_2 and B_2O_3, is shown in Figure 2.4. And the ternary phase diagramme of Na_2O-CaO-SiO_2 is shown in Figure 2.5.

BaO-CaO-SiO_2 slag system has also been used to remove the impurities in silicon.

Method IV, segregation of impurities at the solid/liquid interface. Table 2.8 shows the solid/liquid segregation coefficient of the impurities at the solid/liquid interface when the molten silicon condensates. Because the solid/liquid segregation coefficient of Fe,

Table 2.7 Experimental results of boron removal by slag making unit: ppmw.

Content of impurity	B	Al	Ca
Raw material [ppmw]	9	28	16
Product [ppmw]	1.9	0.1	1.8

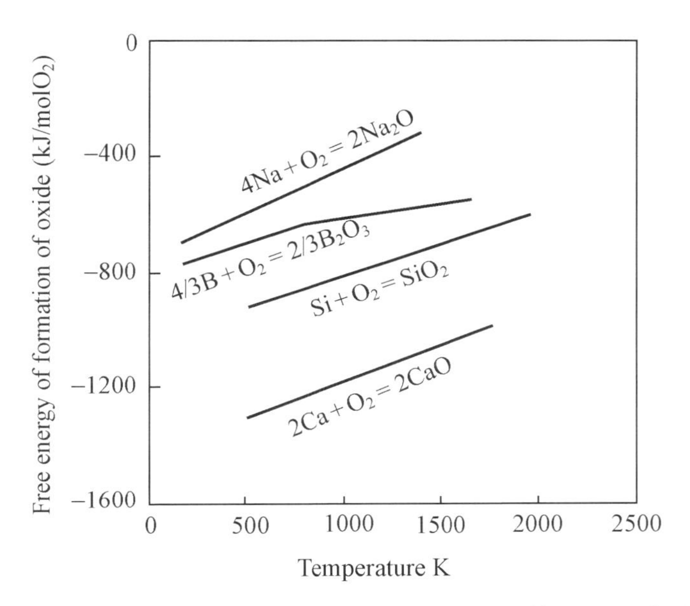

Figure 2.4 Variation relationship of the free energy of formation of oxide with temperature, including Na_2O, SiO_2 and B_2O_3 and so on.

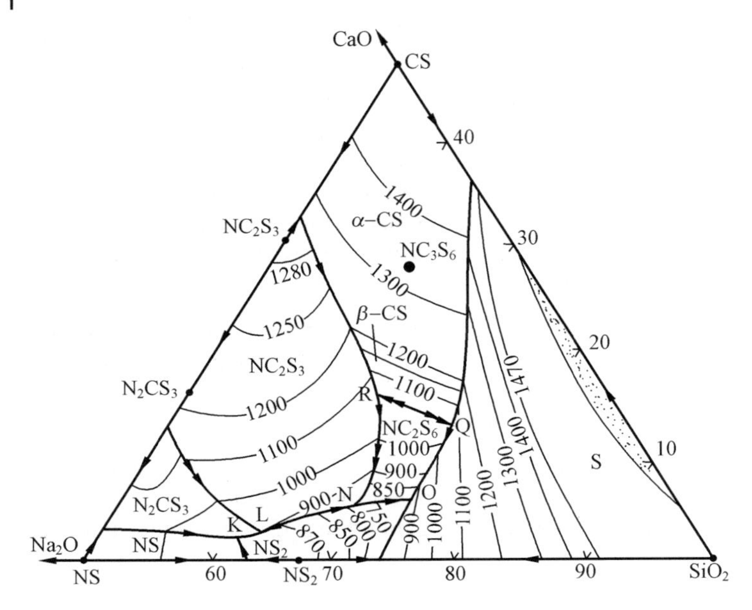

Figure 2.5 Ternary phase diagramme of Na_2O-CaO-SiO_2.

Table 2.8 Solid/liquid segregation coefficients of various impurity elements in silicon.

Impurity	B	P	Al	Fe	Ti	Ca
Segregation coefficient	0.8	0.35	0.0028	0.0000064	0.000002	0.008

titanium, aluminum and calcium is far less than 1, they can be effectively removed during directional solidification. The method can remove some phosphorus but it cannot remove boron effectively.

Method V, gas blowing in silicon molten pool: At the appropriate stage during metallurgical silicon refinement, inject some massive gases (e.g., oxygen, chlorine, argon, argon and water) to the silicon molten pool to make them react with some impurities in the silicon liquid. For example, the oxygen and chlorine can form oxide and chloride with calcium and aluminum in the industrial silicon and then float into the slag; the argon can remove most of the carbon and oxygen; and hydrogen + water (wet hydrogen) can remove boron.

Below is the simple production flow of solar-grade polycrystalline silicon in the metallurgical method: Strictly select SiO_2 ore and the reductant; after SiO_2 is reduced to silicon liquid in the submerged arc furnace, blow gas into the silicon liquid and then agitate and segregate; make slag in the induction furnace to remove impurities and after slag removal, insulate and anneal; finally, carry out directional solidification.

Compared with the high-purity polycrystalline silicon produced by Siemens method, the physical method has the following advantages:

1. The power consumption is only one-third that of the Siemens method;
2. The construction investment is only one-tenth that of the Siemens method;

3. The construction period is only half that of the Siemens method;
4. The environmental pollution is smaller than the Siemens method;
5. The product cost is only a half the Siemens method.

The metallurgical method, however, is facing the following problems:

1. The nonstandard equipment is generally used in the production process due to technical confidentiality;
2. Since the process window is narrow and the production is intermittent (in batches), it must carry out strict on-line monitoring and management to ensure the stability of products;
3. Although the SOG silicon by the metallurgical method can produce the solar cell with PV conversion efficiency in a range of 16–18% and the efficiency degradation problem has been overcome, it still needs some time to overcome the obstacles for entering the market;
4. The theoretical study on metallurgical polycrystalline silicon shall be strengthened to gradually establish the corresponding product standards.

2.1.4 Preparation of Solar-Grade Polycrystalline Silicon by FBR Method

Wacker, Germany, adopts the FBR (FBR furnace) method to replace the bell reduction furnace used by the Siemens method, and uses the polycrystalline silicon of fine particles as the seed crystal to replace the columnar silicon core used by the Siemens method to deposit the polycrystalline silicon, which significantly increases the reduction and deposition area of polycrystalline silicon, raising the deposition rate and reducing the energy consumption and achieving continuous production of polycrystalline silicon. For the reaction gas, similar to the Siemens method, the raw material of FBR is ($SiHCl_3 + H_2$). The reaction temperature, slightly lower, is in the range of 700–1000 °C, and the product is polycrystalline silicon particles. When the fine particle as a seed crystal grows to the polycrystalline silicon grain of certain weight, it will naturally fall out of the furnace. The annual output of single FBR furnace is up to 200–500 t/a (see Figure 2.6 for the production flow of the particle polycrystalline silicon).

Other companies such as MEMC and Hemlock also use the FBR-similar method to produce the low-cost solar polycrystalline silicon, but the reaction gas used is silane SiH_4, and the granular polycrystalline silicon is produced by silane via thermal decomposition.

The FBR method has the following disadvantages: First, the impurities are apt to be introduced during silicon powder pulverisation and fierce frictions between the silicon powder and the furnace wall in the FBR; second, it is apt to catch fire and explode when silane or trichlorosilane gas is in contact with air.

2.1.5 Preparation of Solar-Grade Polycrystalline Silicon by $SiCl_4$ Zinc Reduction Method

The production flow of $SiCl_4$ zinc reduction is to make the metal zinc with purity above 4 N react with the by-product of the Siemens method $SiCl_4$ (purity: 8 N) and produce the silicon of purity above 6 N at 1000 °C, that is, the solar-grade polycrystalline silicon. The by-product in the zinc reduction process, $ZnCl_2$, can generate the metal zinc and release chlorine via electrolysis. The metal zinc can return to the $SiCl_4$ reduction reactor and

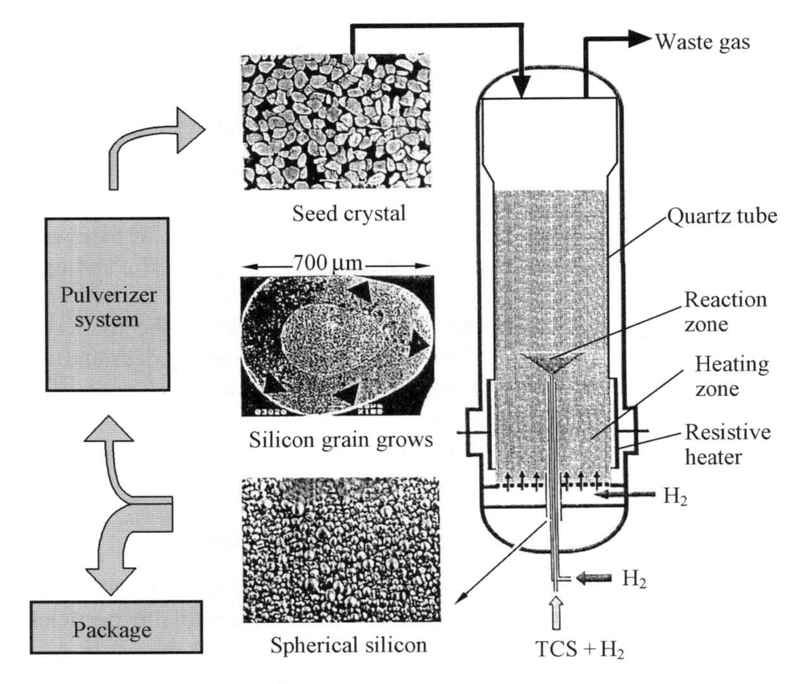

Figure 2.6 FBR furnace of Wacker.

serve as the reductant again; and the chlorine will return to the fluidised bed furnace in the Siemens method for chloridising the unpurified silicon (Formulas 2.1 and 2.2). The '$SiCl_4$ zinc reduction method' can be combined with the '$SiHCl_3$ hydrogen reduction method' of the Siemens method to realise closed circulation in the semiconductor silicon industrial chain, as shown in Figure 2.7.

The zinc reduction reaction of $SiCl_4$ and $ZnCl_2$ is as follows:

$$SiCl_4 + 2Zn = Si + 2\ ZnCl_2 \text{ (endothermic reaction)} \tag{2.1}$$

The electrolytic reaction of $ZnCl_2$ is as follows:

$$ZnCl_2 = Zn + Cl_2 \tag{2.2}$$

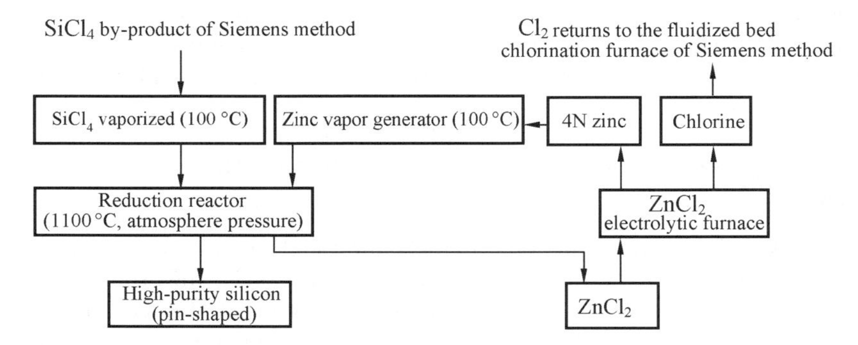

Figure 2.7 Flow chart of $SiCl_4$ zinc reduction method.

The $SiCl_4$ zinc reduction method has the following advantages:

1. Short process flow, less equipment, less investment on fixed assets (one-fifth of that of the Siemens method), and easy operation;
2. Fast deposition speed, low power consumption and short production cycle due to strong chemical activity of zinc;
3. $SiCl_4$, the by-product of hydrogen reduction in the Siemens method, can be reused; $ZnCl_2$ can be recycled, and the product cost is about one-third of that of the Siemens method.

The main problem of the $SiCl_4$ zinc reduction method lies in that the purity of the polycrystalline silicon reduced by zinc is less than that reduced by trichlorosilane hydrogen, and thus it cannot be used for the integrated circuit and other electronic devices and it can only be used for solar cells. In addition, the polycrystalline silicon produced by this method is pin-shaped with big apparent volume and it is generally cast into silicon ingots for use. The method, however, can be used as an associated process with the trichlorosilane hydrogen reduction method to achieve energy conservation and emission reduction.

2.1.6 Preparation of Solar-Grade Granular Polycrystalline Silicon by VLD Method

The vapour to liquid deposition (VLD) method, developed by Tokuyama, Japan, can be used to produce the granular solar-grade polycrystalline silicon (see Figure 2.8 for the experimental equipment). The purified $SiHCl_3$ and H_2 mixture shall be transmitted to the furnace, reduced and deposited on the graphite heater at 1500 °C, forming the liquid silicon. And then, the liquid silicon drips shall fall on a catch tray and generate the granular silicon.

The process has following advantages: The silicon reduction rate at 1500 °C is higher by several times than that at 1100 °C, and the intermittent operation can be turned

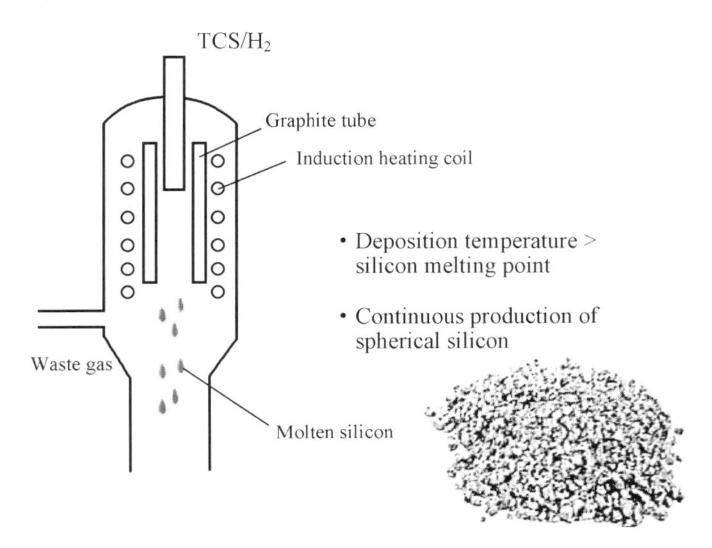

Figure 2.8 VLD experimental equipment of Tokuyama.

to continuous operation, and thus it is easy to scale up. The disadvantage is that the product contains high carbon content. We think if $SiCl_4$ is used to replace $SiHCl_3$, the similar effect can be achieved. Together with the carbon-removal process, it is still a good method to produce the solar cell polycrystalline silicon at low cost.

2.1.7 Hydrogenation of the Main By-Product $SiCl_4$ Produced in the Production Process of Polycrystalline Silicon by the Siemens Method

$SiCl_4$ is a main by-product of polycrystalline silicon by Siemens method, and 14 kg will be generally produced for every kg of polycrystalline silicon, which makes the cost high and also results in severe environmental pollution. How to settle the by-product produced by Siemens method has become the key issue to expand the production scale of polycrystalline silicon. The leading polycrystalline silicon manufacturers in the world have realised closed circulation or comprehensive utilisation of $SiCl_4$ during production of polycrystalline silicon, for example, $SiCl_4$ can be used to produce gaseous phase white carbon black or organic silicon and so on.

$SiCl_4$ shall continue circulating in the Siemens method after hydrogenated to $SiHCl_3$. This is the core technology to improve the Siemens method. It shows the circulation in Figure 2.9.

The author published an article on thermodynamics analysis named 'Exploration on $SiCl_4$ hydrogenation reaction' on the magazine 'Collected works on semiconductor silicon' in 1982 in Shanghai (Formulas 2.3, 2.4, and 2.5). In this article, it shown the relationship between the free energy and the reaction temperature of three hydrogenation reactions, as shown in Figure 2.10.

$$\text{Reaction (1)}\ SiCl_4(g) + H_2(g) = SiHCl_3(g) + HCl(g) \tag{2.3}$$

$$\text{Reaction (2)}\ 3SiCl_4(g) + 2H_2(g) + Si(S) = 4SiHCl_3(g) \tag{2.4}$$

$$\text{Reaction (3)}\ 2SiCl_4(g) + H_2(g) + Si(S) + HCl(g) = 3SiHCl_3(g) \tag{2.5}$$

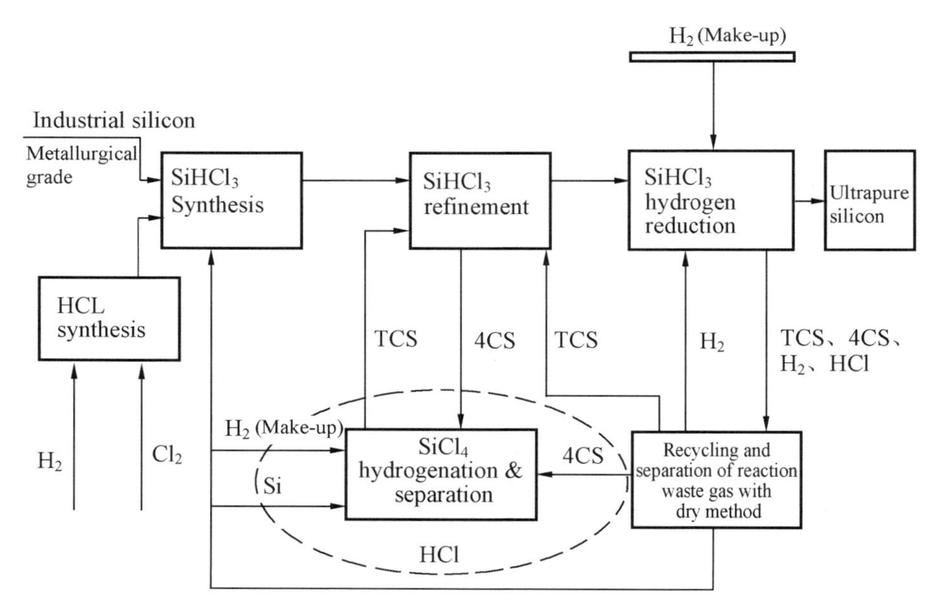

Figure 2.9 Hydrogenation flow chart of the by-product $SiCl_4$ of improved Siemens method.

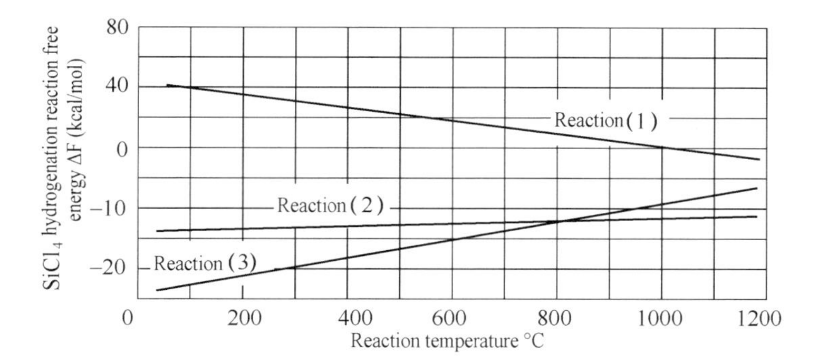

Figure 2.10 Standard free energy-temperature variation curve of hydrogenation reaction.

Figure 2.10 shows Reaction (1) is endothermic reaction (heat-absorbing), and Reactions (2) and (3) are exothermic reactions (heat-producing). The free energy of Reaction (1) becomes negative only above 1000 °C, indicating that the reaction can only happen above 1000 °C; Reaction (2) can happen at lower temperature, below 500 °C, and Reaction (3) can happen even below 300 °C as long as hydrogen chloride is present. Reactions (1), (2) and (3) can be called as 'high-temperature hydrogenation', 'low-temperature hydrogenation' and 'hydrochlorination', respectively. The three hydrogenation reactions, of course, have their own kinetic conditions and associated process equipment.

2.2 Casting Polycrystalline Silicon

2.2.1 General

Based on the working principle of solar cells, when the crystal boundary of the silicon crystal is vertical to the surface of the solar cell component, the directional movement of photo-generated carriers is basically vertical to the P-N junction (cell surface), and the crystal boundary has almost no obstruction to the movement of the photo-generated carrier, that is, the crystal boundary has small influence over the PV conversion efficiency. In 1975, Wacker, Germany, first produced the solar cell with casting polycrystalline silicon in the world. Since the method has the following virtues: unnecessary to crystal pulling, low-cost, high utilisation rate of materials and low energy consumption, it is very attractive. And the casting polycrystalline silicon has been widely used to the newly-built solar cells and material product lines since then, especially after 1990s. It has become one of the most important silicon solar cell materials.

The casting crystalline silicon is refined and profiled by means of directional solidification after molten. Since it is prepared in a square crucible, it is easy to cut into square silicon wafers. The technique has the following virtues: convenient process, easy to produce large-size silicon wafers, easy for automatic growth and control, small material losses and lower energy consumption. More importantly, the tolerance of casting crystalline silicon technique on silicon raw material is wider than CZ silicon. The method has the following disadvantages: more crystal boundaries in the casting crystalline silicon ingot, high dislocation density, high impurity concentration and more micro defects and lower PV conversion efficiency. In recent years, the thermal field technique with flat solid/liquid interfaces and the quartz crucible with silicon nitride coating have been applied to the process of casting crystalline silicon. In addition, the following techniques have also been applied to fabrication of polycrystalline silicon solar cells: silicon nitride and anti-reflection coating on the irradiating surface, hydrogen passivation, and aluminum back electrode impurity gettering and so on. Thanks to these improvements, the PV conversion efficiency of the casting crystalline silicon solar cell has been rapidly raised. In 2014, the maximum conversion efficiency reached 20.3% in the lab. In the actual production, the efficiency of the casting crystalline silicon solar cell can generally reach 17–18%.

2.2.2 Preparation Process of Casting Crystalline Silicon

In the casting crystalline silicon process, the polycrystalline silicon is all molten in the square quartz crucible and then solidified from the bottom of the crucible to the top. The directional solidification technique, based on the impurity segregation principle at solid/liquid interface, can refine the material and the columnar polycrystalline silicon can grow vertical to the liquid. After the crystal growth finishes, it shall be held on for some time at high temperature to anneal the polycrystalline silicon ingot in the original position to minimise the thermal stress and defect density.

In practise, two methods are generally used to make the silicon first cool down at the bottom of the quartz crucible: The first is to make water flow at the bottom and conduct heat exchange between the quartz crucible and the surroundings (heat exchange method, HEM); and the other is to make the quartz crucible gradually move downwards to leave the heating zone (Bridgman Method). Both of them can make the solid/liquid interface flat (see Figure 2.11 for the differences). At present, the polycrystalline silicon ingot furnaces produced in China are mostly based on HEM, that is, there is no relative displacement between the crucible and the heater during silicon melting and solidification, and only water is connected at the crucible bottom for cooling down. Since there is relative displacement between the crucible and the heater during crystal growing by Bridgman method, the ingot furnace has a complicated structure.

The molten directional-solidification casting crystalline silicon has been widely applied to the PV industry. The production process mainly consists of melting, columnar crystal growing, annealing, cooling down and so on. In the whole process, the temperature field is controlled to form a single-direction heat flow (no horizontal

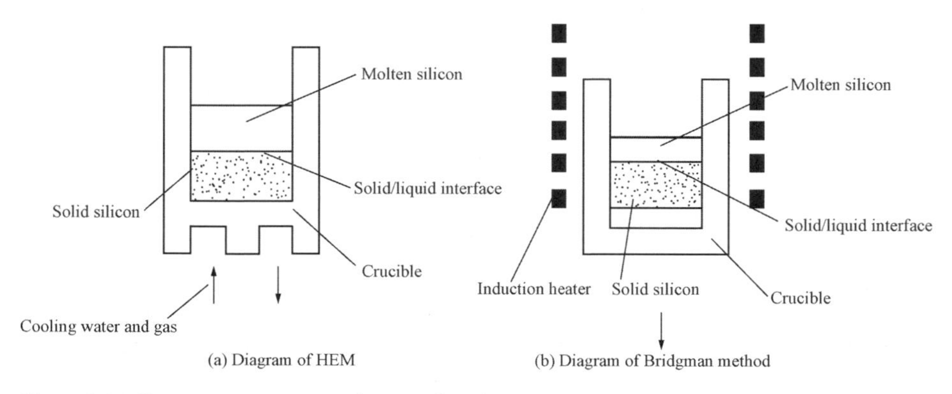

Figure 2.11 Process comparisons of HEX and Bridgman methods.

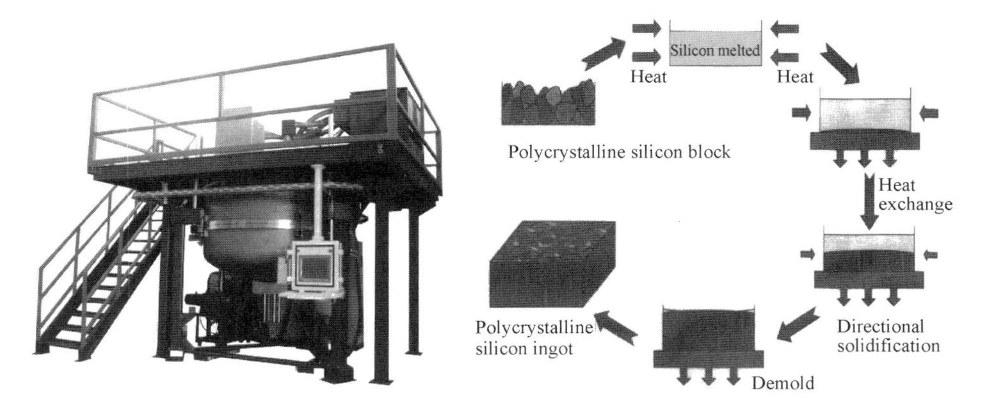

Figure 2.12 Melting furnace and ingot casting diagramme for polycrystalline silicon preparation by directional solidification after melting.

temperature gradient), and the longitudinal temperature gradient at the solid/liquid interface is larger than zero so that the columnar silicon polycrystal can grow directionally, which can meet the requirements of solar cells. Figure 2.12 shows the melting furnace for preparation of polycrystalline silicon by the directional solidification method. When the polycrystalline silicon material is molten in the quartz crucible, the top of the crucible shall be kept at high temperature by the surrounding heaters and the bottom of the crucible shall be gradually cooled down to make the melt at the bottom first crystallised, and the solid/liquid interface shall keep flat and rise gradually, and the grain shall crystallise in the temperature gradient direction until the whole melt is crystallised to crystal ingots.

The casting crystalline silicon grows from the crucible bottom upwards at a rate of 1 cm/hr or so. After casting, the ingot is in cubical. It shows the quartz crucible and the polycrystalline silicon ingots in Figure 2.13. A 800 kg polycrystalline silicon ingot is about 800×300 mm in size.

The qualified casting crystalline silicon ingot shall have smooth surface and be free from any cracks, holes and other obvious defects. For the quality polycrystalline silicon ingot, the crystal grain and the boundary shall be visible from the top, and the grain can be up to about 10 mm in size; and the grain grows in column from the sides, and the main grains grow from bottom to top almost vertical to the bottom, as shown in Figure 2.14. For the casting crystalline silicon ingot by the directional solidification technique, the growth speed is slow, and the quartz crucible is a consumable.

After the polycrystalline silicon ingot is prepared, cut it into the square columns of 125×125 mm, 156×156 mm or 210×210 mm; and then cut it to wafers with a wire-cutting machine (see Figure 2.15).

For the casting crystalline silicon ingot, the quality is poor at the bottom, top and edges in a thickness of several centimetres and thus unusable due to impurity segregation and edge equi-axis crystal.

Figure 2.16 shows the lifetime distribution of the minority carrier on the vertical section of the ingot. Obviously, only the minority carriers in central ingot have a long lifetime.

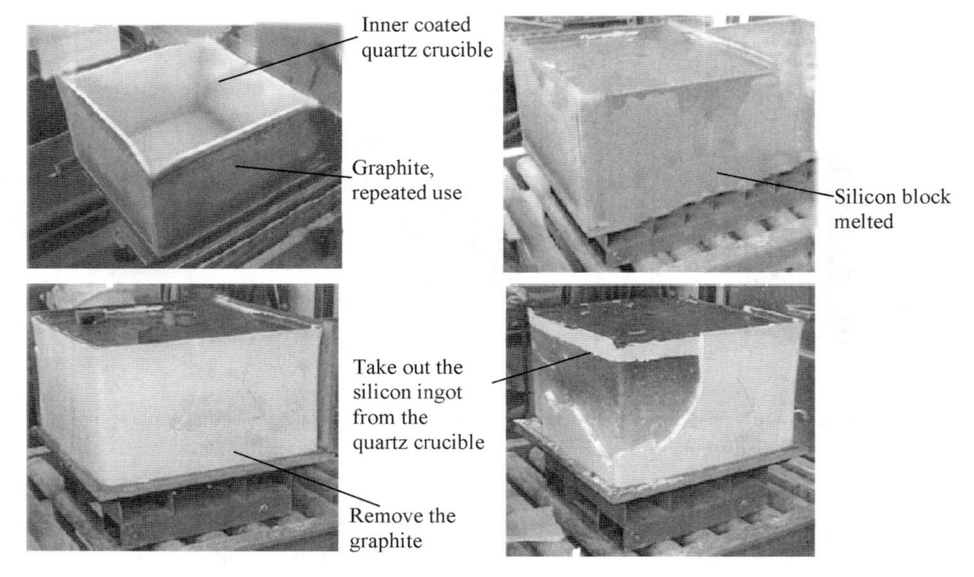

Figure 2.13 Quartz crucible and polycrystalline silicon ingot used in direct melting and directional solidification method.

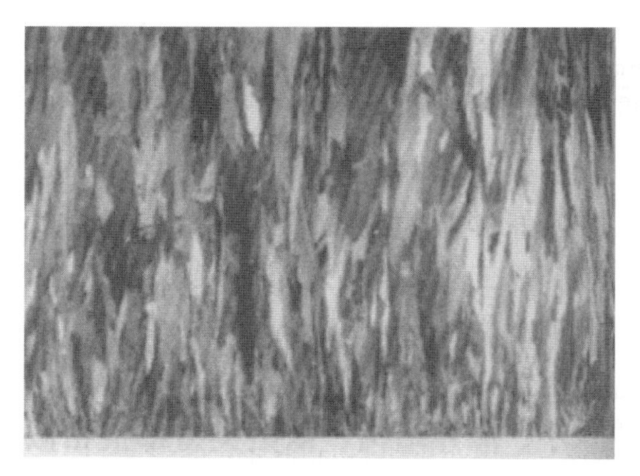

Figure 2.14 Longitudinal section of casting crystalline silicon ingot columns growing in the heat field control conditions.

As for the raw material of casting crystalline silicon, it is more tolerant than that of the pulling and zone melting crystalline silicon, and it can use rejected or scraped materials out of the electronic industry and the low-cost solar-grade polycrystalline silicon. Since the segregation coefficient of boron is close to 1 and it can realise even longitudinal resistivity distribution of the whole ingot, the P-type casting crystalline silicon is often used in solar cell production. The B_2O_3 and the silicon raw material are put into the crucible, and after the silicon is melted, B_2O_3 is decomposed and boron melted into silicon. To meet the requirements of solar cell preparation, the resistivity is usually controlled in a range of 0.1–5 Ω cm, 1 Ω cm if available; and the boron doping concentration is about $2 \times 10^{16}/\text{cm}^3$.

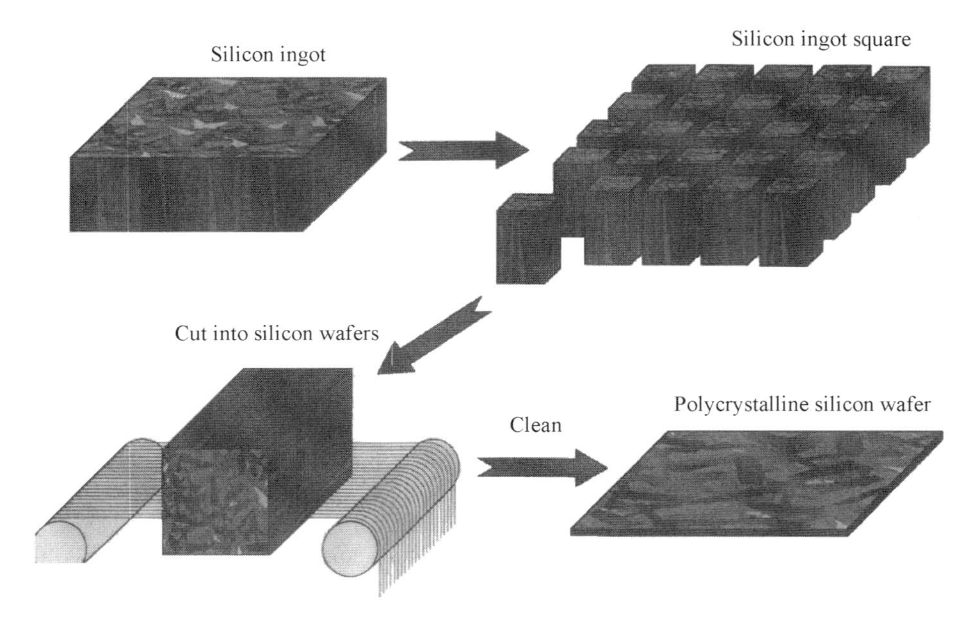

Figure 2.15 Line-cutting diagramme of casting crystalline silicon ingots.

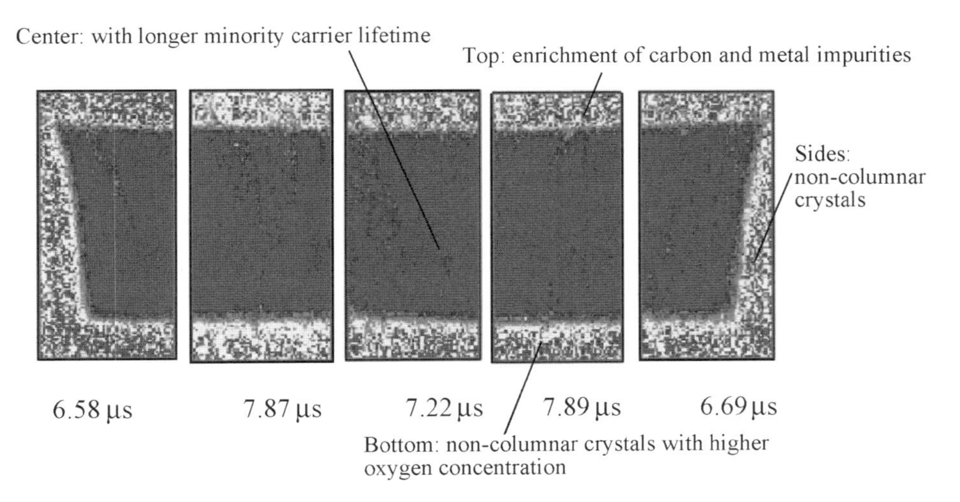

Figure 2.16 Distribution of the minority carrier lifetime in different sections of the ingot.

Since phosphorus has small segregation coefficient, the resistivity of the top is far different from that at the bottom for the casting crystalline silicon crystal doped with phosphorus. As a result, the casting crystalline silicon doped with phosphorus is seldom used for solar cells.

As for the crucible for casting crystalline silicon, since the silicon melt is in contact with the quartz crucible for a long time, it will raise the oxygen content and result in adhesion action. To settle the problems, the crucible is usually coated by Si_3N_4 on the inner wall. It is expected that the application of Si_3N_4 coating technique can also realise repeated use of the quartz crucible.

Below shows a typical growing process of casting crystalline silicon by direct melting and directional solidification:

1. Loading: Put the coated quartz crucible on the heat exchange plate (cooling down plate), add the silicon raw material, and install the heating equipment, insulation equipment and furnace cover, pump down the furnace to 0.05–0.1 mbar and then inject argon as a protective gas and keep the furnace pressure at 400–600 mbar.
2. Heating: Slowly heat by the graphite heater to first vapourise the wet gas absorbed on the graphite components (including the heater, the crucible plate, the heat exchange plate, etc.), the insulation layer, and the silicon raw material; about 4–5 hr later, heat the quartz crucible up to 1200–1300 °C.
3. Melting: Heat the quartz crucible up to 1500 °C at argon, and the silicon raw material will begin to melt and wait until all materials are melted. It will last about 9–11 hr.
4. Growing: After the silicon is melted, reduce the heating power to make the quartz crucible down to 1420–1440 °C (about the silicon melting point); then supply water to the bottom of the quartz crucible to reduce the temperature from the bottom, and the crystalline silicon will be formed first at the bottom and grow upwards; during the growth, the solid/liquid interface should be always kept horizontal until the crystal growth finishes. It will last about 20–22 hr.
5. Annealing: After the crystal grows, the ingot has large stress so that the ingot shall keep at the melting point for about 2–4 hr to make the ingot temperature even and reduce the thermal stress.
6. Cooling down: After annealing, turn off the heater and inject argon to cool down the crystal to the ambient temperature; and the furnace pressure will rise simultaneously until it reaches the atmospheric pressure; finally take out the ingot. It will last about 10 hr.

The followings are important for casting crystalline silicon crystal growth: flat solid/liquid interface, minimal crystal thermal stress, and maximal grain size, and minimal contamination from the crucible and so on. The thermal field distribution and cooling rate play a crucial role.

2.2.3 Impurities and Defects in Casting Crystalline Silicon

2.2.3.1 Non-Metal Impurities in Casting Crystalline Silicon

The non-metal impurities in casting crystalline silicon are mainly oxygen and carbon. Compared with CZ monocrystalline silicon, the oxygen content is generally lower but the carbon concentration is higher.

In casting crystalline silicon, oxygen mainly comes from the raw material and the action of silicon melt with quartz crucible. The segregation coefficient of oxygen in silicon is 1.25, and thus the oxygen concentration at the ingot bottom (about $1.3 \times 10^{18}/cm^3$) is higher than that in the middle and on the top of the ingot (about $3 \times 10^{17}/cm^3$). Since the inner wall of the quartz crucible, however, is coated with silicon nitride and no strong mechanical convection is present (except heat convection) and the silicon melt has weak wash-out on the quartz crucible wall, the oxygen total concentration in the casting crystalline silicon is usually lower than that of the CZ

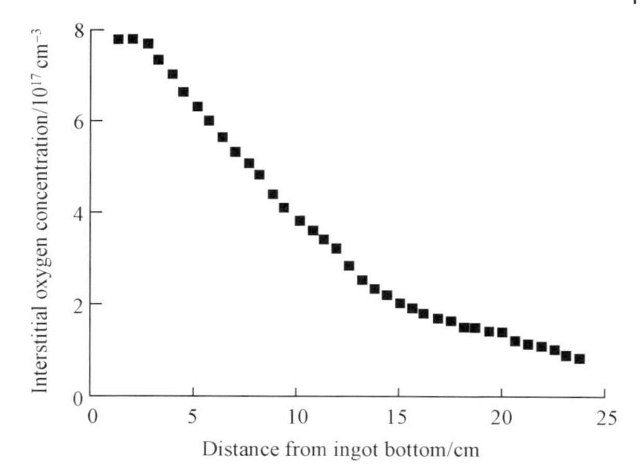

Figure 2.17 Distribution of interstitial oxygen concentration in ingot growing direction.

monocrystalline silicon. The oxygen in casting crystalline silicon is mostly present in interstitial and saturation state, and the interstitial oxygen has no influence on the solar cell's performance. If thermal donor or oxygen deposition occurs at the ingot bottom, it may become the recombination centre or introduce secondary defects, resulting in reduced minority carrier lifetime.

The segregation coefficient of oxygen in silicon melt is generally viewed as larger than 1. Figure 2.17 shows, oxygen is mainly concentrated at the bottom of the silicon ingot and its concentration gradually falls from bottom to top.

In addition to the existence in the raw material, carbon may come from volatilisation of the graphite heater during long time operation. The segregation coefficient of carbon in silicon is 0.07, and the solubility of carbon $C_L(T)$ in the silicon melt can be expressed by Formula (2.6):

$$C_L(T) = 8.6250 \times 10^{-4}T^2 - 2.7643T + 2222.9 \tag{2.6}$$

Figure 2.18 shows the partial Si-C phase diagramme when the SiC particle in the silicon melt sediments at the silicon-enriched zone (diagramme). Since the effective segregation coefficient of carbon in silicon is small, the carbon concentration in the melt will rise during solidification (Formula 2.7). If the carbon concentration exceeds the solid solubility, the enriched carbon will precipitate and have the following chemical reaction:

$$Si + C = SiC \tag{2.7}$$

The study results show, when the carbon concentration in the silicon raw material is larger than $1.26 \times 10^{17}/cm^3$ (the carbon concentration near the ingot top surface is usually larger than this value) and it is larger than the solubility of carbon in silicon, the SiC particles will be present in the top section of the casting crystalline silicon ingot. The SiC sediments will result in serious ohm breakdown in the solar cell, increasing the leak current, forming the new grain core in silicon and further reducing the solar cell conversion efficiency. Generally, the polycrystalline silicon refined by metallurgical method has high carbon content, and thus an attention should be paid to the precipitation process of silicon carbide.

In the casting crystalline silicon, the thermal treatment of carbon has close relations with the nature of oxygen. During the cooling process, carbon impurities in a low-carbon

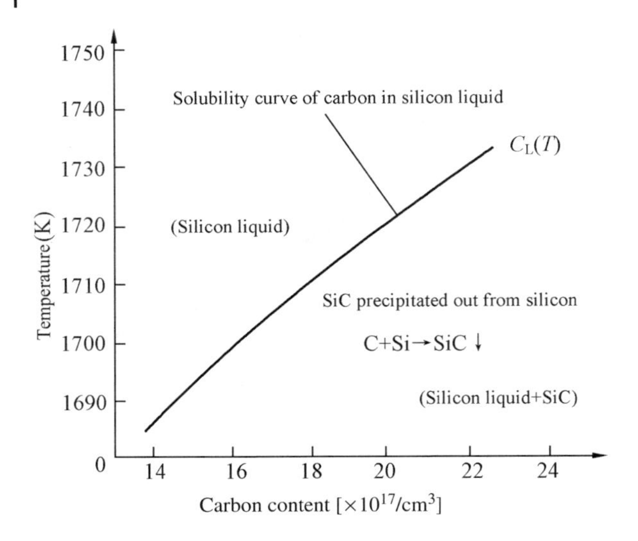

Figure 2.18 Si-C phase diagramme in silicon-enriched zone.

high-oxygen sample can serve as a core of oxygen precipitation; and the carbon impurity of the high-carbon high-oxygen sample can promote generation of a new donor.

2.2.3.2 Metal Impurities and Gettering in Casting Crystalline Silicon

The metal impurities (especially that in the transitional group) element have low solubility in casting crystalline silicon, and most of them are present in sediments. The segregation coefficient of Fe in silicon is $(5 \sim 7) \times 10^{-6}$, and Figure 2.19 shows, the distribution characteristics of the interstitial Fe concentration along the silicon ingot are as follows: The concentration is high at the top and low in the middle, and the distribution is even. In the casting crystalline silicon, a lot of crystal boundaries and dislocations can become the priority sediment place for these impurities. To reduce the influence of these metal impurities on the PV conversion efficiency, the impurity gettering process shall be added during preparation of solar cells with casting crystalline silicon.

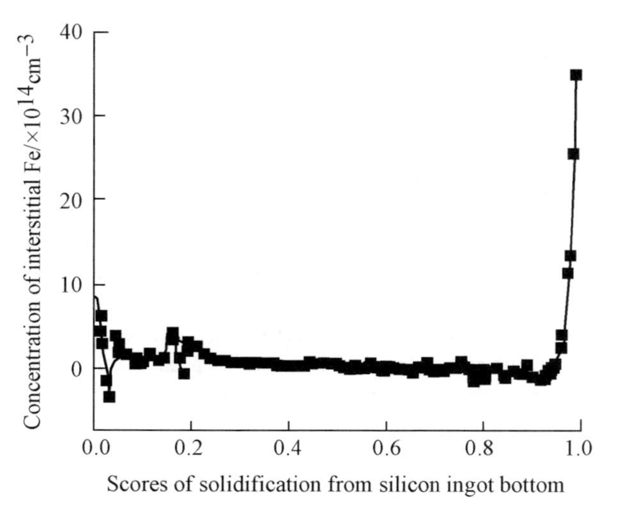

Figure 2.19 Distribution of interstitial Fe concentration along the height direction of silicon ingots.

The phosphorus diffusion method is usually used to produce a P-N junction for the silicon solar cells. The phosphorus impurity gettering is based on that while the liquid-state source of phosphorus oxychloride is used to produce the P-N junction at about 850–900 °C, the phosphorus silicon glass (PSG) will be generated on the front/rear sides of the silicon wafer. Because the phosphorus silicon glass has a lot of micro-defects, which will become the gettering source of metal impurities, the metal impurity atoms will diffuse and precipitate in the PSG layers while the phosphorus diffusion method is used to produce the P-N junction for the silicon solar cell. The chemical reagents such as HNO_3 and HF etc., can be used to remove PSG. In this way, most of the metal impurities in the silicon wafer can be removed, realising metal impurity gettering. As for phosphorus impurity gettering mechanism, it is generally recognised that while the phosphorus diffuses, the metal impurity atoms will diffuse, dissolve and deposit to PSG. The gettering temperature shall not be excessively low; otherwise, it is adverse for metal impurity diffusion and dissolving; but it shall not be excessively high; otherwise, it is adverse for impurity segregation to the impurity gettering zone. As a result, there is an optimum gettering temperature. The experiments show that the phosphorus impurity gettering is not effective for the substitution impurities (e.g., arsenic, stibium, tin, zinc).

In addition to 'phosphorus impurity gettering', the 'aluminum impurity gettering' technique can be also applied to the solar cells, that is, while preparing the aluminum electrodes on the back, the aluminum-silicon alloy will changed to AlSi after 800–1000 °C thermal treatment. While the heavily aluminum doped P-type layer is formed near the AlSi alloy layer, the transition metal impurities in silicon will diffuse to the AlSi layer, which significantly reduces the concentration of the metal impurities. Obviously, the technique can build the aluminum BSF and getter the heavy metal impurities in the silicon wafer. The aluminum impurity gettering technique is based on that the solid solubility of the metal impurities in aluminum is far larger than that in silicon. At 700–900 °C, the solid solubility of many metal impurities in aluminum is larger by 4–10 orders of magnitude than that in silicon.

The phosphorus impurity gettering technique is faster than the aluminum impurity gettering technique but the latter has better gettering capacity and stability than the former. As a result, the two techniques are often jointly applied to the casting crystalline silicon solar cell to improve the gettering capacity of metal impurities.

2.2.3.3 Crystal Boundaries and Dislocations in Casting Crystalline Silicon

The grains of casting polycrystalline silicon grow in a direction basically vertical to the surface of the ingot. The P/N junction of the solar cells produced can be vertical to the crystal boundary of the polycrystalline silicon when the wafers are cut in a direction along the ingot surface. In accordance with the working principle of the solar cell, the photon-generated carriers will move in a direction vertical to the P/N junction when the solar cell is in solar rays. As a result, the crystal boundary will not impede the movement of the photon-generated carriers (as shown in Figure 2.20). The light has almost no influence on the movement of the photo-generated carriers and the crystal boundary has very little influence over the PV conversion efficiency of the solar cells. We also noticed that the grain size will grow gradually (even above 1 cm) via merging the adjacent grains as the height of the casting crystalline silicon rises. Obviously, the crystal boundary is not the main factor restraining the performance of solar cells. On the contrary, if the

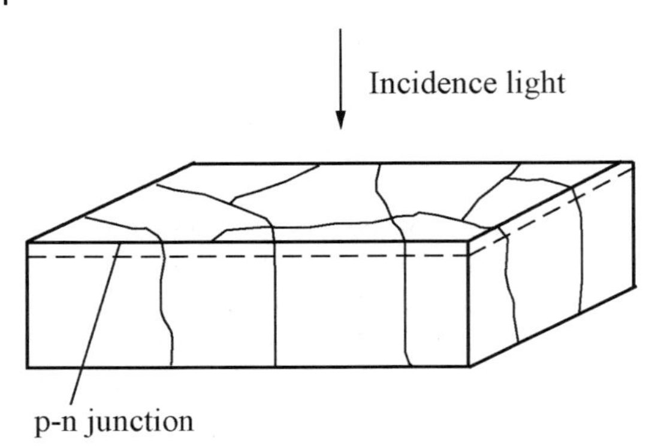

Figure 2.20 When the silicon crystal grain grows in a direction vertical to the P-N junction, the crystal boundary has no influence over the movement of the photo-generated carriers.

crystal boundary is vertical to the incidence light, it will exert great influence over the movement of the photo-generated carriers and thus cannot be used to fabricate solar cells.

The study shows, the pure crystal boundary without metal decoration does not have electrical activity and thus it is not the trapping centre of carriers and it will not affect the electrical performance of the polycrystalline silicon. In addition, the crystal boundary is able to attract metal impurities and a zone with low metal impurity concentration will be present near the crystal boundary, which is the so-called 'crystal boundary impurity gettering'. If the crystal boundary is contaminated by metal impurities, it will have electrical activity and thus affect the diffusion length of the minority carriers and reduce the PV conversion efficiency.

2.2.4 Latest Development of Casting Crystalline Silicon and Wafers

2.2.4.1 Casting of Pseudo-Single Crystal

In March 2011, JA Solar Holdings Co., Ltd., China, successfully prepared pseudo-monocrystalline silicon by the casting method. The process flows are as follows: First, use the monocrystalline silicon wafer (about 5 mm thick) of (100) orientation as the seed and put it on the bottom of the quartz crucible of the casting polycrystalline furnace, and then add the raw polycrystalline silicon blocks; control the heating conditions and the polycrystalline silicon blocks will melt from top to bottom; when the seed crystal in contact with the silicon liquid melts, the operation sequence shall be immediately transmitted to directional crystallisation stage; then, similar to the process of the ordinary casting crystalline silicon, it will solidify from bottom to top to realise directional growth; and then anneal, take out of the furnace and square the silicon ingot. It is found in practise that the middle zone of the ingot is almost all monocrystalline, and the surrounding zones are also monocrystalline and the grains with other orientations are present on the top. Figure 2.21 shows the longitudinal section of pseudo-single crystal.

Compared with the ordinary casting polycrystalline ingots, the average conversion efficiency of the pseudo-monocrystalline ingot is 0.5–1% higher, and the average PV efficiency of the big grains with (100) orientation area larger than 70% is larger than 17.5–18% with maximum above 18.3%. Compared with CZ monocrystalline, it is characterised by low cost and low degradation (the oxygen content of casting polycrystalline is usually lower than that of CZ monocrystalline). For the cell wafers based on

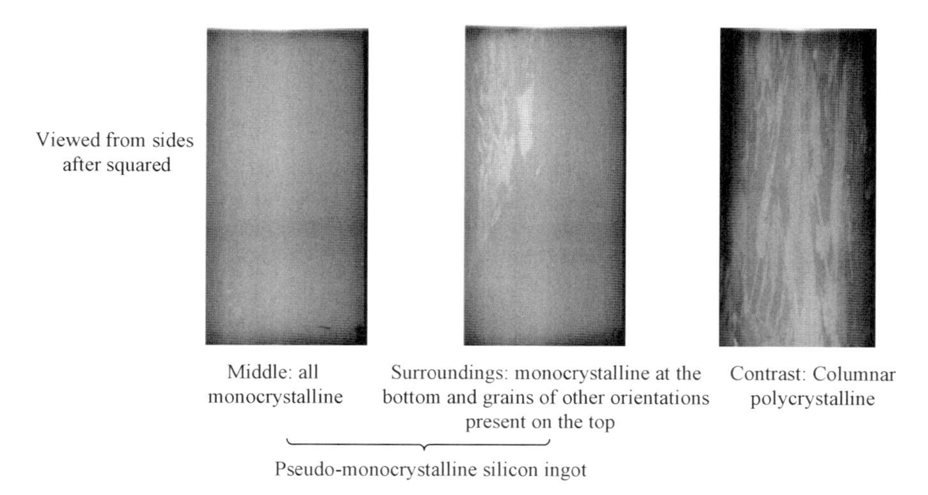

Figure 2.21 Longitudinal section of pseudo-single crystals.

CZ monocrystalline silicon, the degradation rate of PV conversion efficiency is about 2%, and for the cell wafers based on pseudo-monocrystalline ingot, the degradation rate is reduced to 0.5%.

2.2.4.2 Continuous Output Improvement of Casting Crystalline Silicon Furnaces

With technical progress and new equipment development, the output of one casting crystalline silicon furnace was 240 kg in 2003, 400–500 kg in 2007, and 800 kg in 2009. It will exceed 1000 kg/furnace in the upcoming days. As the ingot weight and volume rise, the columnar crystal in the middle zone will take an increasing proportion in the ingot as a whole, which is good for economic benefit. Due to the impurity segregation coefficient has limitation to the longitudinal distribution, it often increases the sectional area of the ingot, instead of its height (generally about 400 mm) to improve the furnace output. The casting furnace capacity will be extended to G8 from the existing G5, G6, G7, and the lateral temperature gradient will become smaller, and so is the equiaxed crystal range near crucible wall, which is good for improvement of quality and throughput. The wall thickness of the quartz crucible, however, shall be correspondingly increased.

As for the polycrystalline silicon casting furnace for solar cells in China, the first one was introduced by Yingli Solar, Baoding from Germany in 2005. At that time, the price of the casting furnace equipment was high. Later, the local production accelerated its pace, and many companies, including 48th Research Institute of China Electronics Technology Group Corporation, Jingyuntong and Hanhong and so on, developed a series of products with good performance/price ratio and then put them into volume production for the market. In August 2009, LDK, Xinyu, Jiangxi announced it had produced 800 kg polycrystalline silicon ingots. In early 2010, Shanghai Hanhong and other companies also successfully developed 800 kg polycrystalline silicon casting equipment, ranked as the frontier in the global PV industry. In addition to improvement of ingot growth techniques and application of impurity gettering technique to cell production, the manufacturing cost of casting crystalline silicon is lower than that of monocrystalline silicon. And thus the conversion efficiency of polycrystalline silicon solar cells has been significantly improved. In the casting crystalline silicon process, the requirements on the purity and mechanical strength of the quartz crucible are very demanding, and the quartz crucible is often cracked due to silicon liquid solidification and volume expansion when the ingots are taken out of the furnace. Since the quartz crucible accounts for about 30% of the production cost, it will dramatically reduce the cost if the quartz crucible can be repeatedly used or the substitute material is used.

2.3 CZ Monocrystalline Silicon

The method to grow the monocrystalline in the melt is also called as CZ method, which is one of the main methods for semiconductor preparation. For example, Ge, Si, GaAs and InP crystals are all grown in the melt. The method requires that the material will be stable near the melting point and will not decompose, sublimate or change phases. The melt-growth is a typical liquid-solid phase conversion process. In the phase change process, the grown atoms or molecules shall change from random to ordered arrangement and release the latent heat of phase change. The phase change is not an integral effect but a process of solid-liquid interface solidifying layer by layer. Most of the monocrystalline silicon solar cells adopt the silicon grown by CZ method as the raw material.

To ensure the grain grows smoothly, the heat and mass transfer problems must be settled. It is known that mass transfer plays an important role in growing grains in gaseous phase and solution while heat transfer plays a more important role when the grain grows in the melt, which has direct influence over the grain growth parameters, morphology of growth interface and integrity of grain and so on.

2.3.1 Heat Flow Continuity Equation of Grain Growth Interface and its Application

Figure 2.22 shows the heat transfer diagramme for growing monocrystalline in CZ furnace, which includes the following seven aspects:

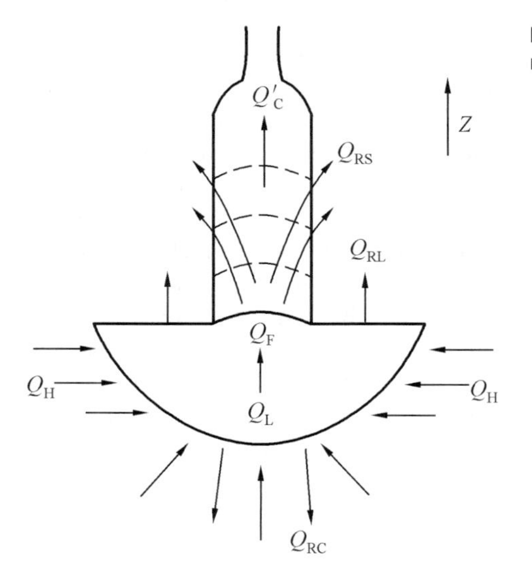

Figure 2.22 Heat transfer of grain growth by CZ method—isotherm → heat flow direction.

1. Q_H: The heat transferred by the heater to the crucible;
2. Q_{RC}: The heat radiated outwards by the crucible (especially at the bottom);
3. Q_L: The heat conducted by the melt to the solid-liquid interface;
4. Q_{RL}: The heat released by the melt;
5. Q_F: The latent heat of phase change released during liquid-solid conversion;
6. Q_{RS}: The heat radiated by the grain surface;
7. Q'_C: The heat conducted by seed crystal

In the above seven heat transfer approaches, the heat flows directly passing the solid-liquid interface are important, that is, Q_L, Q_F, Q_C (Formula 2.8). To ensure smooth grain growth, a necessary condition is to keep thermal equilibrium on the solid-liquid interface, that is, in unit time:

$$Q_L + Q_F = Q_C \tag{2.8}$$

Where, Q_C is the sum of the heat radiated by the grain surface Q_{RS} and the heat conducted by seed crystal Q'_C.

Suppose the sectional area of growing grain is A, the growth rate is f, the latent heat of phase change is $\overline{H}$, the grain density is d, K_L , K_S is the heat conduction coefficient of the melt and the grain, respectively, $\left(\frac{dT}{dZ}\right)_L$, $\left(\frac{dT}{dZ}\right)_S$ is the temperature gradient at the solid-liquid interface of the melt and the grain, respectively, work out the latent heat of phase change released during crystallisation Q_F (Formula 2.9):

$$Q_F = fAd\overline{H} \tag{2.9}$$

The heat transferred from the melt to the interface Q_L (Formula 2.10):

$$Q_L = K_L\left(\frac{dT}{dZ}\right)_L A \tag{2.10}$$

The heat transferred from the solid-liquid interface to the grain Q_C (Formula 2.11):

$$Q_C = K_S\left(\frac{dT}{dZ}\right)_S A \tag{2.11}$$

And the following formula can be obtained (Formula 2.12):

$$AK_L\left(\frac{dT}{dZ}\right)_L + fAd\overline{H} = AK_S\left(\frac{dT}{dZ}\right)_S \tag{2.12}$$

The formula is called as the interface heat flow continuity equation, which is the basic formula of melt-growth grain and can be used to estimate the grain growth rate and prepare the equal-diameter growth process.

1. Estimation of grain growth rate
   Change the interface heat flow continuity equation to the functional relationship of growth rate f (Formula 2.13):

$$f = \frac{K_S\left(\frac{dT}{dZ}\right)_S - K_L\left(\frac{dT}{dZ}\right)_L}{d\overline{H}} \tag{2.13}$$

Since K_S, K_L, d, $\overline{H}$ are all constants, when the growth system is determined, the equation can reflect the functional relationship $f \sim \left(\frac{dT}{dZ}\right)_S, \left(\frac{dT}{dZ}\right)_L$. To maximise the growth rate, $\left(\frac{dT}{dZ}\right)_S$ shall be maximised and $\left(\frac{dT}{dZ}\right)_L$ minimised. If $\left(\frac{dT}{dZ}\right)_S$ is excessively large, it may result in large residual stress and it is also adverse for lattice orientation during cool-down, affecting the integrity of the grain. In addition, $\left(\frac{dT}{dZ}\right)_L$ cannot be excessively low because it may result in uneven solid-liquid interface, big grain defect density, dendritic-web growth and even change to polycrystalline. Obviously, both of them have some limitations (Formula 2.14). Nevertheless, Formula (2.6) can still be sued to estimate the maximum growth rate of a system (limit), that is, the following can be obtained when $\left(\frac{dT}{dZ}\right)_L = 0$:

$$f_{\max} = \frac{K_S}{d\overline{H}}\left(\frac{dT}{dZ}\right)_S \tag{2.14}$$

2. Adjustment of growth rate and heater power to achieve equal-diameter growth
   When the monocrystalline grows, the sectional area of the seed crystal is generally small, and thus the heat conducted from the seed crystal Q'_C can be neglected when the grain grows to the certain height. In this case, the heat consumption of the grain is only the radiation heat on the grain surface, that is, $Q_C = Q_{RS}$. Based on the solid heat radiation law in physics, the heat radiated outwards from the surface at a specific height is $Q_C = 2\pi r\varepsilon\sigma T^4 = B_1 r$ (here, ε is heat emissivity, σ is Stephen-Boltzmann constant, T is grain surface temperature, r is the radius of grain, B_1 is the quantity related to heat conduction), and the interface heat flow continuity equation can be written concerning the grain growth rate (Formula 2.15):

$$f = \frac{B_1}{\pi d\overline{H}r} - \frac{K_L\left(\frac{dT}{dZ}\right)_L}{d\overline{H}} \tag{2.15}$$

   In the equation, the former item is mainly related to the latent heat of phase change and in inverse proportion to the radius. Accordingly, it has relations to the crystallisation amount or the pulling rate. The latter item is mainly in direction proportion to the temperature gradient in the melt, that is, it has relations to the heating power and field distribution of the monocrystalline furnace. To grow the equal-diameter crystal, it is often to change the pulling rate and heating power in practise.

2.3.2 Heat Conduction in the Melt

Based on hydromechanical analysis, two types of liquid flow are simultaneously present in the melt of monocrystalline grown by CZ method: One is the natural convection in the gravity field caused by temperature difference and the other is forced convection caused by rotation of crystal and the crucible. They have great influence over the heat distribution in the melt, the solid-liquid interface shape and even impurity distribution etc.

1. Natural convection
   In the melt of silicon monocrystalline growth, the natural convection is mainly driven by the density difference $\Delta\rho$ caused by temperature difference ΔT. For the natural

Figure 2.23 Natural convection in the crucible.

convection caused by the crucible heated by the outer sides, after the melt near the crucible wall is heated, the density ρ will become small and rise and it will become big after cooling down on the top surface and then fall. The convection is shown in Figure 2.23. It is mainly related to the size of the vessel and the longitudinal temperature gradient $\frac{dT}{dz}$.

The natural convection will result in temperature swing of the melt, and it can be mitigated by reducing the longitudinal temperature gradient $\frac{dT}{dz}$ and increasing the field.

2. Forced convection

The melt is often agitated to accelerate heat and solute transfer and improve uniformity. In the CZ monocrystalline, this is realised via rotation of the crystal and the crucible, and the melt flow artificially generated is called as 'forced convection'. The forced convection caused by crystal rotation will force the liquid to leave the centre and flow outwards under the action of centrifugal force, and the flow at the crucible bottom will rotate upwards along the crystal axis centre. And the forced convection caused by the crucible rotation will only make the liquid flow in spiral shape and the liquid flow direction is similar to the heat convection. If the rotation direction of the crystal ω_s is the same with the crucible ω_c, and $\omega_s > \omega_c$, the melt will rotate to rise along the centre and then separate on the melt top surface; and if $\omega_s > \omega_c$, the melt will rotate to rise from the outside and fall down along the centre.

2.3.3 Temperature Distribution in the Crystal

During monocrystalline preparation by the CZ method, the different parts of the crystal have different temperature, which may result in some thermal stress in the crystal, exerting great influence over the crystal integrity. Figure 2.24 shows the heat flow distribution in the CZ crystal.

In Figure 2.24, suppose it does not take into account the heat conducted by the seed crystal, the crystal is cylindrical, the heat field of the monocrystalline furnace is symmetric and co-axial with the crystal rotation; the crystal homogeneous in all directions. The coordinate origin falls on Point O of the solid-liquid interface centre, and suppose the relative temperature difference $\theta(r.\varphi.z) = T(r.\varphi.z) - T_0$ where T_0 is ambient temperature and T is the temperature of a point in the crystal, r is the radius, φ is angle of circumference and z is the height. The mathematic analysis shows the following characteristics:

(1) The rotation axis of the crystal is the symmetric axis of temperature distribution, and on one horizontal level (the same z-coordinate), the temperature at every point on the circumference with r as the radius is identical.

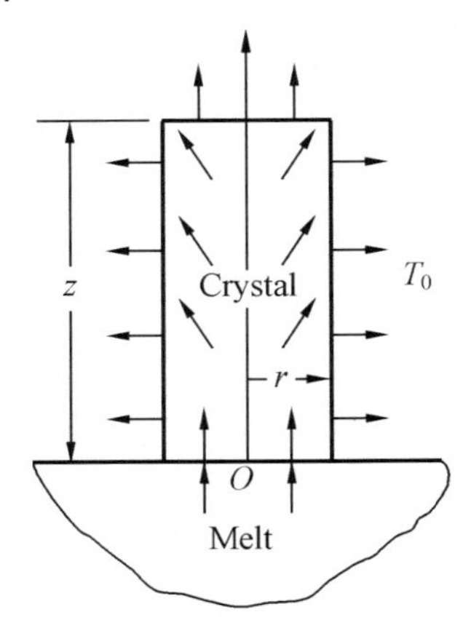

Figure 2.24 Heat flow direction diagramme in the crystal prepared by CZ method.

(2) When r is a constant, the crystal temperature will decrease exponentially with z.
(3) When the ambient temperature is lower than the crystal, the temperature will decrease with r. In this case, the isotherm inside the crystal is recessed toward the melt, and the axial component of the temperature gradient $\frac{\partial\theta}{\partial z}$ will decrease with r. When the ambient temperature is higher than the crystal, the temperature will increase with r. In this case, the isotherm inside the crystal is convex toward the melt, and $\frac{\partial\theta}{\partial z}$ will increase with r.

Although the above analysis on heat distribution in the crystal has some approximate hypotheses during inference, including not considering the impact of agitation on the temperature distribution in the melt, and the impact of crucible wall temperature on heat conduction and so on, no other more accurate model is available up to now. The above conclusions are similar to the test results of monocrystalline grown by CZ method.

2.3.4 Impurity Segregation Between Solid and Liquid

During monocrystalline growth, the impurities (solutes) have different concentrations in liquid and solid, which is impurity segregation. The followings are defined: $k_o = \frac{C_s}{C_1}$ is the equilibrium segregation coefficient, C_s, C_1 is the impurity concentrations in solid and liquid when the growth rate is infinitely slow (i.e., the solid and liquid are in equilibrium), respectively. If $k_0 < 1$ (It is the case for most impurities in semiconductor), the impurities will be continuously enriching to the melt during growth, as shown in Figure 2.25.

The equilibrium segregation coefficient of impurities in silicon is listed in Table 2.9.

Table 2.9 shows, the segregation coefficient of most impurities is less than 1, which means that most impurities will stay in the melt during growth. As a result, when the crystal grows, the impurities will be increasingly in the melt. For most semiconductor crystals, some N-type or P-type impurity is doped on purpose into the high-purity raw material to achieve the desired electrical performance. For silicon, boron and phosphorus are the most common dopant of N-type or P-type materials, respectively.

For the existing solar cells, the silicon monocrystalline is generally produced by the CZ method with diameter of 6″ and 8″ and then processed into 125×125 mm and

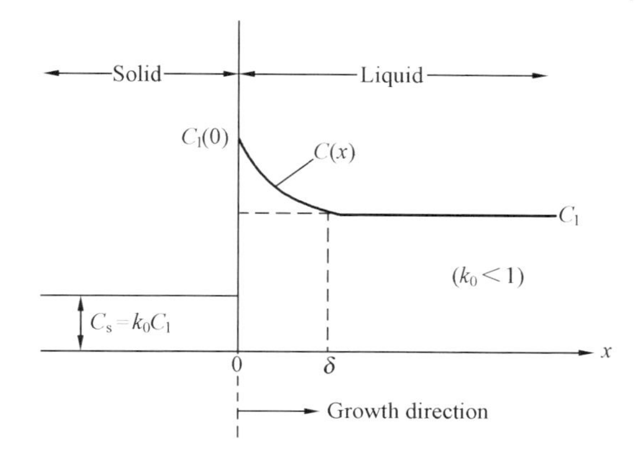

Figure 2.25 Impurity concentration distributions near solid-liquid.

Table 2.9 Equilibrium segregation coefficient of impurities in silicon.

Impurity	k_0	Type
B	8×10^{-1}	P
Al	2×10^{-3}	P
Ga	8×10^{-3}	P
In	4×10^{-4}	P
O	1.25	N
C	7×10^{-2}	N
P	0.35	N
As	3.0×10^{-1}	N
Sb	2.3×10^{-2}	N
Te	2.0×10^{-4}	N
Li	1.0×10^{-2}	N
Cu	4.0×10^{-4}	X
Au	2.5×10^{-5}	X

156 × 156 mm square cell type. Figure 2.26 shows TDR-105 silicon monocrystalline furnace produced by Bai Er Te Optoelectric Co., Ltd., Jiangsu, China. The furnace is high efficient and energy-saving. It can produce 2 × crystals at one time and the output of one furnace reaches up to 350 kg and the effective crystal production rate is above 85% and the cost reduces by 30%. The monocrystalline silicon wafer produced has achieved good conversion efficiency both for N- and P-type silicon cells.

At present, the CZ monocrystalline silicon is developing towards high purity, high integrity, high uniformity and big diameter. For refinement effect concerns, the pulling growth rate is usually slow and the cost is high, also it is apt to introduce impurity oxygen. As a result, it shall be considered comprehensively. To further control the evenness of impurity defects in silicon monocrystalline and raise the production efficiency, the magnetic-field-applied CZ method (MCZ) and the double-crucible CZ method have been developed.

Figure 2.26 TDR-105 silicon monocrystalline furnace for solar cells.

2.4 Nature of a-Si/μC-Si Thin Film

2.4.1 Nature of a-Si Thin Film

The absorption layer thickness of the crystalline silicon is generally about 100 μm to adequately absorb most of the sunlight. The a-Si thin-film cell has far thinner absorption layer thickness compared with the crystalline silicon, and several hundred nm in thickness can absorb adequate solar energy, which is because the crystalline silicon has different electron transition structure from a-Si. The crystalline silicon energy band has an indirect transition structure while a-Si with direct transition structure. As a result, a-Si can absorb photons in a more effective way than the crystalline silicon. In the range of visible light, a-Si has the light absorption coefficient about one to two orders of magnitude larger than that of the crystalline silicon, and the intrinsic absorption coefficient can reach up to $10^5/\text{cm}^{-1}$. Accordingly, the absorption layer of the a-Si thin-film cells is only one percentage of the crystalline silicon solar cells in terms of thickness.

2.4.1.1 Basic Nature of a-Si Thin Film and its Application to PV Sector

The a-Si thin film has the following basic characteristics: The atoms are arranged in order in the short distance but out of order in the long distance. That is to say, for an independent silicon atom, similar to the silicon atom in the monocrystalline silicon, its surrounding is the covalent bond composed of four silicon atoms where the bond length is similar but the bond angle is deviated. In the tetrahedral structure of the crystalline silicon, the angle of a bond with the other is 129.5 ° while the bond angle of a-Si will deviate by about ± 10 °. The atoms near a-Si are arranged in order but those far away have no rule. For a-Si, the order of the near atoms correlates the amorphous material with the corresponding crystal materials but those atoms far away are out of order, which makes them different in nature.

Since a-Si has different atomic order from the crystalline silicon, their energy band structures are different, too. As a result, the band structure theory of the crystal cannot be simply applied to the a-Si thin-film semiconductor, and some corrections shall be madeb (see Figure 2.27 for the energy band diagramme of a-Si). Unlike the crystalline silicon whose energy band is separate, the defects in the band gap are continuous and in the a-Si band, the valence band top and the conduction band bottom have developed to the 'band tail structure' and extended to the deep gap. In a-Si, the carrier mobility is two to three orders of magnitude less than that of the crystalline silicon. The material is not suited to produce high-speed micro-electronic components but it can be applied to the solar cell with simple structures.

There are many energy band theories about the a-Si semiconductor, amongst those, Mott-CFO model is very successful. The Mott-CFO model holds that in the nearest zone of a-Si, the crystalline semiconductor band theory can be applied, and in the near and far zones, the atomic arrangement deviation can be viewed as a micro disturbance, resulting

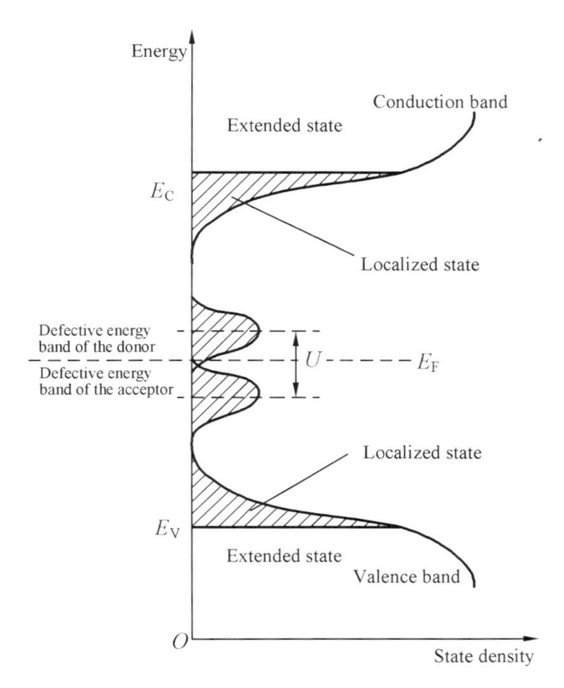

Figure 2.27 Energy band diagramme of a-Si.

in the band tail structure. The amorphous defect concentration is high, which can reach up to about $10^{17}/cm^3$, widening the defect energy level and forming defect band. In the amorphous semiconductor band, no accurate valence band top or conduction band bottom is present, but the 'mobility edge' (still expressed by E_c and E_v), which has similar meaning, is existent. For a-Si, the energy band consists of two parts: One part is called as the extended state ($E > E_c$ or $E < E_v$ zone), and the other part as the localised state ($E_c > E > E_v$ zone). In the extended state, the amorphous semiconductor has low electron and hole mobility. In the localised state, the carrier can only leap between localised energy levels in thermal excitation or tunnel effect.

a-Si, which is deposited by physical vapour deposition such as sputtering and so on, contains a lot of silicon dangling bond, resulting in pinning of Fermi level. The thin film is insensitive to doping and it is difficult to form P-type and N-type via doping and thus it cannot be applied in practise. As a result, the a-Si film is mainly produced by such chemical vapour deposition methods as plasma-enhanced chemical vapour deposition (PECVD), hot wire chemical vapour deposition(HWCVD), photo chemical vapour deposition (photo-CVD) and so on. These chemical vapour deposition methods use hydrogen to passivate the silicon dangling bond to obtain the hydrogen-contained a-Si film (a-Si:H), which can significantly improve the performance of the a-Si film. It is worth noting that when the hydrogen bond is introduced, the energy band structure of a-Si is simultaneously changed, and the energy gap of a-Si:H can be raised to about 1.7 eV in width.

Compared with the crystalline silicon, the a-Si film has following basic characteristics:

1. The a-Si atoms are in order in short distance in the nm range;
2. The a-Si covalent bond has continuous random network structure;
3. The physical nature of the monocrystalline silicon is heterogeneous in all directions; and the polycrystalline silicon and μC-Si are multi-directional; and a-Si is homogeneous in all directions;
4. The electron transport property of the a-Si energy band is greatly different from that of the crystalline silicon, where the pseudo-vertical transition conduction mechanism is present, the light absorption coefficient is larger than the crystalline silicon but the carrier mobility is small due to existence of dangling bonds and other defects;
5. After hydrogenation, the a-Si band is about 1.7 eV in width, which is larger than that of the crystalline silicon (1.12 eV). The band gap width of a-Si can be also continuously adjusted by various alloys in a range of 1.4–2.0 eV, which is good for development and optimisation of new materials with better performance;
6. A-Si is metastable in thermodynamics. Under appropriate heat treatment conditions, a-Si can be converted to polycrystalline silicon, μC-Si or nanometer silicon.

Thanks to the virtues of a-Si film, including the unique physical performance, simple preparation technique, low-cost and easy to develop large-scaled continuous production, it has attracted the attention in the industrial sector. The a-Si film has the following advantages in the solar cell sector:

1. Low material and production cost, which is mainly reflected in three aspects: First, the cell can be produced on low-cost substrate material, for example, glass, stainless steel, plastic and so on; second, the absorption layer of the a-Si film is only 1%

of the crystalline silicon in thickness; and third, the a-Si film can be produced at 100–300 °C, thus the energy consumption is lower;
2. Simple preparation technique and suited to produce large-sized film, full automation available to improve productivity;
3. Integrated cell and module production, and the products for various demands can be designed and developed according to the requirements of the user on power, output voltage and output current;
4. Apt to realise light and flexible cells and integrated to the buildings.

Compared with the crystalline silicon, however, the energy band of a-Si has band tail due to existence of dangling bonds and other structural defects. Since the structural defect and band tail will work as recombination centres, which will affect significantly the carrier transport, the efficiency of the amorphous thin-film silicon solar cell is relatively low, and the PV conversion efficiency has some attenuation in the sunshine for a long time. Accordingly, the focus shall come to improve the performance of the a-Si film and the structural design of solar cells, accurately control the thickness of each layer, and improve the interface state between layers. In recent years, the new thin-film silicon solar cells have been developed, including the a-Si stacked layers and so on. And the new a-Si alloy is also a key approach, for example, a-SiGe and a-SiC alloy.

2.4.1.2 Fermi Level Pinning and Efficiency Degradation Mechanism for a-Si Thin-film Cells

In the crystalline silicon, the phosphorus and boron atoms are usually in substitution position, serving as the donor or the acceptor. The doped atoms in a-Si are, instead of the substitution position, easy to react with the network defects and hydrogen ions contained in a-Si, so that some doped atoms cannot provide the electron or hole for the a-Si base, and cannot serve as the donor or acceptor. The Fermi level representing the carrier concentration distribution shows, the doping seems insensitive to Fermi level variation, which is called as 'Fermi level pinning'. Since the number of the doped atoms capable of supplying carriers is different from the actual number of the atoms doped into a-Si, the ratio of them is defined as 'activation coefficient of doped atoms'. It is mainly dependent on the defect density in the thin film. For a-Si grown in PECVD (plasma enhance chemical vapour deposition), the activation coefficient is usually 30% for the phosphorus atom and much lower for the boron atom.

The studies over years show, the hydrogen-contained a-Si has the defect that can result in efficiency degradation. For the solar cell produced by a-Si, the light conduction and dark conduction will be simultaneously reduced after radiated in the sun for some time and then keep stable, resulting in reduction of PV conversion efficiency of amorphous solar cells. The early degradation of the amorphous solar cells is called as Staebler-Wronski effect (S-W effect), and its attenuation rule is described below:

(1) The a-Si solar cells will degrade exponentially in the early state, which mainly happens on the early stage (about three months to one year). It is caused by the metastable defect resulted by hydrogen atoms moving in the lattice;
(2) Generally, the degradation is about 24% for the single-junction a-Si solar cell, and about 15% for the double-junction a-Si cell, which shall be deducted by the seller when the solar cell is sold;

(3) If a-Si is stacked with μC-Si to produce the solar cell, the degradation will be dramatically reduced.

The study discovers that the a-Si film produced by PECVD generally contains 10–15% hydrogen, which, on one hand, can passivate the silicon dangling bond, reducing the defect density, and on the other hand, if the hydrogen content is far larger than the silicon dangling bond density, the excessive hydrogen will be in a state with low activation energy in a-Si, including SiH_2 and SiH_3 groups, hydrogen-related micro-holes and so on, and take on various forms and metastable. That is to say, the hydrogen in large amount in a-Si can make the dangling bond saturated to form inactive Si-H bonds and form other hydrogen bonds. These hydrogen bonds have various binding energies in a-Si, and they will decompose or have other reactions after long-time sunshine, which may result in hydrogen atoms diffusing and mobbing in the film and generating new metastable defect centres to reduce the lifetime of the photo-generated carriers and the efficiency of the solar cells.

As a result, the intrinsic a-Si layer cannot contain excessively high hydrogen content. In the preparation techniques, helium or argon or the like is used during annealing, and how wire (HW) chemical vapour deposition is also used to reduce the S-W effect. In the a-Si:H film produced by inert gas annealing, the hydrogen content is less than 9%; and in the a-Si:H film produced by hot wire chemical vapour deposition, the hydrogen content is only 1–2%. These techniques are all aimed to reduce the excessive hydrogen and the defect density in the a-Si film to make the Si-Si and Si-H bonds stable.

2.4.2 Nature of μC-Si Thin Film

The 'silicon grains' in the thin-film silicon solar cell are mainly composed of μC-Si or the nanometre silicon of locally crystallised and fine grains embedded in a-Si. Generally, the grain with size in 10–30 nm is called as the μC-Si film (μc-Si); and the grain with size below 10 nm is called as the nanometre silicon film (nc-S) but no strict definitions are found in the literatures. The μC-Si films can work independently or combine with a-Si to form various new thin-film silicon solar cells. For example, the μC-Si cells can be used as the bottom cell to absorb the red light, which can be stacked with the a-Si top cell that can absorb the blue and green light to improve dramatically the efficiency of the solar stacked cells. μC-Si has been applied to the thin-film solar cell sector. In this section, the nature of μC-Si used to the thin-film solar cells will be discussed.

The μC-Si film is based on the uneven growth mechanism where a layer of a-Si in certain thickness (about dozens of nanometres) is first deposited on a substrate surface, which then forms into cores for growth of polycrystalline silicon. Figure 2.28 shows the diagramme of μC-Si incubated from a-Si and the grain-growing process.

The μC-Si (μc-Si:H), a mixed-phase material, contains amorphous matrix and grains where the grains are embedded in the a-Si network. In the grain, the silicon atoms have similar bonding structure to the monocrystalline silicon, and the μc-Si material with large-sized grains has similar band gap with the crystalline silicon (1.12 eV), and does not have direct transition characteristics. Compared with a-Si, it has higher conductivity, higher doping efficiency and lower light absorption coefficient. As a result, the μC-Si is often used with a-Si, which can significantly extend the spectrum response range of the thin-film cell to the long wavelength and improve the PV conversion efficiency. In addition, the μC-Si is in good order than a-Si, and thus it can dramatically improve the

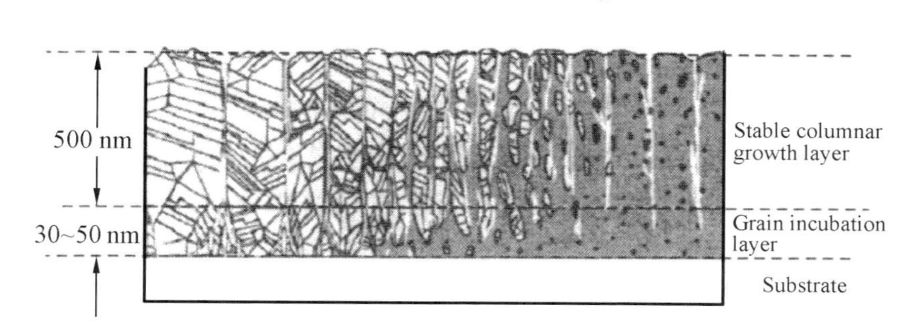

Figure 2.28 Diagramme of μC-Si incubated from a-Si and grain growth.

stability of the thin-film cell. Since the a-Si cell has good short wavelength spectrum response and the μC-Si cell has good long wavelength spectrum response, together with compatible deposition techniques of μC-Si and a-Si, the solar cell can adopt the stacked layers of a-Si/μC-Si to improve the cell efficiency and stability as well as the economic benefit. Since μC-Si is a semiconductor material with indirect band gap and small light absorption coefficient, it must have an absorption layer (1–2 μm) thicker than a-Si in order to fully absorb the solar rays.

The phase change from amorphous to microcrystalline is not sudden, and there is a complex phase zone where the amorphous and microcrystalline phases coexist, and the microcrystalline grains can be formed in the amorphous network. The a-Si containing grains and in the phase change threshold is good for producing the absorption layer of the solar cell. For the μC-Si material with good grain size and crystalline phase ratio, its band gap is similar to the crystalline silicon (1.1 eV), and thus it is suited to produce the bottom cell of the stacked cells of a-Si/μC-Si.

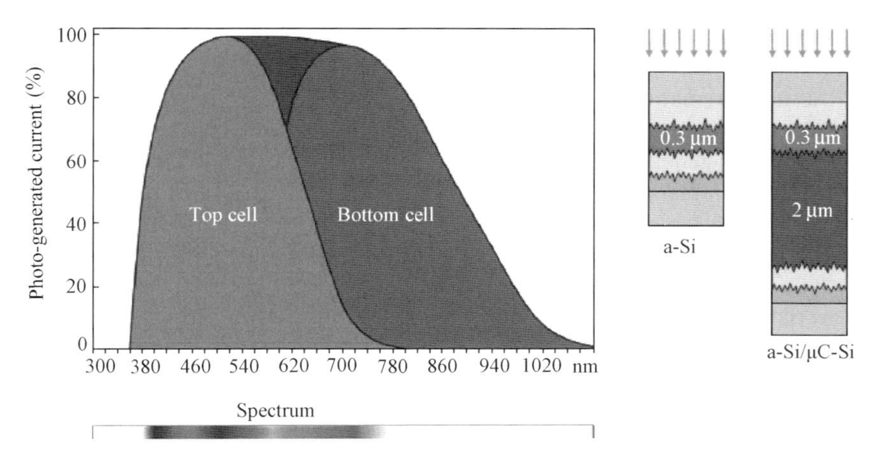

Figure 2.29 Solar cells with stacked layer of a-Si/μC-Si to extend the absorption light wavelength range.

See Figure 2.29 for the solar cell with stacked layer of a-Si/μC-Si. Since the μC-Si bottom cell can absorb the red light and the a-Si top cell can absorb the blue and green light, it is a thin-film silicon cell with high efficiency and good stability.

2.5 Preparation Methods of a-Si/μC-Si Film

2.5.1 A Main Raw Material for Silicon Film Preparation—Silane

The silane was mainly used for polycrystalline silicon and integrated circuit manufacture. In late twentieth century, the thin-film transistor liquid crystal display (TFT-LCD) was industrialised at large scale. In recent years, the thin-film silicon solar cell (TFSSC) has seen rapid development, which has dramatically changed the global commercial silane market structure. Table 2.10 shows, the global commercial silane output was 2680 t in 2008 (except those directly for producing polycrystalline silicon), and those used for PV industry will account for above 50% soon.

Since silane can be used to produce the TFSSC, TFT flat-panel displays and quality electronic-grade monocrystalline, achieving localisation of silane production is has important significance for China's new energy industry and information industry.

There are mainly three techniques to produce silane:

1. **Chlorosilane disproportionation:** In plant-producing polycrystalline by the Siemens method, the by-product $SiCl_4$ shall be first used as the raw material and hydrogenated to $SiHCl_3$, and then go through disproportionation, separation and cryogenic rectification refinement process to achieve high-purity silane SiH_4. The reaction principles are given below (Formulas 2.16, 2.17, and 2.18):

$$2SiHCl_3 = SiH_2Cl_2 + SiCl_4 \text{----------------Reaction (1), catalyst} \quad (2.16)$$

$$2SiH_2Cl_2 = SiH_3Cl + SiHCl_3 \text{----------------Reaction (2), catalyst} \quad (2.17)$$

$$2SiH_3Cl = SiH_4 + SiH_2Cl_2 \text{----------------Reaction (3), catalyst} \quad (2.18)$$

The most important technique to produce SiH_4 lies in selection of the catalyst for the above three reactions. In 1987, the author participated in a R&D project on SiH_2Cl_2 preparation organised by Shanghai Institute of Metallurgy, CAS, and found the catalyst and the transformation from trichlorosilane to dichlorosilane was realised at 60–80 °C. The disproportionation process from trichlorosilane, dichlorosilane and monochlorosilane is generally used across the globe (see Figure 2.30 for the disproportionation reaction plant).

The boiling point of chlorosilane is given below: $SiCl_4$: 57.6 °C; $SiHCl_3$: 31.5 °C; SiH_2Cl_2 : −8.3 °C; SiH_3Cl : −30.4 °C; SiH_4 : −111.5 °C. Obviously, all of them are

Table 2.10 Silane consumption growth with PV industry development in the world in recent years.

	IC	TFT	PV	Others
2008	37%	33%	26%	4%
2010	21%	19%	58%	2%

Figure 2.30 Chlorosilane disproportionation plant.

flammable and explosive. Accordingly, the cryogenic rectification process shall be used for refinement, and the liquid method shall be used for storage and transport. Special concern shall be given to safety during silane preparation.

2. **Fluoride method:** First, the silicon tetrafluoride (STF) can be produced by the low-cost calcium fluoride, sulphuric acid, quartz or hydrogen fluoride, and the triethylaluminium and the aluminum powder shall be mixed in the flux bath to synthesise silicon tetrahydride (SAH) in the agitated reactor; then, generate the silicon tetrafluoride (STF) and the silicon tetrahydride (SAH) to SiH_4 by steps (Formulas 2.19 and 2.20).

$$5SiF_4 + 5NaAlH_4 = Na_5Al_3F_4 + 2AlF_3 + 5SiH_4 \quad (2.19)$$

$$SiF_4 + NaAlH_4 = SiH_4 + NaAlF_4 \quad (2.20)$$

The silane produced by fluoride is cheap but it shall prevent environmental pollution during production.

3. **Reaction of magnesium silicide and ammonia:** Magnesium silicide can react with ammonia to generate silane (SiH_4): The technique was developed by the Zhejiang University and Sichuan Emei Semiconductor Plant in the 1970s when it was mainly used to produce quality electronic-grade polycrystalline silicon (Formulas 2.21 and 2.22). The chemical reaction and principle are shown below:

$$2Mg + Si = Mg_2Si \text{ (in the synthesis sintering furnace)} \quad (2.21)$$

$$Mg_2Si + 4NH_4Cl = SiH_4 + 2MgCl_2 + 4NH_3 \text{ (in the reaction vessel)} \quad (2.22)$$

$MgCl_2$ is solid sediment, NH_3 is subjected to cryogenic separation and SiH_4 can be refined to electronic grade by the special molecular sieve. The production process, however, is intermittent, and the product quality is apt to fluctuate. Besides, the output is small, and it is difficult to achieve massive production of a-Si thin-film solar cells and thin-film transistor liquid crystal display (TFT-LCD).

2.5.2 Introduction to Silicon Thin-film Growth Methods

The main raw material is silane (SiH_4) in the TFSSC industry. Silane is gaseous at normal pressure and temperature, and the boiling point is $-111.5\,°C$. And it is not corrosive, flammable in the air and apt to decompose at high temperature. The chemical vapour reaction of silane in silicon deposited on the substrate is the thermal deposition of silane $SiH_4(g) \rightarrow Si(s) + 2H_2(g)$. The reaction is basically irreversible although it can deposit a-Si and polycrystalline silicon at the temperature down to 600 °C. To curb the trend of silane gas thermally decomposed to particles in gaseous phase, the hydrogen shall be used to dilute silane gas in most epitaxial growth processes and reduce the reaction system pressure.

The silane (reaction gas) and the carrier gas (hydrogen) can become plasma by means of glow discharge in vacuum conditions, which can make silane thermally decomposed at lower temperature (about 300 °C) and enable some hydrogen atoms entering the a-Si film to passivate the dangling bond in a-Si. This is plasma enhance chemical vapour deposition (PECVD). In addition to PECVD, the hot wire chemical vapour deposition (HWCVD) can also use thermal decomposition of silane to make a-Si deposited on the substrate.

Since the solar cell with a-Si/μC-Si stacked layers requires that μC-Si should be nanometre in thickness and the traditional PECVD techniques can grow the silicon thin film at a rate about 0.3 A/s, it generally takes several hours to grow the μm-level-thick μC-Si absorption layer. To reduce the production cost, the deposition rate must be raised, and other methods can be used to improve the deposition rate, including very high frequency (VHF) PECVD, hot wire (HW) CVD, very high frequency (VHF) PECVD and radio frequency high pressure depletion (RF-HPD)PECVD and so on, and see the next section for descriptions. In addition, the hydrogen content in a-Si and the transformation from a-Si to μC-Si are also the issues of great concerns. When the μC-Si film is produced on glass or other substrates, the a-Si film is generally first deposited and then transferred to the μC-Si film.

For the proportion of μC-Si in a-Si (i.e., crystalline fraction X), the Raman spectrum can be used for measurement and then it can be figured out by the following formula (Formula 2.23):

$$X = (I_C + I_{GB})/(I_C + I_{GB} + I_A) \tag{2.23}$$

Where, I_C, I_{GB} and I_A refer to the scattering intensity of crystalline state, the interface and the amorphous state, respectively, which respond to 516.8 cm^{-1}, 507.5 cm^{-1} and 480.0 cm^{-1} of the diffraction spectrum.

2.5.3 Preparation of a-Si/μC-Si Thin Film by PECVD Method

PECVD is short for 'plasma enhance chemical vapour deposition'. The silane gas and hydrogen will first have ionisation and chemical reaction in the plasma reaction chamber and then deposit silicon thin film on the substrate. Generally, when the hydrogen concentration is low, a-Si will be deposited on the substrate. As the H_2 concentration rises, the crystalline proportion of the silicon film will increase. The PECVD method can be also used to grow the μC-Si film but the growth rate is lower since it needs the hydrogen of higher concentration for dilution during crystallisation of μC-Si.

The crystallisation proportion of μC-Si is mainly dependent on the hydrogen concentration in the reaction gas and the temperature of the substrate. Generally, the high hydrogen concentration and the high temperature of the substrate are good to generate μC-Si.

In the PECVD technique, the ion energy in the plasma is also a key factor for a-Si to transform to μC-Si. Generally, it is apt to grow a-Si when the ion energy is high, and otherwise, it is apt to grow μC-Si. For example, when the very high frequency (VHF) PECVD is used to produce μC-Si film, it can reduce the energy of the high-energy ion and decrease the bombardment of the high-energy ion on the growth surface and raise the ion current density reaching the substrate surface. Compared with 13.56 MHz, it is better for generation of μC-Si, crystalline fraction and deposition rate of silicon film.

It is reported, the growth rate of μC-Si film can reach above 3 nm/s when the VHF-PECVD method is used to deposit μC-Si at high pressure (about 10 Torr), high power density (about 1 W/cm^2), spray air and small electrode spacing (about 1 cm).

In addition, the electrodes in the plasma reaction chamber can be also designed to change the quantity of the high-energy ions in the plasma. For example, a negative electrode can be added between the traditional electrodes, which can make the high-energy ion escape from the plasma and thus fail to reach the surface, reducing the bombardment and influence on the substrate.

2.5.4 Preparation of a-Si/μC-Si Thin Film by HWCVD Method

HWCVD is short for 'hot wire chemical vapour deposition'. The silicon thin film can be grown in the following processes by HWCVD method: The hot tungsten wire up to 1500–2000 °C shall be installed near the glass substrate, and when silane or hydrogen or other source gas passes the hot wire, the molecular bond will break to form various neutral groups and deposit a-Si on the substrate after chemical reaction. The tungsten wire, generally 0.3–0.7 mm in diameter and shaped in disc or flat plate, is about 3–5 cm away from the substrate in the reaction chamber. The big current shall be connected to heat up the tungsten wire. Silane or other gas source will be decomposed under high temperature catalysis of the wolfram wire while flowing to the substrate, which is good for the silicon atom to form the μC-Si film on the substrate. The hot wire μC-Si film has high deposition rate with maximum of 3–5 nm/s, and the hot wire has low heat radiation to the substrate, and the deposition temperature is 175–400 °C. The μC-Si film grain produced by HWCVD method is about 0.3–1.0 μm, and it has columnar structure and it is oriented towards (110) crystal. It can be applied to PV modules. The a-Si film produced by HWCVD method has far less hydrogen content than that produced by PECVD method. See Figure 2.31 for the diagramme of the equipment in the reaction chamber for specially designed hot wire (HW) CVD where the silane gas flow enters the reaction chamber via the shower heads and the hydrogen or the ammonia gas enters via the bypass. The hot wire is used for heating, which can also catalyse the reaction.

If the distance between the hot wire and the substrate is shortened or the turn of the hot wire is increased, it can raise the deposition rate of the thin film. The pressure in the

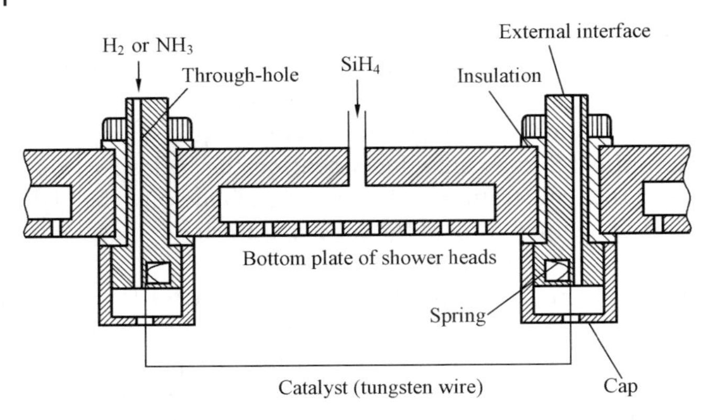

Figure 2.31 HWCVD reaction chamber diagramme.

reaction chamber is also a key factor when the HWCVD method is used to produce the a-Si film. Generally, the growth rate of the thin film will increase with chamber pressure.

In the HWCVD technique, the temperature of the hot wire is recognised as the leading factor for the quality of the μC-Si film. The study shows, if the hot wire is in a range of 2000–2100 °C, it can produce smooth μC-Si film, but if it is lower than 1500 °C, it can only produce the a-Si film because excessively high temperature is an adverse environment to growing the hydrogenated silicon-based film. NREL, USA, carried out a systematic study on the HWCVD technique where the hot wire is controlled at 1900 °C, and the substrate is fixed at 360–380 °C. In this way, the a-Si film of high quality was obtained. The film has low hydrogen content, only 1% or so. In contrast, in the general a-Si deposited by the plasma glow, the hydrogen content is at least 10–15%. Excessively low hydrogen content will result in high defect density but excessively high hydrogen content may result in metastable defects.

For the μC-Si film produced by the HWCVD technique, the grain size can reach above 1 μm and the electron mobility of μC-Si can reach up to 20 cm^2/Vs, and the deposition rate can reach 4 nm/s.

Obviously, the HWCVD technique has some advantages in producing μC-Si film. In addition, it is also a key technique to produce the a-Si film. Regardless of the substrate temperature and H_2 concentration, the a-Si film produced by the HWCVD technique usually has larger growth rate than that by the ordinary PECVD technique. The hot wire is generally made of wolfram or the wolfram and tantalum alloy. The material will affect the maximum temperature of the hot wire. It should be pointed out, since the heating temperature of the wolfram wire is high, it has high demands on some equipment and may cause some contaminations to the product. In addition, since it is difficult to install the hot wire in the reaction chamber, it is not suited to produce large-scaled even film on the glass plate. This is also the reason why the HWCVD technique cannot be widely used in practise.

2.5.5 Growing Silicon Thin Film by Other Methods

2.5.5.1 Direct Preparation of μC-Si Thin Film by LPCVD Technique

LPCVD is short for 'low-pressure chemical vapour deposition'. Different from the ordinary CVD techniques, the LPCVD technique reduces the pressure in the reaction chamber to 0.05–5 Torr by means of mechanical pumps and vacuum pumps. In this case, the reaction temperature is relatively low (550–800 °C), and silane serves as the gas source. In this technique, it can grow the μC-Si film with good uniformity. The growth rate is about 0.1–1 μm/min, and it is suited to epitaxial growth of thin layers. LPCVD has good control effect on auto-doping and the impurity distribution is sharp with good uniformity. Compared with the ordinary PECVD techniques, the mobility of minority carriers is larger, and the stress in the grain is lower. Since the growth rate is low and the deposition temperature is relatively high, the selection of substrate is limited.

2.5.5.2 a-Si Crystallised to Polycrystalline Silicon Thin Film by SPC Technique

When subjected to long-time heat treatment above 600 °C in certain protective atmosphere, the a-Si film can be transformed to polycrystalline silicon thin film in the condition lower than the silicon crystallisation temperature to achieve solid phase crystallisation (SPC). In the temperature range, the grains will be combined with each other due to movement of the crystal boundary and the small grains will gradually grow to big ones. Generally, the grain will increase in size with temperature. SPC can obtain the silicon grain in large size and the carrier mobility is large. Since the crystallisation temperature is generally above 600 °C, however, it has some requirements on the substrate material. In addition, it is also a disadvantage that the SPC process shall take a long time.

2.5.5.3 a-Si Crystallised to Polycrystalline Silicon Thin Film by Metal-Induced Method

In the metal-induced crystallisation technique, some metal atoms (aluminum, nickel, gold, etc.) are introduced into a-Si for induction, and it can realise continuously induced crystallisation for a-Si and obtain continuous polycrystalline silicon on the glass substrate. As a result, the polycrystalline silicon is also called as continuous grain silicon (CGS).

The metal-induced crystallisation technique for producing polycrystalline silicon thin film lies in the metal category and the crystallisation temperature, and it has nothing to do with the structure of a-Si or the metal layer thickness and so on. As a result, the requirements on a-Si are not demanding, and it can simplify the preparation processes of a-Si film and reduce the production cost.

The metal-induced crystallisation technique can make a-Si crystallised to polycrystalline silicon at 200–350 °C in a period of 30 min or so. Although it has been widely applied to the thin-film transistor display sector, it has a fatal disadvantage—the metal atom for induction may affect the electrical performance of the semiconductor silicon film or even result in short circuit of the cell as a whole.

2.5.5.4 a-Si Crystallised to Polycrystalline Silicon Thin Film by RTP Technique

The rapid thermal process (RTP) refers to the heat treatment technique where the material shall be rapidly heated up to above 1000 °C by light in dozens of seconds and then rapidly cooled down. In the RTP system, the iodine tungsten lamp is generally used to heat up. Since the spectrum of the iodine tungsten lamp covers the range from ultraviolet to infrared, it, on one hand, can heat up the material, and on the other hand, the high-energy photon in the short wave of the iodine tungsten lamp can promote diffusion. In the RTP technique, there are thermodynamic effect and light effect, especially high-energy photon effect. For example, after produced at about 250 °C, the a-Si film can be rapidly crystallised at 700 °C by the RTP technique. The RTP technique can be used to produce intrinsic polycrystalline silicon thin film and heavy doping thin film, and the polycrystalline silicon thin film produced has been applied to polycrystalline silicon emitter and field-effect transistor (FET) of the solar PV and integrated circuit. Compared with traditional heat treatment, the RTP technique significantly reduces crystallisation heat and time but the grain of polycrystalline silicon thin film grain is much smaller and thus higher heat stress will be existent and the repeatability and evenness are also poor.

2.5.5.5 a-Si Crystallised to Polycrystalline Silicon Thin Film by Linear Laser Technique

The linear laser crystallisation refers to making local area of a-Si film rapidly rise to a certain temperature and crystallised under the action of pulse laser. The laser beam is extended to linear laser with even intensity distribution by means of optical methods, and when the linear laser scans on the a-Si film, it can obtain the crystallised and even polycrystalline silicon thin film. The XeCl laser is mainly used for this purpose, that has the following parameters: wavelength: 308 nm; pulse width: generally, 15–50 nm; and light absorption depth: only dozens of nm. Since laser is featured by short wavelength, high energy and shallow optical absorption depth, it can make a-Si rise to the crystallisation temperature in a range from dozens of seconds to hundreds of nanosecond (ns) and rapidly transform into polycrystalline silicon while the substrate has small temperature rise. In this technique, the substrate has low temperature so that it is not demanding for the substrate materials. Since the pseudo-molecular laser pulse has short time (10–30 ns) and the wavelength falls in the UVB range, the absorption happens mainly in the depth of 6 nm underneath of the material surface. As a result, it is an ideal tool to crystallise silicon thin film on the glass substrate. The laser crystallisation technique, however, has its own weakness: complex equipment and high production cost. Although it has been applied to the TFT display sector with high additional value, it is difficult to realise massive industrial application in the solar cell industry.

2.5.6 Comparisons of Various Silicon Film Growth Methods

The above silicon film growth methods can be divided into high-temperature techniques (> 600 °C) and low-temperature techniques (< 600 °C). Most of them first form a-Si of certain thickness on the substrate surface and then, generate cores in the special conditions and gradually grow to polycrystalline silicon (μC-Si) with big grain. The formation of polycrystalline silicon (μC-Si) belongs to uneven growth mechanism.

For the low-temperature deposition (< 600 °C) techniques, the selection range of substrate is wide (e.g., glass) but the deposition rate is slow with small throughput and poor film quality. Its advantage lies in low cost. PECVD, one of the main low-temperature deposition techniques, is often used to deposit the a-Si film of thin thickness and low crystallisation requirements. The SPC process takes a long time, which is adverse for industrialisation; and the linear laser technique is rapid but the equipment is complicated and expensive. For the low-temperature techniques, it is a key direction for the study how to raise the film growth rate in condition of polycrystalline silicon (μC-Si) quality. At present, the joint application of PECVD, SPC and MIS has come into existence. And even the liquid phase Epitaxy (LPE) technique has been developed where the growth rate can reach 4 μm/min and the minority carrier diffusion length can reach 60 μm.

For the high-temperature deposition (800–1200 °C) techniques, for example, APCVD and RTCVD, the grains deposited are larger and the growth rate is higher compared with the low-temperature techniques, but it is demanding on the substrate. To prevent the impurities diffusion out of the substrate, a layer of silicon dioxide or silicon nitride is usually deposited on the substrate to serve as an insulation layer. In addition, hydrogen is apt to diffuse outwards in the high-temperature techniques, and to achieve hydrogen passivation, it shall need low-temperature treatment.

2.6 Compound Semiconductor Materials

The band gap of semiconductor material for solar cell applications usually falls in a range of 1–2 eV. The theoretical study shows, the semiconductor materials with band gap about 1.4 eV can achieve the maximum conversion efficiency. Although wide band gap is good to improve the open-circuit voltage, excessively wide gap will make the absorption spectrum narrow, reducing excitation of carriers and photo-generated current.

Although silicon is deemed as the most important and matured PV materials nowadays, the band gap of the crystalline silicon is 1.12 eV wide, which is not in good match with the solar spectrum. In addition, it is an indirect transition material, and thus the light absorption coefficient is low. From these points, silicon is not the best PV material. Accordingly, the human being has always been looking for other PV materials.

For the compound semiconductor materials, the forbidden band gap is generally very wide and most of them are direct transition materials with large light absorption coefficient. It only takes several micrometres of the material in thickness to produce high-efficient solar cells. In addition, the compound semiconductor solar cell has obviously better anti-radiation performance and working temperature range than that of the silicon solar cell.

2.6.1 GaAs and Other Semiconductor Materials

Although GaAs, InP and III-V group compounds can be all used as the material of crystal solar cells, only GaAs and its ternary compounds have been widely used for cost, preparation difficulty and material performance concerns.

GaAs, whose forbidden band gap is 1.43 eV wide, is a direct band gap material. Compared with the light absorption coefficient of Si and Ge, for III-V group compound semiconductor material, the light absorption coefficient of GaAs will rise sharply up to above 10^4/cm, 1 order of magnitude higher, when the wavelength is below 0.85 μm. This is the strongest part of the solar spectrum. As a result, for the GaAs solar cell, as long as the thickness reaches 3 μm, it can absorb about 95% energy of the solar spectrum.

Nowadays, the GaAs solar cell has developed from simple single P-N junction to multi-junction stacked layers in structure, for example, AlGaAs/GaAs, GaInP/GaAs, GaInP/GaAs/Ge, GaInP/GaAs/GaInAs/Ge, GaAs/GaSb and so on, as well as the GaAs cell and the concentrator cells prepared on the low-cost Si and Ge substrate.

It is reported, the PV conversion efficiency of GaAs multi-junction cells has reached 32% and even up to 42%. Since III-V-group compound solar cell has high cost on materials and production, it is mainly used to special applications. The concentrator technique can further raise the power generated and reduce PV generation cost.

The preparation techniques of GaAs materials and cells include: monocrystalline growth and diffusion, vacuum evaporation, liquid phase epitaxy (LPE), organic metal oxide chemical vapour deposition (MOCVD) and molecular beam epitaxy (MBE) and so on, which will be described in Chapter 5.

2.6.2 CdTe and CdS Thin Film Materials

For CdTe polycrystalline thin film, the band gap is 1.45 eV, and the theoretical conversion efficiency of solar cells is about 29%. Obviously, it is a low-cost material of high efficiency and stability for thin film solar cells, and it is easy to realise massive production. Accordingly, it is currently the one with maximum output of thin film solar cells in the PV sector.

CdS, also an important material for the solar cells, has its band gap of 2.4 eV. It is mainly used as the material of N-type window of the thin film solar cell and it is generally combined with CdTe, $CuInSe_2$ thin film materials to form the heterogeneous junctions of good performance.

At present, the high-efficient CdTe thin film solar cell is basically N-CdS/P-CdTe in structure, that is, a layer of 'window material' CdS shall grow on the CdTe surface. Compared with the silicon solar cell, the CdTe solar cell is characterised by simple process and low cost. Although the compound CdTe is nontoxic at ambient temperature, Cd and Te are toxic. In addition, the Cd and Te resources are limited, especially Te.

2.6.3 $CuInSe_2$ and $CuInS_2$ Thin Film Materials

$CuInSe_2$ (CIS), the material with direct gap and of great importance for solar cells in addition to GaAs and CdTe, has its band gap in a continuously adjustable range of 1.04–1.68 eV. The PV conversion efficiency of the solar cell can reach 25–30% theoretically; the thin film of 1–2 μm in thickness can absorb over 99% of the solar rays, significantly reducing the cost of the solar cell. Based on $CuInSe_2$ (CIS), many PV materials have been developed, including $CuIn_xGa_{1-x}Se_2$ (CIGS) and $CuInS_2$, where Ga and S substitute some rare elements such as In and Se. The production lines of Cu (In, Ga) (S, Se) have been developed in the globe. At present, the CIS thin film solar cell materials are mostly doped with Ga, and the CIGS thin film is used as the absorption layer, which can dramatically improve the efficiency of solar cells. The sample in the lab has reached above 21.7% in 2015, and CIGS, similar to CdTe, has become the most important material of the compound thin film solar cell and the hot point of PV power sector in recent years.

$CuInS_2$, also a material with direct band gap and forbidden band gap of 1.50 eV or so, is better suited to the solar cell than $CuInSe_2$ (CIS) thin film materials. Its optimum forbidden band gap is 1.45 eV, and it may generate higher open-circuit voltage and smaller temperature coefficient. $CuInS_2$ can produce the P/N-type thin film of high quality and it is easy to produce homogeneous junctions and thus suited to massive production. Theoretically, the PV conversion efficiency can exceed 30% for the $CuInS_2$ homogeneous junction solar cell, which also features simple production, stable performance and low cost. Accordingly, deemed as a promising material for solar cells it is expected to replace the $CuInSe_2$(CIS) thin film, becoming the new solar material. $CuInS_2$, doped with Ga, can also improve the PV conversion efficiency of the solar cell. At present, the actual efficiency of the CuInS thin film solar cell, however, still has a big gap with the theoretical one and new techniques are under exploration with hope to achieve breakthroughs.

2.7 Analysis on Impurities in Semiconductor Materials

2.7.1 Glow Discharge Mass Spectrometry (GDMS) Analysis

GDMS, short for glow discharge mass spectrometry, is suited to chemical analysis of most elements in the periodic table. It can cover almost all the elements in the periodic table except for H,O,C,N with detection limit of 6 N (99.9997%). It is very suited to analyse the metal impurity element in 6 N UMG silicon. GDMS adopts direct sampling, and the sampling can be directly analysed (or subjected to simple mechanical processing) and no chemical method (acid dissolution) is necessary. The sample can be the solar-grade silicon in block, particle, chip, power or other shape.

GDMS has been used for ultra-pure metal analysis over 30 years. It can analyse 73 elements at one time and it can meet the analysis requirements on the silicon material of the solar cell. The technique is globally recognised. SEMI PV committee defined GDMS as the PV silicon standard testing method in October 2008 (SEMI PV 1–0309).

The analysis accuracy of GDMS has close relations with the sample. For the low content, the analysis error falls in a range of 10–25%. Table 2.11 shows the limits of GDMS method.

GDMS is not suited to Siemens silicon (>8N), and the analysis effect is not so good for electronic-grade silicon.

Table 2.11 GDMS impurity analysis limits (Unit: ppm).

Element	Limit	Element	Limit	Element	Limit	Element	Limit	Element	Limit
Li	<0.001	Cr	<0.01	Nb	<0.01	Pr	<0.005	Hf	<0.01
Be	<0.001	Mn	<0.005	Mo	<0.05	Nd	<0.005	Ta	<100
B	<0.001	Fe	<0.05	Ru	<0.01	Sm	<0.005	W	<0.05
F	<1	Co	<0.005	Rh	<0.01	Eu	<0.005	Re	<0.01
Na	<0.1	Ni	<0.01	Pd	<0.01	Gd	<0.005	Os	<0.01<
Mg	<0.005	Cu	<0.01	Ag	<0.01	Tb	<0.005	Lr	<0.01
Al	<0.01	Zn	<0.05	Cd	<0.05	Sm	<0.005	Pt	<0.01<
Si	Matrix	Ga	<0.05	In	<0.01	Eu	<0.005	Au	<0.1
P	<0.01	Ge	<0.05	Sn	<0.01	Gd	<0.005	Hg	<0.01
S	<0.1	As	<0.05	Sb	<0.01	Tb	<0.005	Tl	<0.01<
Cl	<0.05	Se	<0.01	Te	<0.01	Dy	<0.005	Pb	<0.01
K	<0.05	Br	<0.01	I	<0.01	Ho	<0.005	Bi	<0.01<
Ca	<0.05	Rb	<0.01	Sc	<0.001	Er	<0.005	Th	<0.005
Sc	<0.001	Sr	<0.01	Ba	<0.01	Tm	<0.005	U	<0.005
Ti	<0.005	Y	<0.01	La	<0.01	Yb	<0.005		
V	<0.005	Zr	<0.01	Cs	<0.005	Lu	<0.005		

Another method is ion coupled plasma mass spectrometry (ICPMS), which is mainly suited to material surface metal analysis. For the impurities inside, the sample shall be dissolved and then it can be used to most elements. Although the sample can remove silicon by acid dissolution to work out the mean impurity content in the sample, it may reduce the impurity in the sample or introduce foreign impurities to the sample, resulting in deviations if the sample processing is inappropriate. In addition, ICPMS, similar to GDMS, is not suitable to oxygen, carbon and nitrogen analysis.

2.7.2 Secondary Ion Mass Spectrometry (SIMS) Analysis

SIMS, short for secondary ion mass spectrometry analysis, can analyse almost all the elements, including boron, phosphorus, oxygen, carbon, Fe and nitrogen. And for most elements, the detection limits fall in a range of 8–11 N, and it is suited to impurity element analysis for electronic-grade silicon and the PV silicon of high purity. It has high accuracy (all element analysis dependent on the standard sample). The SIMS adopts direct sample (or via simple mechanical treatment, and thus it is free from disturbance of sample preparation, and the sample can be in blocks, particles, chips or powder. SIMS can offer the optimum detection limit for almost all elements. For the impurity of low content, the analysis scheme shall be developed and the instrument conditions shall be optimised. The method is better applicable to analyse several elements and a given element. In case of many elements, the cost will be high. The SIMS method can carry out profile distribution analysis on the impurities in the sample (Figure 2.32). The SIMS analysis features high accuracy and precision with error less than 10%.

SIMS has been applied to electronic-grade silicon analysis for more than 30 years, and it has become the indispensable analysis means in the semiconductor sector, including the silicon sector. It can carry out quantitative analysis for impurities with detection limits from 8 N to 11 N. To achieve the optimum detection limit (see Table 2.12), however, one to three elements can be only analysed in one analysis condition.

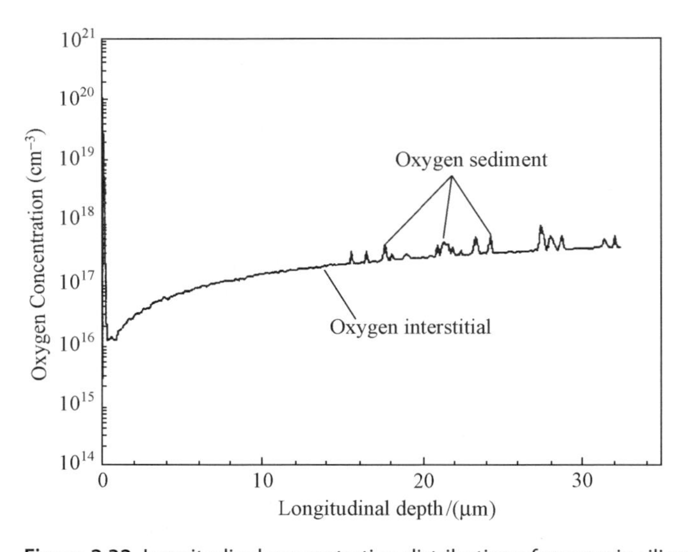

Figure 2.32 Longitudinal concentration distribution of oxygen in silicon by SIMS, EAG.

Table 2.12 SIMS impurity analysis limits (Unit: /cm^{-3}).

O_2^+ primary ion beam, for positive ion				Cs^+ primary ion beam, for negative ion	
He	1E17	Cr	3E11	H	5E16
Li	5E11	Mn	5E12	C	2E15
B	1E12	Fe	1E13	N	5E13
Na	5E11	Ni	1E14	O	5E15
Mg	1E12	Cu	1E14	F	1E14
Al	5E12	Zn	1E15	P	1E13
K	2E12	Mo	1E14	S	2E14
Ca	5E12	In	1E13	Cl	5E14
Ti	1E12	W	5E13	As	1E13
				Ge	5E13
				Sb	1E13
				Au	1E13

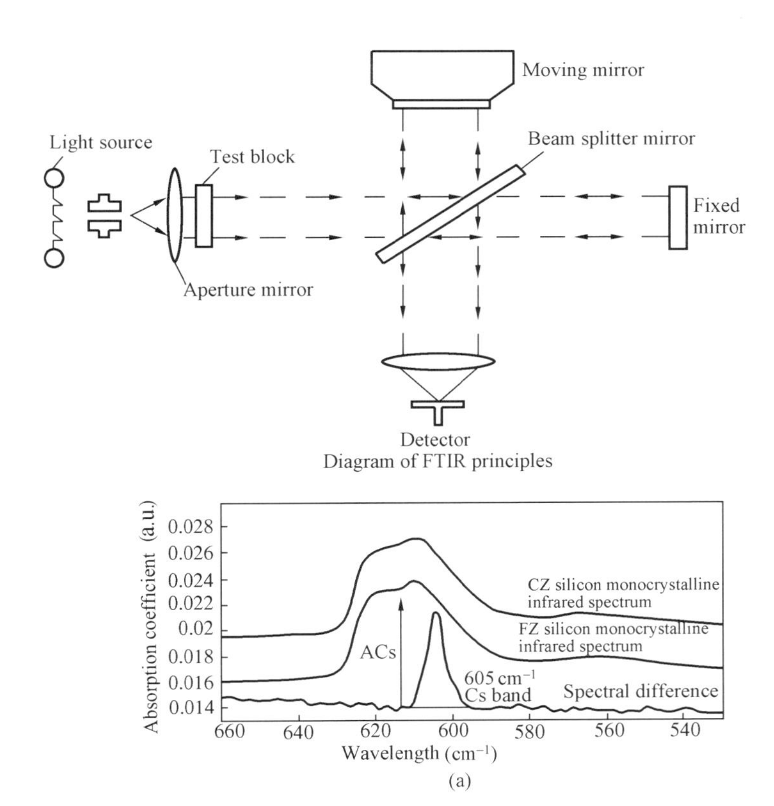

Figure 2.33 Detection principles of carbon/oxygen content in silicon wafer infrared spectroscopy.

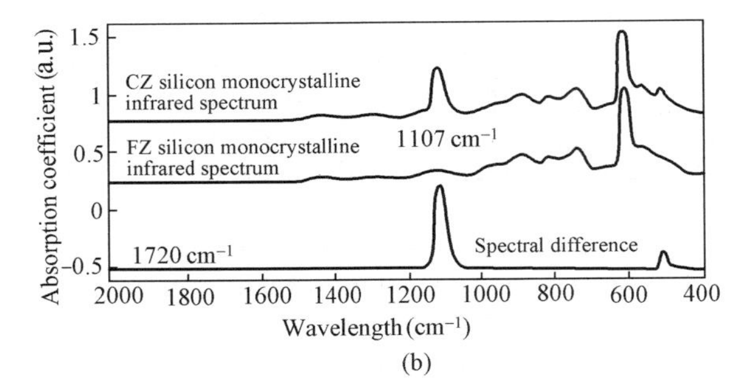

Figure 2.33 (*Continued*)

2.7.3 Infrared Spectroscopy to Detect the Carbon and Oxygen Contents in Silicon Wafer

The carbon atoms in silicon often exist in the form of substitution position, and the substitution carbon in silicon has direct relations with the absorption peak at 607.2 cm^{-1} in the infrared spectrum. The absorption peak can be measured by a double-beam infrared spectrophotometer or a Fourier transform infrared spectroscopy (FTIR), and then the content of carbon at substitution position can be figured out.

The oxygen atoms of the silicon crystal are basically present at the interstitial position of the silicon atoms. The interstitial oxygen in silicon has relations with the absorption peak at 1107 cm^{-1} in the infrared spectrum. The absorption peak can be measured by a double-beam infrared spectrophotometer or a Fourier transform infrared spectroscopy (FTIR), and then the content of interstitial oxygen can be figured out. The measurement method and sample preparation requirements are basically similar to the interstitial carbon in silicon except the spectrum scan range, that is, the position of absorption peak (see Figure 2.33 for the diagramme).

3

Preparation Methods of Crystalline Silicon Solar Cells

3.1 Preparation Process Flow of CSSCs

3.1.1 Basic Structure of CSSCs

For industrial preparation of monocrystalline silicon solar cells, it is generally to diffuse an N-type dopant on the P-type silicon wafer of (100) orientation with depth about 0.3–0.5 μm to form the P-N junction. When the cell is in the solar rays, a lot of negative charges (electrons) will be accumulated on the irradiating surface of the solar cell (N-type zone) and a lot of positive charges (holes) will be accumulated on the bottom surface (P-type zone) under the action of the built-in field of the P-N junction. The wafer surface shall be textured and an anti-reflection film shall be coated to reduce reflection of the solar rays. Then the metal electrode can be printed and sintered on the top/bottom surface of the cell and the conductor and load connected to form a solar cell. The following actions, including surface texturing, light trapping and passivation of anti-reflection film, and carrier reflection of back surface field (BSF) and so on, can reduce carrier recombination and light losses and improve the conversion efficiency of the solar cell.

Figure 3.1 shows the structural diagramme of the traditional crystalline silicon solar cells (CSSCs).

At present, most of the solar cell manufacturers use the P-type silicon as the crystal solar cell substrate, which is because boron—the doping agent P-type dopant of silicon—has solid/liquid distribution coefficient close to 1 and has even resistivity distribution in CZ monocrystalline and casting polycrystalline, and the yield is high. For silicon monocrystalline or quasi-monocrystalline, it usually selects (100) orientation.

To reduce the series resistance with the metal gate, an N^+ layer of high concentration (usually an N^+-P junction) shall grow on the irradiating surface of the solar cell. The depth of the N-layer junction is generally no larger than 0.5 μm. If the junction is too deep, the minority carriers generated in the N layer of the irradiating surface may fail to reach the junction zone. If the depth of the N-layer junction reduces to 0.1–0.2 μm, it can increase the spectral response of the short wavelength range. The cell is called as violet cell.

The N zone of the irradiating surface has shallow junctions and light doping, which is good for improvement of cell conversion efficiency, as shown in Figure 3.2. All in all, the junction depth and doping concentration shall be optimised.

If the surface recombination speed is high, it may reduce the quantity of the minority carriers reaching the junction zone. Figure 3.3a and b show the influence of carrier

Technology, Manufacturing and Grid Connection of Photovoltaic Solar Cells, First Edition. Guangyu Wang.

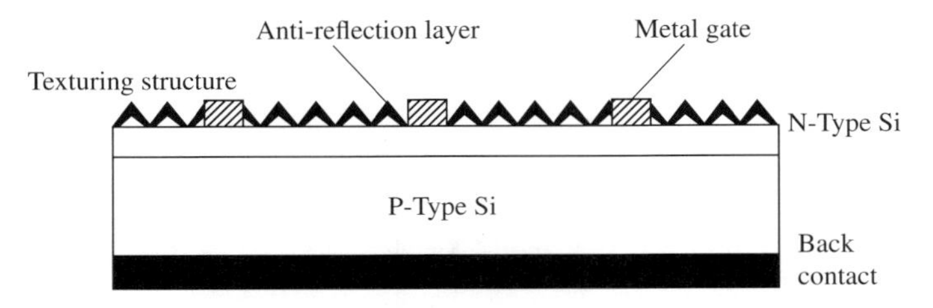

Figure 3.1 Structural diagramme of crystalline silicon solar cells.

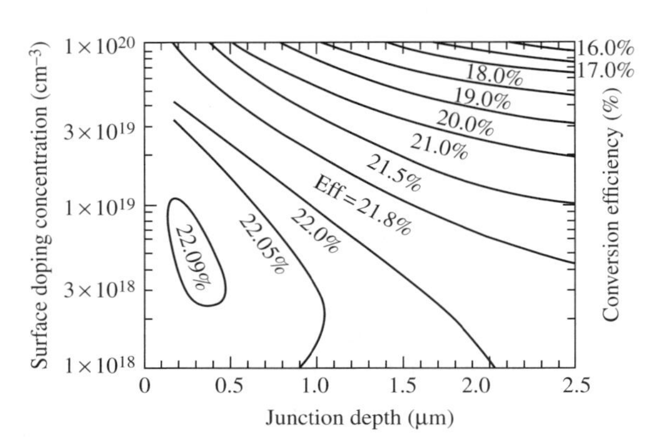

Figure 3.2 Relations of junction depth, cell efficiency and doping concentration of N zone.

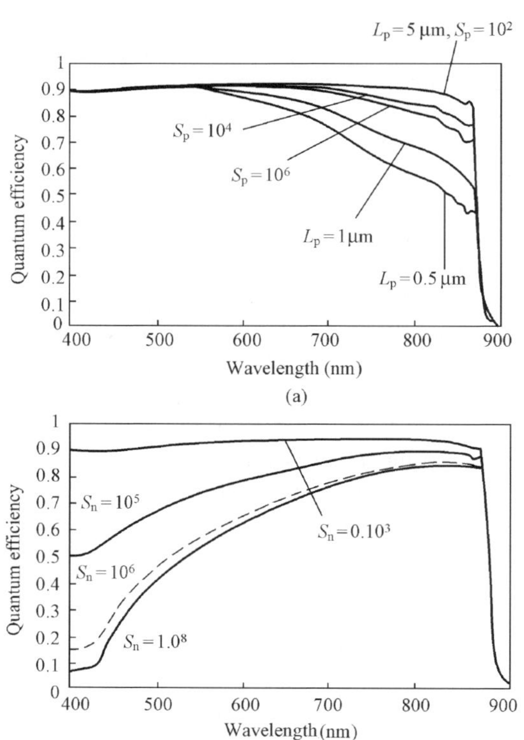

Figure 3.3 Influence of carrier surface on quantum efficiency of PN junction solar cell (a) back surface; (b) irradiating surface,

recombination of the back surface and the irradiating surface on the solar quantum efficiency. Figure 3.3a shows the carrier recombination on the back surface will mainly affect the long wavelength response while Figure 3.3b indicates the carrier recombination on the irradiating surface will mainly affect the short wavelength response.

3.1.2 Production Flow of CSSCs

Figure 3.4 shows the traditional production process steps of CSSCs.

The CSSC flow and equipment will be introduced in the following sections. It shall point out that the production processes may vary greatly in the cell manufacturers due to the following factors: The sources of crystal silicon are different, the equipment

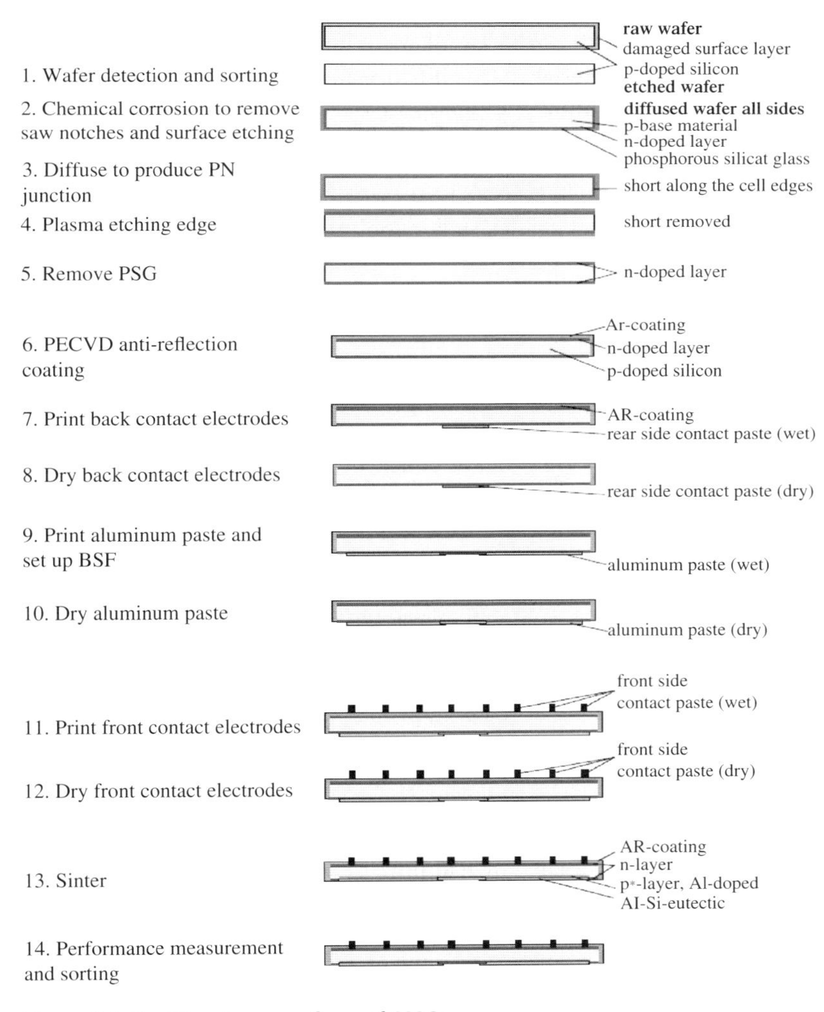

Figure 3.4 Traditional process flows of CSSCs.

adopted is in different conditions, the low-cost solar-grade silicon has been developed, the polycrystalline silicon solar cell has been produced and some high efficient cells of new structure have been developed in recent years. The solar cells of new structure will be described in Chapter 5. The diversification of raw materials and technical advance of cells are all dependent on improvement of process flows and upgrade of equipment.

3.2 Performance Detection and Sorting of Raw Silicon Wafer

A solar cell production line with annual output of 25 MW needs 2500–3000 silicon wafer per hour to produce the cell substrate. For a silicon wafer manufacturer with annual output at 1000 t level, hundreds of million pieces of silicon wafer in 125 × 125 mm or 156 × 156 mm shall be produced every year to meet the demand of the 100 MW-level cells. To control the quality of raw silicon wafer and consistency of cell parameters and raise the yield of cells, both the silicon suppliers and the cell manufacturers shall detect and sort the CSSC raw material (bare silicon wafer) one by one according to the conduction type, resistivity, lifetime of minority carriers, thickness, saw marks, surface contamination and micro-cracks and so on. Obviously, the original silicon wafer tester for semiconductor production cannot meet the demands, and thus the high-speed multi-function wafer tester has been developed by the equipment manufacturer to meet the demand of solar cell production. First, let's take a look at the principles of these main performance detections.

3.2.1 Measurement of Silicon Wafer Conduction Type

There are many methods to measure the conduction type of monocrystalline silicon, and the most common and easiest one is cold/hot probe method. For the silicon wafer of solar cells, the hot probe is 40–60 °C while the cold one keeps at the ambient temperature. When the cold/hot probes are simultaneously in contact with the target single crystal, the hot probe contact will generate a lot of electron-hole pairs excited by heat. And the minority carriers will be swept into the hot probe by the potential difference caused by contact of metal with semiconductor, and the majority carriers have concentration gradient at the cold/hot positions and thus diffuse towards the cold probe contact. In this case, the potential differential will be established between the two probes. If the tested sample is N-type semiconductor, the majority carrier generated at the hot probe contact will be electrons, which will diffuse towards the cold contact, resulting in rise of the hot probe potential and fall of the cold probe potential. And in the external circuit, a current from the hot probe to the cold probe via galvanometer with optical point will be generated. And the galvanometer with optical point will deviate towards one direction. If the tested sample is P-type, the majority carrier generated at the hot probe contact will be holes, and the potential of the cold probe will rise and the potential of the hot probe will fall. And a current will flow from the cold probe to the hot probe via galvanometer with optical point and the galvanometer with optical point will deviate towards the other direction. The deviation direction of the galvanometer can be observed to determine the tested sample—N-type or P-type. As a result, the voltage (+/−) between the two probes can be used to determine the conduction type of the wafer. Figure 3.5 shows the measurement principles of silicon wafer conduction.

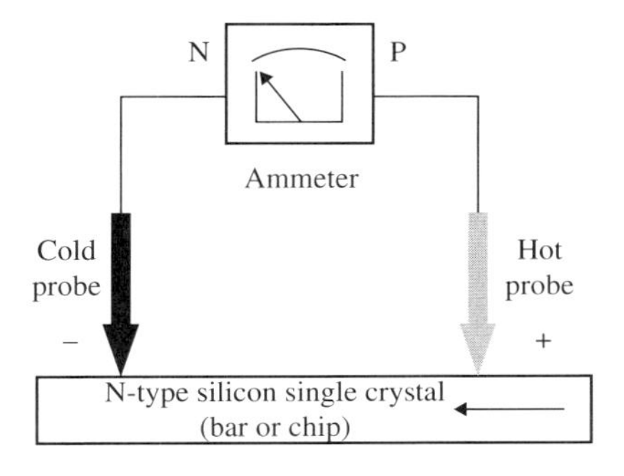

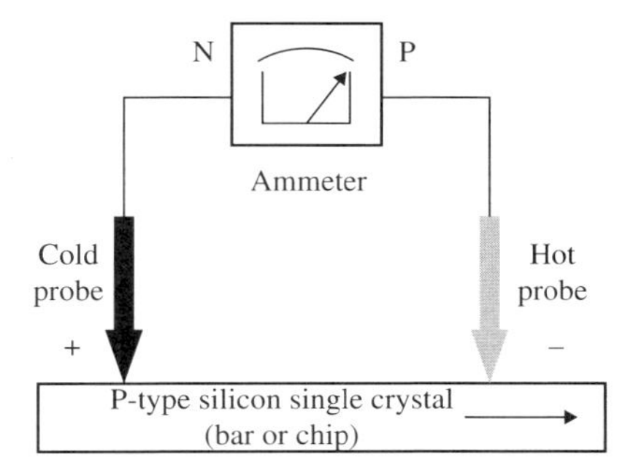

Figure 3.5 Principle diagramme of hot probe method to measure the silicon wafer conduction type.

3.2.2 Measurement of Silicon Wafer Resistivity and Thin Layer Square Resistance

The ordinary objects are usually measured by an AVO meter for the resistance, but the AVO meter cannot accurately measure the resistance of semiconductor because the contact resistance between the metal and the semiconductor is often very high and a big voltage drop will be generated when the current flows through the contact position. In addition, the minority carriers are likely injected to the semiconductor at the contact position under the action of the externally applied voltage, changing the resistivity of the semiconductor. One pair of probes is only applied with current and the other pair is used to measure the voltage drop. It will introduce the direct-row four-probe method in the following sections, including the principle, equipment and operation steps. See Figure 3.6 for its principle diagramme.

In Figure 3.6 four probes are pressed vertically on the surface of the silicon wafer. The probes, made of wolfram, tungsten carbide or high-speed steel, is built in a conic shape on the tip. The insulation between any two probes or between any probe and any part is larger than $10^9\ \Omega$. The tips of the four probes are in linear arrangement with equal clearance (Formula 3.1). When I mA current is applied on the outer two probes (1, 4), the potential difference V mV will be generated between the two central probes (2, 3). Based on it, the silicon wafer resistance can be worked out:

$$\rho = kV/I \tag{3.1}$$

Figure 3.6 Principle diagramme of four-probe method for resistance measurement.

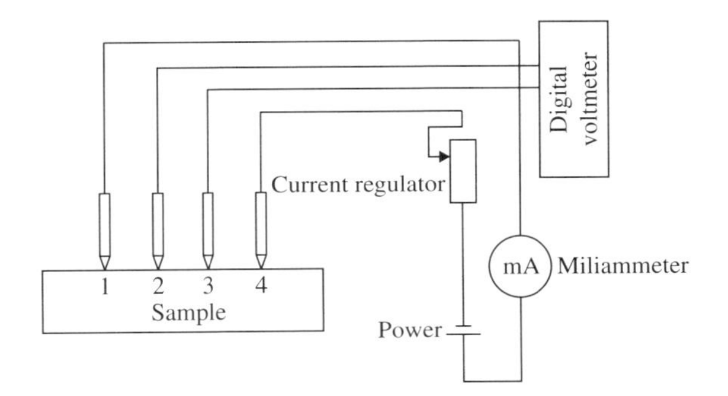

Where, ρ is silicon wafer resistivity; V is the potential difference between probes 2 and 3; I is the current applied between probes 1 and 4; and k is the correction coefficient of the probe.

The direct-current two-probe method can be also used to measure the circular, square or rectangular monocrystalline ingot with even sectional area but the two ends of the sample shall be first metallised to form good ohm contact with the metal electrode.

Obviously, the operation is complicated. And the four-probe method can be applied to measure the block sample with irregular shape.

According to the principle of four-probe method as shown in Figure 3.6, it can be used to measure the thin layer square of semiconductor. The resistance of thin layer square R_S can be viewed as the resistance between the two ends of the square thin layer on the silicon wafer. It has relations to the resistivity and thickness of the thin layer instead of the size of the square thin layer. It is measured in Ω/cm. The tool for square resistance measurement is still direct-row four probes (The clearance between probes S shall be less than the linearity of thin layer). When the current I is applied on the outer two probes (1, 4), the potential difference V between the two central probes (2, 3) can be measured. In this way, it can avoid the complicated treatment to thin-film contact resistance.

The resistivity of the sheet square has the following relationship with the current and voltage of the four probes (Formula 3.2):

$$\text{Sheet resistivity } \rho_S = 2\pi S\ V/I(\Omega\text{cm}) \tag{3.2}$$

As a long as the area of the thin layer is far larger than the linearity of the probe, the size of the thin layer can be viewed as infinitely large, and in this case, the above formula can be corrected and the follow formula of the square resistance can be obtained (Formula 3.3):

$$\text{Square resistance } R_S = 4.53V\big/I(\Omega\big/\square) \tag{3.3}$$

Where, 4.53 is the correction coefficient, which can be adjusted to various values if the area of the thin layer is not infinitely large.

The measurement principle of four-probe method shows, it can be used to the square resistance of the intermediate products in CSSC technique (e.g., the diffusion layer) and the surface thin layer square resistance of the intermediate products in the production techniques of thin film silicon solar cells (e.g., ITO).

Since the four-probe method is in contact, it is likely to damage the silicon wafer. And a non-contact method has been developed in recent years for the semiconductor industry where the high-frequency current of the coil will produce eddy current in the conductive thin layer beneath the coil. The energy losses generated by eddy current can come down to the negative resistance effect generated by the conductive film resistance. The final variation of resistance can be used to work out the square resistance of the measured thin film.

3.2.3 Measurement of the Minority Carrier Lifetime

The lifetime of nonequilibrium minority carriers, a key parameter of the semiconductor materials, plays a crucial role in the PV conversion efficiency of solar cells. Figure 3.7 shows, the key parameters of the solar cell, including open-circuit current, short-circuit current and filling factor, all rise with the growth of the minority carrier lifetime in

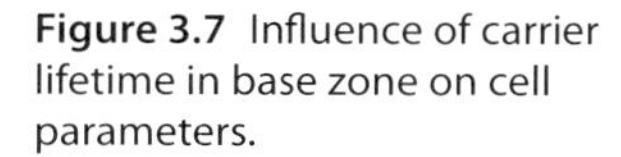

Figure 3.7 Influence of carrier lifetime in base zone on cell parameters.

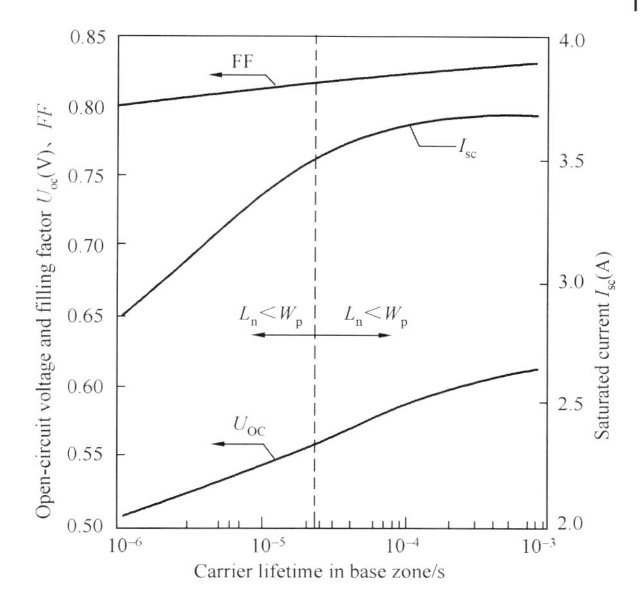

the base zone. The analysis was conducted on the variation rule of the minority carrier lifetime in the solar cell preparation techniques such as raw material silicon wafer, phosphorous diffusion impurity gettering, silicon nitride passivation, which can play an important role in cell production process optimisation, cell efficiency improvement and finished product rate increase.

The photoconduction attenuation method uses the pulse light source to generate the nonequilibrium minority carriers in the silicon wafer, generating additional photoconduction. Once the light stops, the additional photoconduction will attenuate exponentially, and an exponential attenuation curve will be shown on the oscilloscope where the time constant τ is the lifetime of nonequilibrium minority carriers (see Figure 3.8 for the measurement principle).

It shall be pointed out that the light of different wavelength has different penetration depth in the silicon monostylline. The two minority carrier lifetime testers available on the market have two pulse light wavelength: One is Semilab Rt. WT1000, Hungary, with

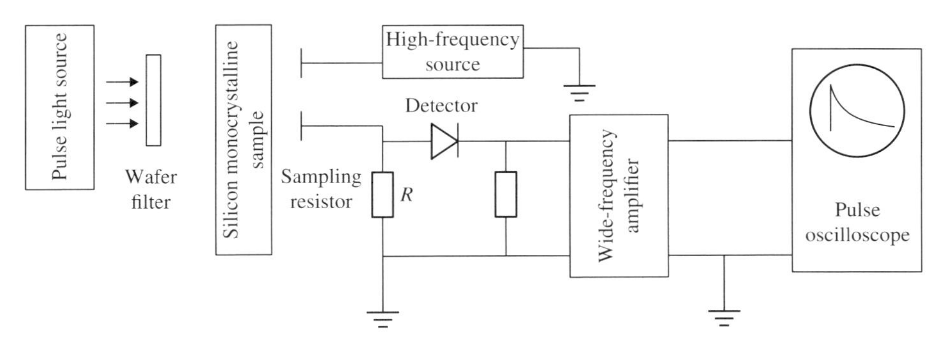

Figure 3.8 Measurement circuit diagramme of the minority carrier lifetime by microwave photoconduction attenuation method.

pulse wavelength of 0.904 μm and working frequency of 30 MHz and penetration depth of 30 μm; and the other is LT-1 Dr, China Kunde, with pulse wavelength of 1.07 μm and working frequency of 10 GHz and penetration depth of 220 μm. Obviously, since the silicon wafer for solar cells is thin, the former is more appropriate; but for the block silicon ingot, the latter is better.

In addition, the wafer surface impurity and damage will result in recombination of carriers, affecting the measurement of lifetime. For the naked silicon wafer, since the chip surface has some damage layers of certain depth and contamination, which forms the recombination centre on the minority carrier surface, the measured lifetime of the minority carriers is under great influence of the lifetime of the surface. The mixed acid of HF and HNO_3 with volume ratio of 1:6 can be used for 5–7 min to remove the damaged layer and pollutants and smoothen the surface so as to improve the measurement accuracy of minority carriers.

After the silicon wafer surface is passivated to reduce surface recombination, the lifetime of the silicon wafer minority carrier can be dramatically improved.

3.2.4 Measurement of Silicon Wafer Thickness

The silicon wafer thickness and homogeneity are key data to production of CSSCs. Based on the principle that the capacitance of a plate capacitor is in direct proportion to the area of two plate electrodes and in inverse proportion to their distance, the silicon wafer to be tested can be put between two plate electrodes and the area of the plate electrodes can be fixed to work out the thickness of the silicon wafer via the measured capacitance. Figure 3.9 shows the silicon wafer thickness tester of MTI, USA.

3.2.5 High-Speed Multi-Purpose Silicon Wafer Testers

FORTRIX, Korea, produces a high-speed bare silicon wafer tester, which is in modular design and can detect simultaneously the silicon wafer thickness, surface saw notch, resistivity, lifetime of minority carriers, pollution and micro-cracks and so on with resistivity, thickness and minority carrier lifetime testing measured by SEMILAB standard modules. It can detect more than 3000 pieces of silicon wafer per hour. The wafer damage rate is less than 2% during testing. The instrument is compatible with 125 × 125 mm

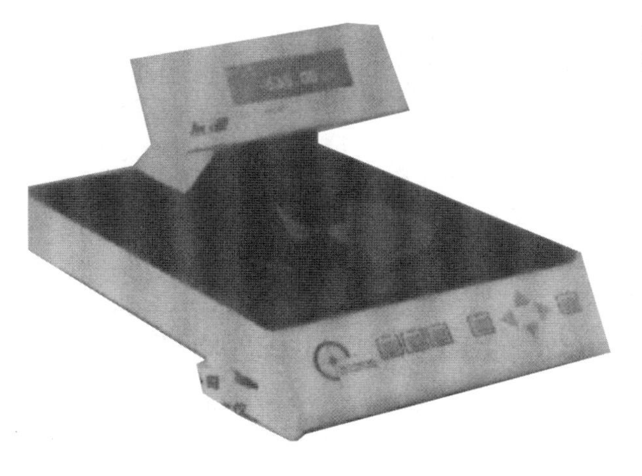

Figure 3.9 Silicon wafer thickness tester of MTI, USA.

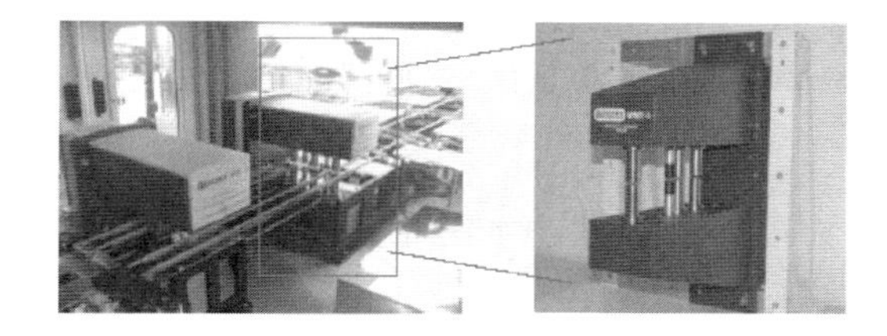

Figure 3.10 Outline of FORTRIX silicon wafer tester (4800 mm × 1900 mm × 1900 mm).

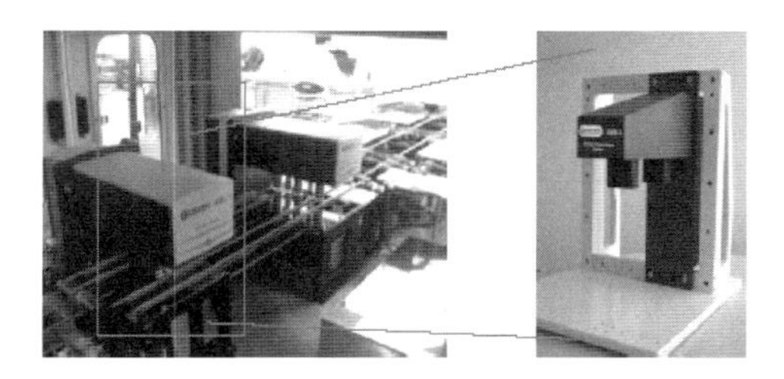

Figure 3.11 Sorting line of FORTRIX silicon wafer tester.

and 156 × 156 mm wafer. It can put the wafer in cassette or stacking manners. The tested wafer can be classified by specifications into 12 types. Figures 3.10 and 3.11 show the overall outline and sorting flow line pictures of the instrument.

3.3 Silicon Wafer Surface Cleaning and Texturing

3.3.1 Principles of Chemical Cleaning and Texturing

The process is aimed to remove the cutting damages on the silicon wafer surface and etch the texture on the surface by corrosive liquid so as to reduce light reflection.

For the P-type monocrystalline wafer, NaOH (or KOH) shall be first used to remove about 30 μm damaged layer of the chip, and then the NaOH (or KOH) solution shall be used for surface texturing. The chemical reaction of texturing is given below (Formula 3.4):

$$\mathrm{Si} + 2\mathrm{NaOH} + \mathrm{H_2O} = \mathrm{Na_2SiO_3} + 2\mathrm{H_2} \tag{3.4}$$

Since the (100) orientation monocrystalline silicon wafer has orientation optimum texturing performance during chemical corrosion, many (111) facets in 'pyramids' texture, that is, the texture with light trapping function will be formed on the wafer surface. It can reduce the reflectivity of the wafer surface down to below 9% (and the reflectivity can be further reduced to below 2% if an anti-reflection coating is added). See Figure 3.12 for the scan electron microscope picture of monocrystalline texture structure.

For polycrystalline wafer, since the surface orientations are inconsistent, the alkaline corrosion has no good texturing effect on polycrystalline wafer surface. As a result, the corrosive liquid is used to the mixed acid solution of HNO_3 and HF (HNO_3 : HF = 1 : 12, volume ratio). In this case, the effect is good. Figure 3.13 shows the scan electron microscope picture of casting polycrystalline surface texture structure

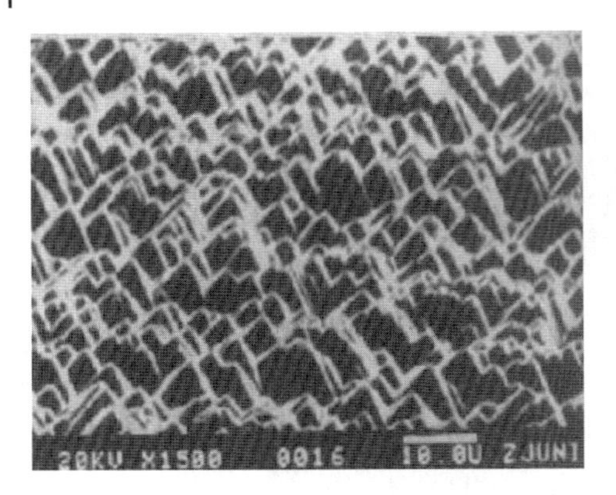

Figure 3.12 Scan electron microscope of monocrystalline texture structure.

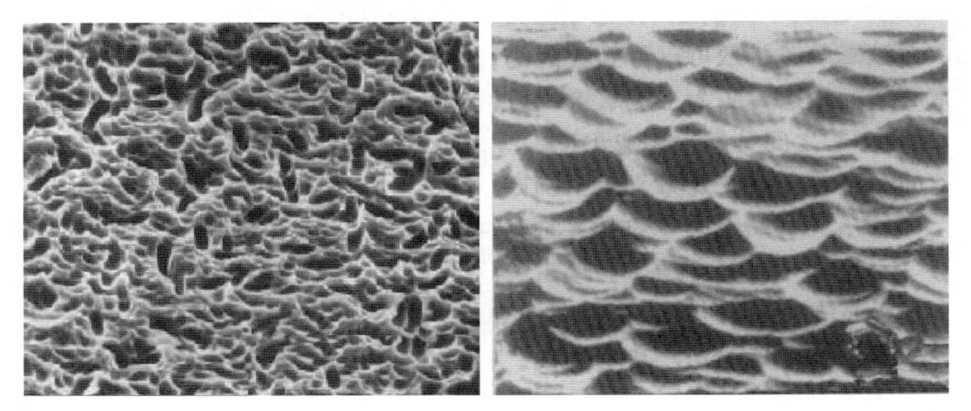

Figure 3.13 Scan electron microscope picture of casting polycrystalline surface texture structure.

(Formulas 3.5 and 3.6). And the chemical reaction is shown below:

$$3Si + 4HNO_3 = 3SiO_2 + 2H_2O + 4NO \tag{3.5}$$

$$SiO_2 + 6HF = H_2SiF_6 + 2H_2O \tag{3.6}$$

3.3.2 Production Equipment and Process of Chemical Corrosion Texturing

The typical cleaning and texturing liquid is KOH for monocrystalline silicon wafer (see Figure 3.14 for the texturing process).

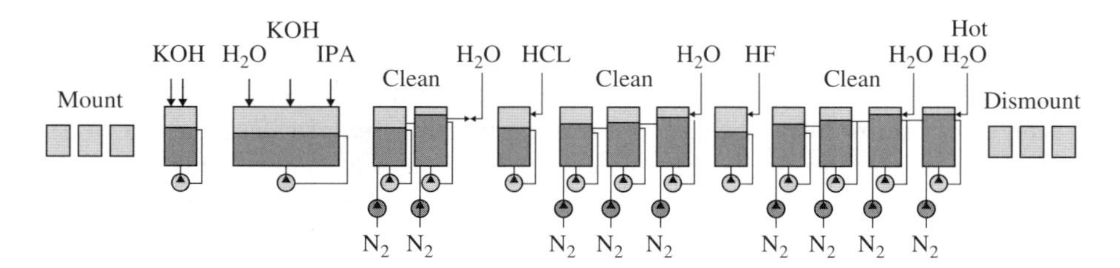

Figure 3.14 Flow chart of typical bath-type chemical texturing for monocrystalline silicon wafer.

Basic steps: Mounting→KOH corrosion texturing→DI water cleaning+nitrogen bubbling + spraying → HCl neutralisation → DI water cleaning + nitrogen bubbling + spraying→HF corrosion→DI water cleaning+nitrogen bubbling+hot water spraying→ dismounting→drying by a centrifugal machine (off-line). In Figure 3.14, IPA (isopropanol) is a chemical additive, which is good for the bubble desorption from the silicon wafer. Refer to Table 3.1 for silicon wafer cleaning and texturing process and the parameters.

For the polycrystalline silicon wafer, the texturing liquid is HF + HNO3 and the neutralisation liquid is KOH, and the rest are similar to monocrystalline silicon wafer.

At present, there are mainly two types of chemical texturing equipment: the bath-type and the chain-type. In china, the bath-type equipment has been widely used, which is provided with full PLC and touch screens. To reduce cleaning cost and pollutant discharge, the even liquid and air duct design and sealing and isolation techniques have been employed. The cleaning function can be provided in the bath to remove the particles polluted by chemical liquid and left on the silicon wafer. The robot is generally used in the bath with process time adjustable. The acid/alkaline bath is provided with environmentally friendly cleaning and texturing liquid where the automatic liquid make-up equipment is provided. The hot nitrogen drying technique has replaced the original dryer, which can significantly reduce the broken chips. The equipment is suited to 125 × 125 mm-156 × 156 mm silicon wafer with output of 1500–3000 pcs/hr. It is provided with unloading robots, air draft system, and electric control gear and so on. It shall be provided with the running water, pure water, nitrogen and compression air system.

To meet the requirements on quality and environmental protection of the chemical cleaning and texturing product, raise the productivity and reduce the cost, the bath-type chemical corrosion equipment must be provided with proper corrosion liquid concentration, corrosion temperature and time and so on. In addition, the automatic make-up and agitation shall be available according to the real-time concentration variation of the cleaning and texturing liquid to save water and liquid and reduce the cost and pollution. For example, the frequency control pump can be used to reduce the turbulence on the silicon surface; the multi-spraying technique can be employed to reduce water consumption; and the near infrared process analysis instrument can be used to monitor the concentration of the chemical agent in real time to reduce chemical consumption and automatically make up the chemical agent.

RENA has developed the chain-type cleaning and the texturing equipment where the silicon wafer is directly put on the roller of the conveyor instead of the basket, and then passes the various chemical agent bathes. Compared with the bath-type equipment, it can work continuously with higher productivity and lower broken chips. See Figure 3.15 for the conveyor chain-type cleaning and the texturing equipment developed by RENA.

3.3.3 Laser Texturing and Reactive Ion Etching (RIE) Techniques

Since the chemical texturing techniques on silicon wafer surface has low cost, it has been widely applied to the production process of solar cells. For the polycrystalline silicon wafer, however, the anti-reflection effect of the surface after chemical texturing still has a big gap with the monocrystalline silicon wafer. Accordingly, some equipment manufacturers are still seeking new surface texturing methods.

Table 3.1 Steps and Process Parameters of Monocrystalline Silicon Wafer.

	Step	Auxiliary	Liquid	Time	Temperature	Material	Air draft		
1	Auto-mounting treatment								
2	Remove the damaged layer	Bubbling	KOH	7 min	60–90 °C	SUS316	Yes		
3	Warm water isolation	Bubbling	DI water	7 min	50 °C	PP	/		
4	Monocrystalline texture	Bubbling	Alkaline, Liquid IPA	25 min	60–90 °C	SUS316	Yes		
5	Monocrystalline texture	Bubbling				SUS316	Yes		
6	Monocrystalline texture	Bubbling				SUS316	Yes		
7	Monocrystalline texture	Bubbling				SUS316	Yes		
8	Spraying	Bubbling	DI water, Acid liquid	7 min	50 °C	PP	Yes		
9	Cleaning	Bubbling				PP	Yes		
10	HCl neutralisation	Bubbling		25 min	RT	PP	Yes		
11	Cleaning	Bubbling				PP	Yes		
12	HF treatment	Bubbling	HF	7 min	RT	PP	Yes		
13	DI water spraying	/	DI water	7 min	RT	PP	/		
14	DI water cleaning	Bubbling	DI water	7 min	RT	PP	/		
15	Pre-dehydration	/	DI water	7 min	RT	PP	/		
16	Drying	/	Hot air Nitrogen	15 min	110°C	SUS316	Yes		
17						SUS316	Yes		
18	Automatic dismounting								

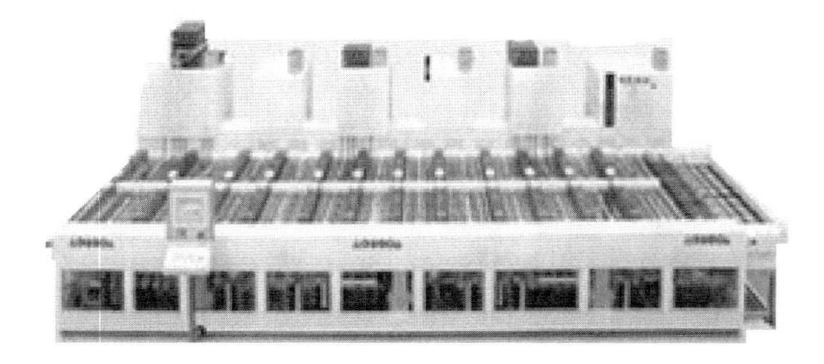

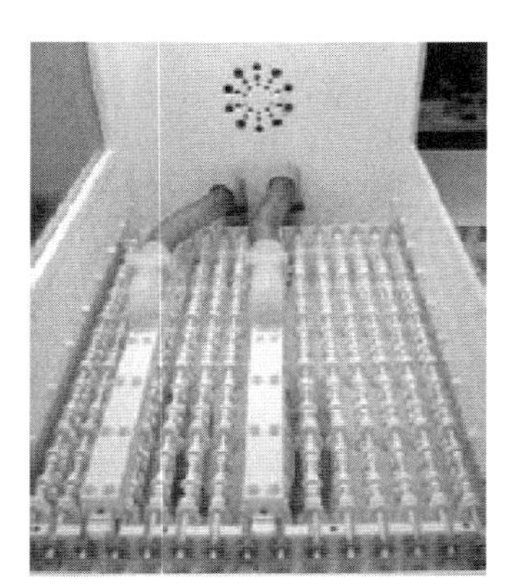

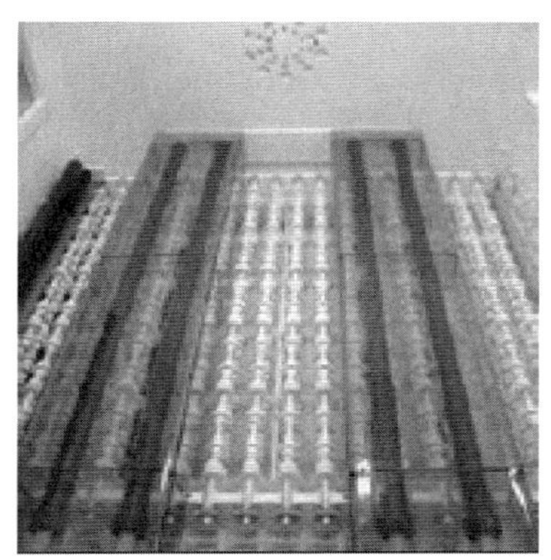

Figure 3.15 Conveyor chain-type cleaning and texturing equipment, RENA.

Recently, some companies of Japan and Korea use laser texturing and RIE techniques to replace the chemical wet texturing to achieve dry chain-type production. These methods can reduce chemical consumption, improve the environment and PV conversion efficiency as well as reduce the silicon wafer damage rate. Below we will introduce the laser texturing and RIE techniques.

3.3.3.1 Laser Texturing

The laser has three functions to the silicon material: laser annealing, laser zone melting and recrystallisation and laser texturing. Laser annealing means that the laser shining zone temperature will rise after heated but not up to the melting point and it is used for annealing. Laser zone melting and recrystallisation means that the laser shining zone temperature will rise up to the melting point to melt the material after heated, and then fall to recrystallise during cooling down. Laser texturing means that the laser irradiating zone temperature will rise sharply after heated to ionise the material into plasma and then vapourise, forming a sunken uneven zone on the irradiating surface.

The silicon wafer surface laser texturing, one type of laser texturing, can texture the polycrystalline surface to one with homogeneous orientations, playing a role similar to an 'excavator'. The principle is that the high-energy laser pulse irradiation can make the local temperature of the silicon wafer surface rise sharply, melt and gasify, forming an uneven surface texture on the irradiation zone (Figure 3.16). It offers a good means to anti-reflection treatment of polycrystalline silicon. It shall be pointed out, however, that

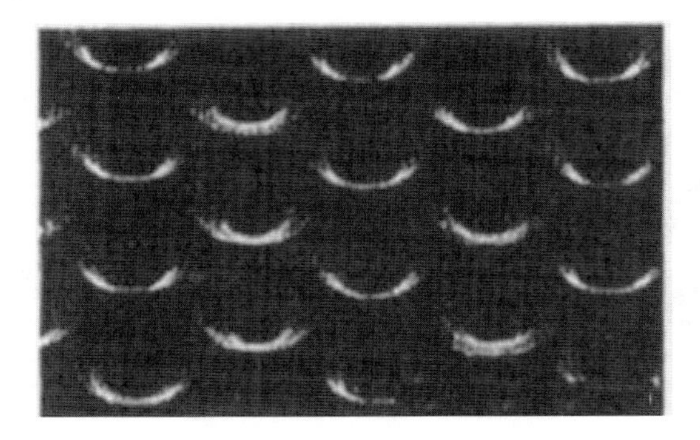

Figure 3.16 Silicon wafer surface texture after laser texturing.

the effect of silicon wafer surface texture after laser texturing is obviously better than that of chemical corrosion and even better than that of the ordinary monocrystalline pyramid texture, but the experiments show that the lifetime of minority carriers of the laser texturing sample is obviously less than that of the pure chemical treatment, which is because the residuals and damage layers on the silicon wafer surface after laser treatment have not been completely removed and it needs the comprehensive treatment of acid/alkaline corrosive liquid.

3.3.3.2 RIE

RIE is to make an active substance (e.g., SF_6, N_2O) react with silicon wafer surface to generate gaseous phase substance in the condition of plasma, and then be drawn by a pump for treatment.

Figure 3.17 shows the picture of RIE system, IPS, Korea. It mainly consists of 3 × RIE chamber and 8 × 8 bracket, covering an area of 12 × 10 m^2. It can handle 1500 pieces of silicon wafer in the size of 156 × 156 mm. Figure 3.18 shows the silicon wafer surface and anti-reflection reduction after RIE process is used.

After cleaning and texturing, the scan electron microscope can be used to observe the silicon wafer surface texture and measure the reflectance to check the texturing effect.

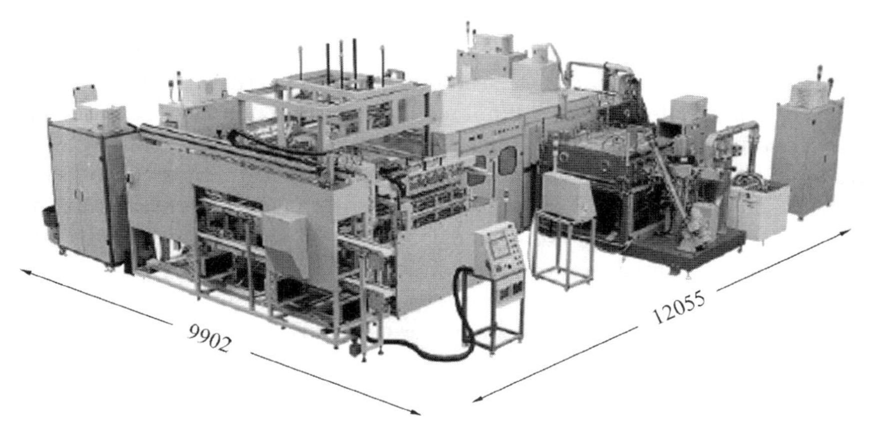

Figure 3.17 RIE system, IPS.

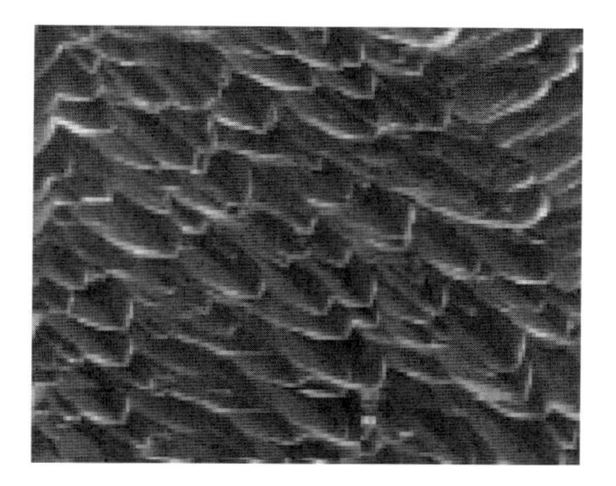

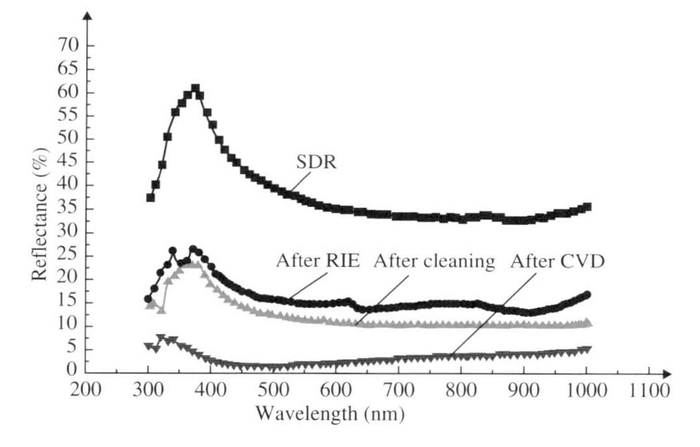

Figure 3.18 Silicon wafer surface and anti-reflection reduction after RIE process is used.

3.4 Junction Preparation by Diffusion

3.4.1 Principles

The diffusion junction preparation is aimed to generate the P-N junction by doping phosphorous element to the P-type silicon wafer. The following methods can be used to prepare P-N junctions: thermal diffusion method, ion implantation, epitaxy method, laser method and high-frequency injection method and so on. At present, it mainly uses the thermal diffusion method to produce the silicon P-N junction of crystal solar cells and this method can be further divided into liquid source diffusion, coating source diffusion and solid source diffusion. In production, $POCl_3$ or H_3PO_4 is generally used as the diffusion source, and phosphorous is usually diffused in quartz tube or chain-type diffusion furnace at the diffusion temperature of 850–900 °C.

As for $POCl_3$ diffusion junction preparation method, the liquid $POCl_3$ shall be first gasified and then sent to the diffusion furnace. In the furnace, the chemical reactor below will occur (Formula 3.7):

$$4POCl_3 + 3O_2(\text{excessive}) = 2P_2O_5 + 6Cl_2(\text{gas}) \tag{3.7}$$

As for H_3PO_4 diffusion junction preparation method, the liquid H_3PO_4 shall be first evenly coated on the silicon wafer in the coating machine and then sent to the diffusion furnace. In the furnace, the chemical reactor below will occur (Formulas 3.8 and 3.9):

$$2H_3PO_4 = 2P_2O_5 + 3H_2O \tag{3.8}$$

$$2P_2O_5 + 5Si = 5SiO_2 + 4P \tag{3.9}$$

To get qualified solar cells, the requirements on P-N junction diffusion include appropriate junction depth and diffusion square resistance. Generally, the shallow junction has small dead layer and good cell short-wave response but it is apt to result in electric leakage of P-N junctions in the follow-up electrode sintering process. If the junction is too deep, the dead layer will be obvious. On the other hand, if the diffusion concentration of phosphorous is too small, the series resistance may rise, and thus the density of gate electrodes shall be increased, raising the process difficulty. If the diffusion concentration is too large, the heavy doping effect may occur, reducing the cell open-circuit voltage and short-circuit current. In practise, the junction depth is generally controlled in a range of 0.3–0.5 μm, and the mean square resistance is 50–70 Ω/□. If the square resistance deviation is less than 10%, the lifetime of minority carriers shall be larger than 10 μsec.

In addition to produce P-N junctions, the diffusion process is also aimed to achieve phosphorous impurity gettering so as to improve the lifetime of minority carriers. The good diffusion process can achieve good impurity gettering effect and raise significantly the lifetime of minority carriers to further improve the solar cell efficiency. The adjustment of the minority carrier lifetime can optimise the diffusion technique and explore the optimal impurity gettering condition.

Since the junction depth of the solar cell is at the submicron level, special attention shall come to junction protection after diffusion. It is often found in practise that the reasons for low cell conversion efficiency and finished product rate mainly lies in careless operations after junction preparation (e.g., mechanical scratches and inappropriate ultrasonic cleaning), which may damage the texture and the P-N junction of the irradiating surface. As a result, great care shall be paid to the operations.

In the CSSC production process, if the phosphor silicate glass (PSG) layer is too thin, it is apt to result in uneven junctions. If the coating layer is too thick, it is also difficult to control. As a result, it shall control the thickness of the PSG diffusion layer. The common method is to weigh the silicon wafer before and after diffusion and then the weight of the coating layer can be figured out and finally the thickness of PSG can be estimated. For example, if the PSG deposited weight is 35 mg after the silicon wafer (156 × 156 mm) is textured, the PSG is about 40–80 nm thick.

3.4.2 Process and Equipment

There are two techniques for diffusion junction preparation: One is $POCl_3$ gaseous diffusion in the tubular furnace, and in the other technique, the silicon wafer is first coated by a dilute phosphoric acid coating machine and then sent to the chain-type diffusion furnace for diffusion.

3.4.2.1 Gaseous Diffusion of $POCl_3$ in Tubular Furnace

The home-made diffusion equipment is mostly based on $POCl_3$ gaseous diffusion in tubular furnace, and the output has been greatly improved in recent years, and the control method has been changed from the original manual control to PLC semi-auto control to the existing computer-based fully automatic control. The gas velocity control has been developed from analog to digital MFC. The original open-tube diffusion has been changed to environmentally friendly, energy-saved, safe and fully closed diffusion. The homogeneity of single silicon wafer and the tube as a whole has been raised from the original 7% to 4%.

The 48th Research Institute of CETC adopts $POCl_3$ closed-tube furnace (high-temperature diffusion/oxidation system) as shown in Figure 3.19. The closed-tube diffusion process is completely isolated, and the slots are made on the rack in the quartz boat. Each tube can be mounted with up to 400 pieces of silicon wafer. The exhaust is collected by directional manner. The production course is clean. The diffusion temperature is controlled in a range of 400 – 1100 °C; the zone of constant temperature is 1080 mm long, and the temperature stability of single point is $\pm 1\ °C/hr$. The purification level of the purifier worktable is Level 1000 (Environmental Level: 10,000). The silicon wafer is transported by the silicon carbide cantilever slurry automatic transmission system. The gas flow, which can be displayed in real time on the computer monitor, is controlled by the mass flowmeter, and 100 working curves can be stored. The temperature of the diffusion source is controlled by the thermostatic controller with accuracy of $\pm 0.2\ °C$. It adopts the hot input and output of silicon wafer, heat dissipation and air exhaust at the furnace outlet with furnace outlet below 150 °C.

$POCl_3$ is brought by inert gas bubbling, and the tubular diffusion furnace shall keep even temperature distribution inside. The typical diffusion temperature is 850 °C and it

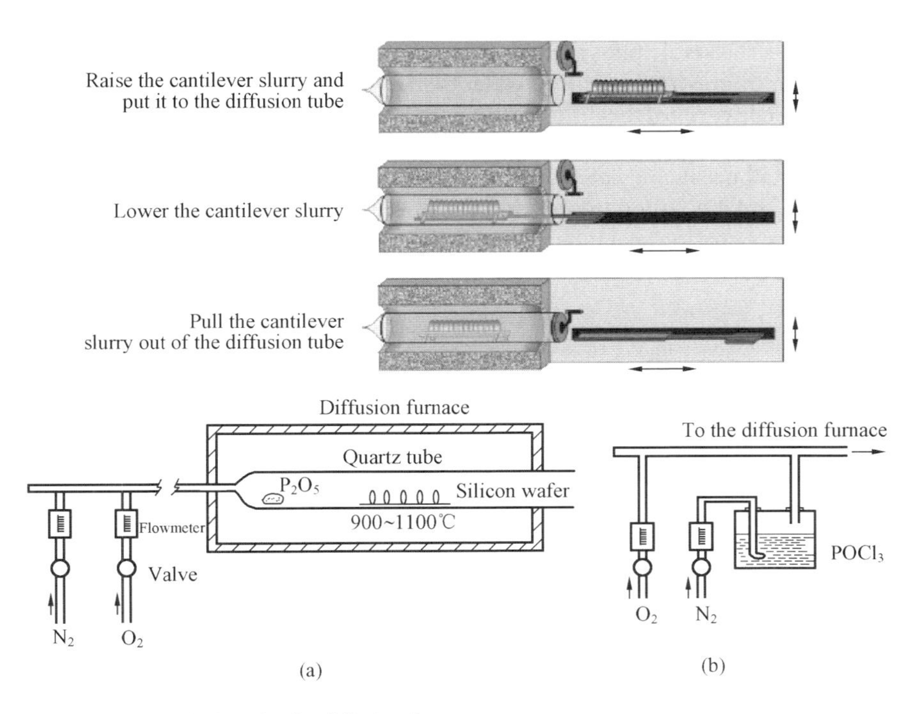

Figure 3.19 $POCl_3$ closed-tube diffusion furnace.

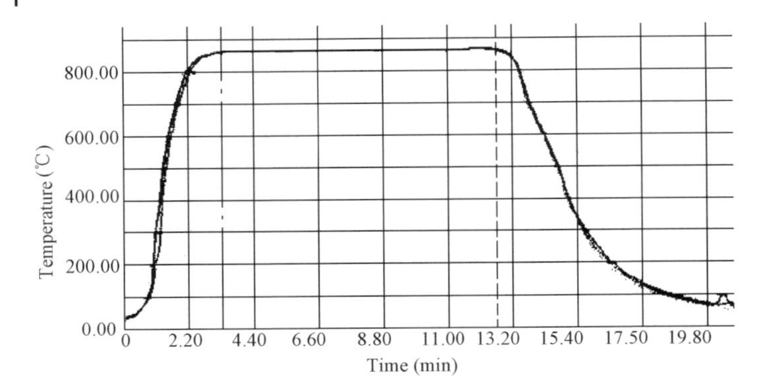

Figure 3.20 Temperature distribution curve of diffusion furnace.

shall be held on for 16 min or so. The diffusion furnace is the porcelain furnace without metal on the inner wall. It shows the temperature distribution curve of the diffusion furnace in Figure 3.20. The tubular diffusion furnace is generally heated by resistance wire, and the cantilever intermittent mounting is used. The furnace wall is hot.

3.4.2.2 Dilute Phosphoric Acid Doper and Chain-Type Diffusion Furnace

SCHMID uses the dilute phosphoric acid as the dopant. Before the silicon wafer enters the diffusion furnace, a phosphorus coating machine is installed on the chain-type production line (see Figure 3.21). The doping process is described below: First, the dilute phosphoric acid shall be atomised by ultrasonic to obtain very small drips and then the silicon wafer passes the stable phosphoric acid fog, and the drips will be coated evenly on the silicon wafer by gravity. The silicon wafer then enters the diffusion furnace via the worktable, and the conveyor of the diffusion furnace is driven by the porcelain roller. The method can carry out even phosphoric doping for both the polycrystalline silicon wafer and the monocrystalline silicon wafer. The phosphoric coating operation does not need manual operation and it is suited to very thin wafer and it has low damage rate.

In the phosphorus coating machine of SCHMID, the phosphoric acid of 85% concentration shall be mixed by 1:3 to get dilute phosphoric acid, and then pumped to the

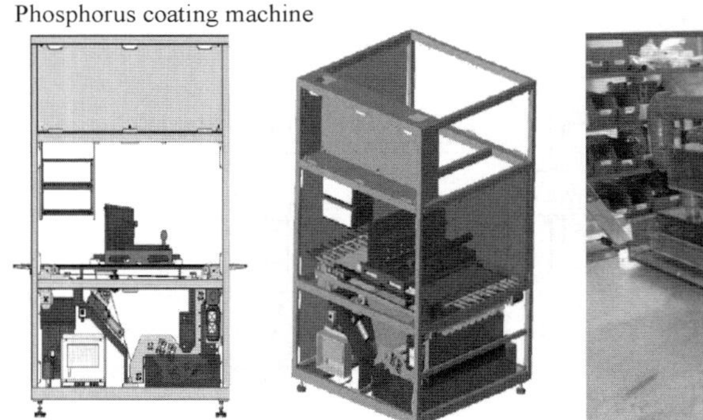

Figure 3.21 Phosphorus doping machine, SCHMID.

atomisation chamber by a magnetic peripheral pump. The concentration of H_3PO_4 is 27%; the doping temperature is 35°C. Ten ultrasonic generators are installed in the atomisation chamber to generate phosphoric acid fog. When the silicon wafer in horizontal arrangement enters the phosphorus coating machine, the dilute phosphoric acid liquid will be evenly coated on the silicon wafer surface. The dense and small phosphoric fog can ensure the wafer surface can be evenly wetted by the phosphoric acid liquid. Next, the silicon wafer will be directly sent to the next procedure—the diffusion furnace to produce P-N junctions. SCHMID phosphorus coating machine, can turn over the silicon wafer to coat on both sides. And the phosphoric fog can coat evenly and smoothly simultaneously on four to five pieces of silicon wafer.

After the wafer is coated, the dilute phosphoric acid can enter the chain-type diffusion furnace to produce P-N junctions at 870°C.

The chain-type diffusion furnace is heated by halogen tungsten lamps. Since the wall is cold, the temperature can rapidly rise or fall. The diffusion chamber is easy to clean, and the square resistance has better evenness than the tubular one. It is applicable to the thin and large silicon wafer. It features small silicon wafer deformation and high productivity, automatic and continuous mounting/dismounting, free from artificial pollution and thus free from silicon wafer contamination by metaphosphoric acid. Accordingly, it has caught great attention. It needs the automation equipment and the volatile of metaphosphoric acid may condense on the lamp tube surface (see Figure 3.22 for the equipment outline and diagramme).

In Table 3.2, it shows the main data and production capacity of BTU chain-type diffusion furnace.

3.4.3 Measurement of Diffusion Layer Thickness (Junction Depth)

The above discussion shows, the thickness of the diffusion layer (junction depth) plays an important role in the efficiency of solar cells. The junction depth is mainly measured by SIMS and spreading resistance probe (SRP) methods.

3.4.3.1 Measurement of the Longitudinal Distribution of the Phosphorus Concentration on the Diffusion Layer by SIMS Method

The secondary ion mass spectrometer is based on the following principle: The accelerated ion in the field bombards the silicon wafer surface and crashes or sputters other ions (secondary ions), which shall be collected and analysed by a mass spectrometer in the vacuum chamber to identify the type and concentration of the impurities in the silicon wafer. SIMS technique is destructive, and it needs ultrahigh vacuum (about 10^{-10} – 10^{-12} torr). Figure 3.23 shows an example where profile analysis of the phosphorus concentration on the diffusion layer of the solar cell by SIMS method.

3.4.3.2 Measurement of PN Junction Depth by SRP

In this method, two probes in accurate alignment are installed, which can move in steps on the slope of the silicon wafer, and the resistance between them shall be measured for every step (see Figure 3.24). As the probes moves through the P-N junction, the variation of conduction types can be measured. The method can plot the sectional view of the depth of very shallow P-N junctions. Its advantages lie in that it can carry out accurate doping concentration measurement with limitations to junction depth and its

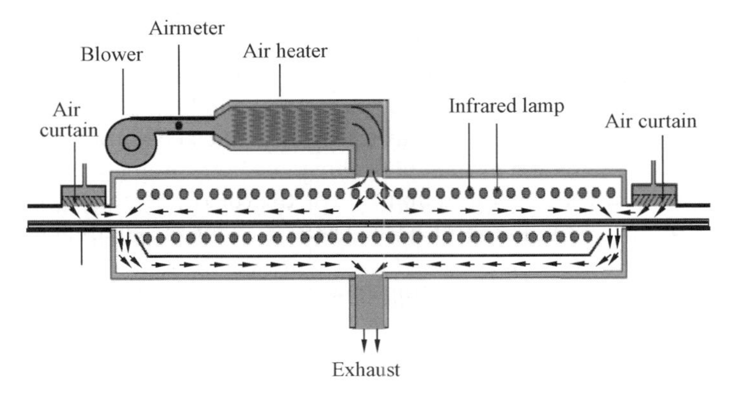

Figure 3.22 Conveyor and internal structure of BTU chain-type diffusion furnace.

disadvantages include that the operator shall be of rich experience and skillful on sample preparation and it has damages to the sample.

3.4.4 CSSC Phosphorus Impurity Gettering

For the P-type silicon solar cell at the phosphorus diffusion stage, a phosphorus silicon glass (PSG) layer will be formed on both sides of the silicon wafer while the p-n junctions are generated. The PSG layer contains a lot of micro defects, which can getter the heavy metal impurities in the silicon wafer, and make them diffuse and deposit in the PSG layer. If the PSG layer can be removed by such chemical means as HPO_3, HNO_3 and HF, the metal impurities thereof will be also removed, achieving the goal of metal impurity gettering.

For the casting and metallurgic polycrystalline, the phosphorus impurity gettering plays a more important role. For the metal impurities in the interstitial state, for example, Fe, Cu, Co, Cr and Ag and so on, the impurity concentration in the silicon wafer after gettering all falls to some extent, showing obvious effect. This is because the metal in interstitial state diffuses in interstitial manner at high diffusion speed, and it is easy to getter. The reduction of the metal impurity concentration is good to improve the lifetime of the minority carriers in the casting polycrystalline silicon.

Table 3.2 Main data and production capacity of BTU chain-type diffusion furnace.

	TQ161-8-240-N48 S	TFQ202-15-360N66GT	TFQ361-16-384A108GT
Maximum check temperature	1050 °C	1050 °C	1050 °C
Temperature zone	8	15	16
Heating length	240 inch	360 inch	384 inch
Width of conveyor	16 inch	20 inch	36 inch
Material of mesh belt	Nichrome V	Nichrome V	Nichrome V
Atmosphere	N2/Ar	N2/Ar	N2/Ar
Material of passage wall	Fused quartz, Muffle	Fused quartz plate	Fused quartz plate
Cooling	48 inch	66 inch	66 inch
Speed of conveyor	22 inch/min	20 inch/min	25.6 inch/min
Size of silicon wafer 210 × 210	142 pc/hr 1, at an interval of 25 mm	130 pc/hr 1, at an interval of 25 mm	496 pc/hr 3, at an interval of 25 mm
Size of silicon wafer 156 × 156	370 pc /hr 2, at an interval of 25 mm	336 pc /hr 2, at an interval of 25 mm	1083 pc/hr 5, at an interval of 25 mm
Size of silicon wafer 125 × 125	445 pc /hr 2, at an interval of 25 mm	600 pc /hr 3, at an interval of 25 mm	1568 pc /hr 6, at an interval of 25 mm

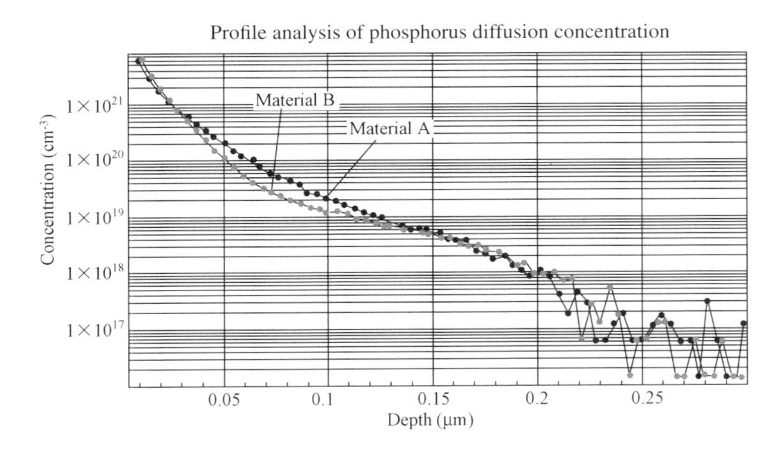

Figure 3.23 Measurement of diffusion layer thickness of solar cells by SIMS method.

The impurity gettering effect of polycrystalline has relations with the gettering temperature and time. Generally, the gettering temperature is 900 °C with duration of 90 min or so. The effect becomes better as the treatment time prolongs and the temperature rises.

Generally, it takes three steps to getter the metal impurities: First, the metal deposits in the silicon wafer are dissolved; second, the metal atoms are diffused to the gettering position; and third, the metal impurities are redeposited at the gettering position.

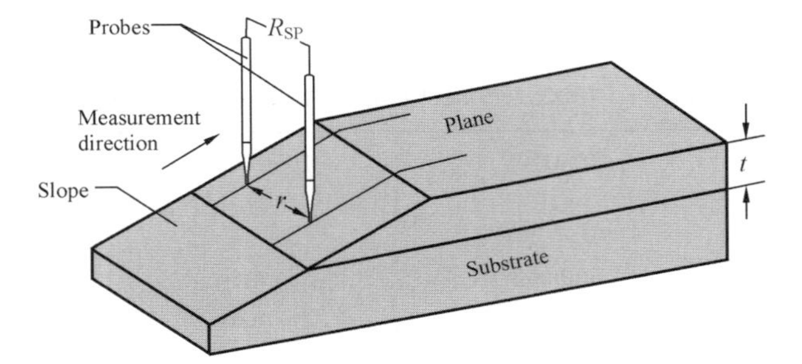

Figure 3.24 SRP diagramme.

Two gettering mechanisms can explain the principles: One is the relaxation mechanism, which needs a lot of defects outside the active zone to work as the gettering points and the metal impurity shall reach over-saturation; and the other is segregation mechanism, which prepares an gettering layer of high solid solubility outside the active zone of the component, and the metal impurities will diffuse from the crystal silicon of low solid solubility to the gettering layer (over-saturation is unnecessary), gettering and removing the metal impurities. In principle, both of the mechanisms can explain why the concentration of the metal impurities in the crystal silicon reduces to a low level. As for the mechanism, some researchers hold that the metal impurity in PSG has a solid solubility far larger than that in the crystal silicon and thus more metal impurities will deposit in the PSG.

3.5 Plasma Corrosion and Laser Edging Isolation

3.5.1 Objectives and Means of Edging Isolation

During diffusion, it is inevitable to form a PSG diffusion layer around the silicon wafer, which may result in the short-circuit fault in the cell. The procedure is aimed to remove the diffusion layer which may cause the short-circuit fault of the cell top/bottom electrodes.

Plasma etching is viewed as the leading etching technique so far. In the technique, the diffused silicon wafer of identical area shall be first closely stacked as shown in Figure 3.25, and automatically pressed and then put into plasma etcher to prevent the corrosive gas from entering the silicon wafer gap. The power electrodes of the plasma shall be connected to the RF power supply and the earthing electrodes to the side wall of the etcher, which shall be earthed. Apply the reaction gas CF_4 etc., and use the glow discharge to make CF_4 separate into fluorinion and form the plasma active base group under excitation of RF. It shall have chemical reaction with PSG at the edge of the cell in the action of the field to form the volatile reaction product and then fall off the etched surface and finally drawn out of the reaction chamber by the vacuum system.

Below shows the chemical reaction for the PSG diffusion layer on the silicon wafer edge to generate volatile product SiF_4 (Formula 3.10):

$$CF_4 + O_2 + Si = SiF_4 + CO_2 \tag{3.10}$$

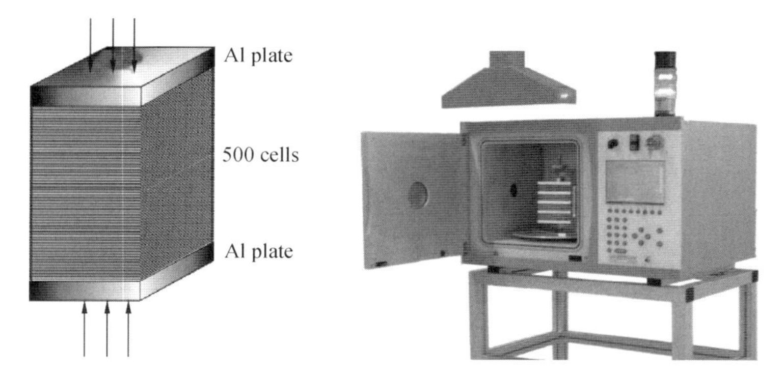

Figure 3.25 Mounting and equipment of plasma edging isolation.

During edging isolation by the plasma etcher, the silicon wafer must be automatically pressed and the rack shall rotate to reduce the undercutting risk and raise the etching homogeneity. The edging isolation, however, cannot be operated in chain and it is apt to break the silicon wafer.

3.5.2 Principles and Equipment of Plasma Etching

Plasma etching is dry corrosion. And the plasma is a physical form electrically excited to make gas glow. The plasma etcher can ionise the gas molecule in the glow at the vacuum chamber by RF. Plasma etching is the process where the plasma has chemical reaction with the substance on the silicon wafer surface to generate gaseous product. In glow, fluorine and oxygen will act alternately to silicon to remove the PSG on the edge of the silicon wafer. The plasma etcher and the plasma etching physical process are shown in Figure 3.26.

The plasma dry etching system consists of the following device: the etching reactor, the RF power supply, the gas flow control system, and the vacuum system to remove the etching products and gas. Figure 3.27 shows the plasma etching system.

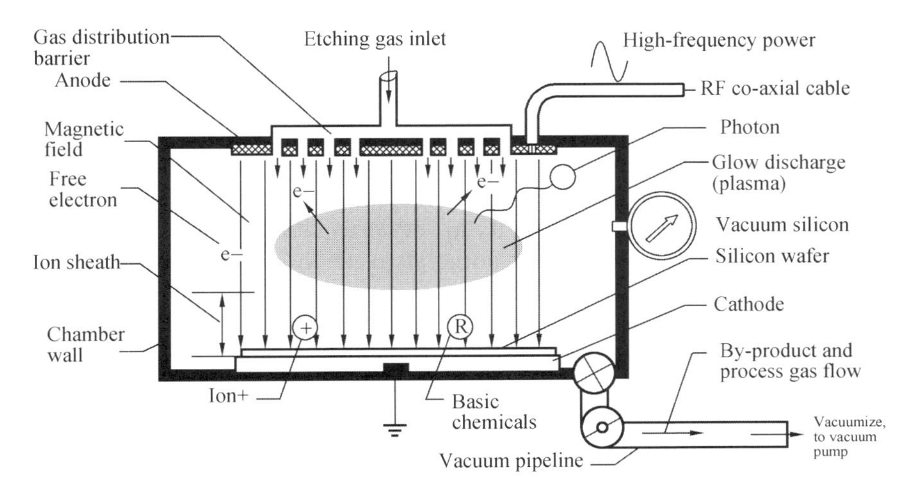

Figure 3.26 Diagram of plasma etcher.

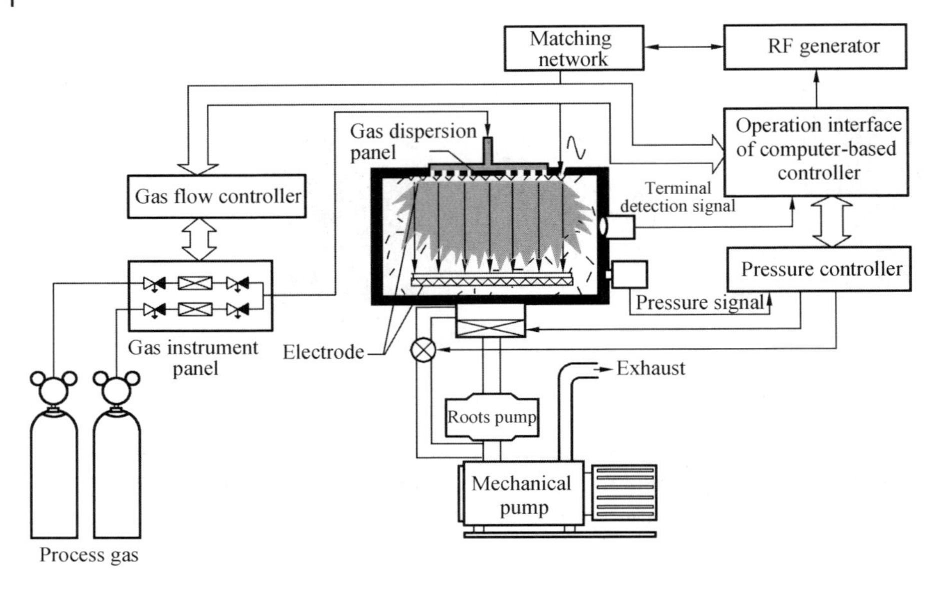

Figure 3.27 Diagramme of plasma etching system.

In the plasma etching system, the etching is mainly dependent on the chemical reaction between the substance and the chemical gas. To achieve high selection ratio (the bottom material to be etched with minimal chemical reaction), the gas entering the chamber (generally containing chlorine or fluorine) shall be selected with care. During etching, the vacuum system plays a crucial role because it may affect the plasma parameters concerning the gas flow and pressure. In addition, the vacuum system must be capable of handling the etching product generated during etching because the chlorine, fluorine and the etching product for etching must be drawn in time. The vacuum pump for etching process is generally tubopump and Roots pump and so on.

The process parameters during etching process can be determined by process optimisation, for example, power frequency, pressure, temperature, air flow rate. The quality indexes of etching process include etching rate, selection ratio, homogeneity, and residuals and so on.

The etching gas for crystal silicon is traditionally fluorine-based, including CF_4, CF_4/O_2, SF_6, C_2F_6/O_2 and NF_3. The fluorine atom plays a crucial role in etching. The etching generated by fluorine is homogeneity in very orientation. Different from wet etching, the plasma etching has generally no good selection ratio for the following materials, and the terminal detection shall be used to monitor and stop the etching process to prevent over-etching. As a result, the plasma etching terminal detection is a must where the process parameter variation, including variation of etching rate, variation of corrosive product removed and the variation of reaction gas or product during gas discharge, shall be measured (see Figure 3.28).

The common terminal detection method is optical emission spectroscopy (OES), which is based on that the base group excited will transmit the light of certain length during glow discharge. For the CF_4/O_2 system, the reaction product is SiF_4 with

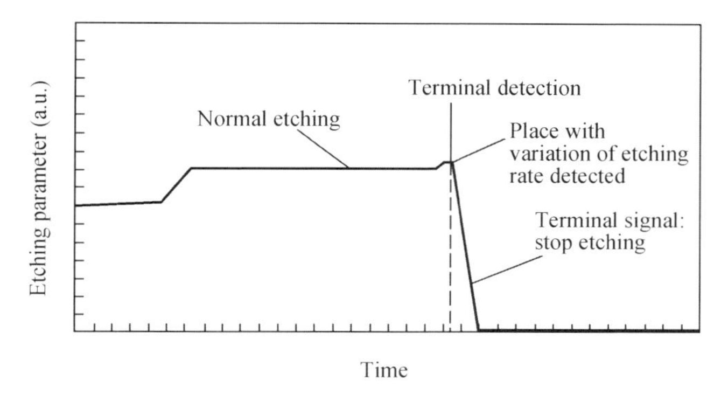

Figure 3.28 Plasma etching terminal detection.

characteristic wavelength of 440 and 777 nm. The OES can be used to analyse and identify the etched material. Based on the relative concentration relationship of the emitted light concentration and the associated element, the terminal detector can detect when to stop etching to prevent over-etching on the bottom material.

After etching, the multimeter and cold/hot probes can be used to judge the etching status of the silicon wafer edging isolation.

3.5.3 Laser Edging Isolation

When glow discharge is used to remove the PSG on the edge of the silicon wafer, the wafer shall be closely stacked together, which is difficult and easy to break the wafer. In recent years, the laser edging isolation technique has been developed by other countries, which can realise chain-type production and be designed with fast, environmentally friendly, low damages and accurate edging isolation functions but the procedure is set as the final one in solar production.

Compared with the traditional plasma corrosion edging isolation technique, the laser edging isolation technique has the following advantages:

- Since it does not need chemicals and special gases, it is free from waste water, acid/alkaline treatment, or environmental pollution;
- The silicon wafer is transported on the conveyor, and no stacking is required and thus the broken rate is low;
- The operation and maintenance cost is low;
- The productivity is high, 1500 pc/hr.

The laser edging isolation technique transports the wafer to the laser processing machine via the conveyor where the wafer shall be sucked by the vacuum system. The camera system will detect the actual position, shape and size of the wafer. The laser beam controlled by the computer programme shall move around the silicon wafer to gasify the PSG and the residuals produced will be drawn away by the exhaust system. During edging isolation, the user can set the special laser parameters to realise automatic continuous production.

The light source for laser edging isolation is $Nd:YVO_4$ (wavelength: 532 nm) laser, the maximum laser power is 10 W, the scan accuracy is ± 25 μm, and the edging isolation process time is about 2.4 sec. The laser edging isolation machine of MANZ,

Figure 3.29 Diagramme of laser edging isolation processing.

Germany, can process the silicon wafer in the size of 125 × 125 mm, 156 × 156 mm and 210 × 210 mm (as demanded). The silicon wafer is 150 – 350 μm thick. Figure 3.29 shows the diagramme of laser edging isolation processing.

At present, SCHMID and BACCINI can supply the laser edging isolation system for CSSC production. It shall be noted, however, that in the manufacture procedures of CSSC, the traditional plasma edging isolation procedure is generally set behind the phosphorus diffusion procedure while the laser edging isolation procedure is often set behind contact alloying (sintering). The cells after laser edging isolation can be subjected to final detection and sorting.

In addition, RENA has developed the chain-type PSG-removal chemical corrosion equipment, which can meet the edging isolation demand simultaneously. The wet edging isolation process can shorten the flow and achieve good effect (please refer to Section 3.6 in this chapter).

3.6 Removal of PSG

3.6.1 Principles and Processes of PSG Removal

During phosphorus diffusion, a PSG layer (SiO_2 doped with P_2O_5) will be formed on the top/bottom surface of the silicon wafer. The layer is loose with many defects and it is easy to absorb humidity. As a result, it may result in performance attenuation and reduction of the solar cell and exert adverse influence over the follow-up procedures of the solar cell. As a result, it must be removed. To remove the PSG, the silicon wafer is generally soaked in the fluoride acid solution to have chemical reaction and generates the soluble complex hexafluorosilicic acid (Formula 3.11). The chemical reaction is shown below:

$$SiO_2 + 6HF = H_2SiF_6 + 2H_2O \tag{3.11}$$

See Figure 3.30 for the process procedure for PSG removal:

The hydrophobic nature of the tested silicon wafer surface can be used to identify whether the PSG has been completely removed. The silicon wafer completely removing the PSG shall be free from water on the surface when taken out from the solution.

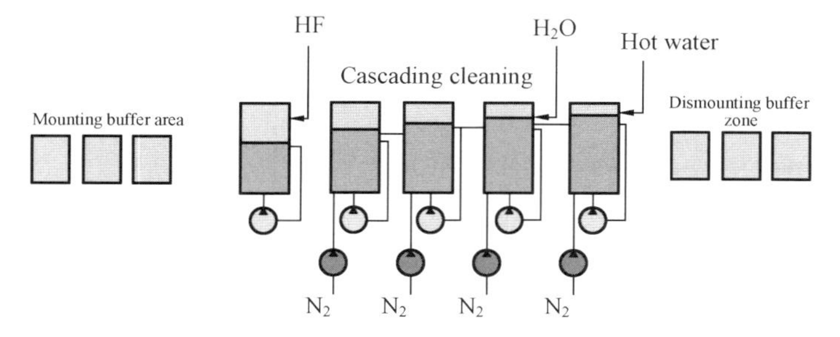

Figure 3.30 Diagramme of process procedure for PSG removal.

3.6.2 Equipment and Production Line for PSG Removal

The HF acid solution is generally used for chemical corrosion to remove PSG, and the production line consists of two types: the bath-type and the chain-type.

3.6.2.1 Bath-Type PSG-Removal Integrated Production System

The bath-type PSG-removal system generally consists of the series cleaning bath, the unloading robots, the exhauster, the servo drive system, the electrical control system and the automatic acid mixing system. The main power sources include hydrofluoric acid, nitrogen, compression air, pure water, hot exhaust and so on. The system is designed with PSG removal, cleaning and drying functions.

When the PSG is soaked in HF acid, the conductivity of the corrosion solution can be controlled to add HF acid. In addition, the multi-jet technique can reduce significantly the water consumption; and the gas jet technique can achieve such goals as rapidness, dirt-free and dry. It lists the procedures and process parameters for PSG removal in Table 3.3 for reference.

Table 3.3 Procedure and process parameters for PSG removal.

	Procedure	Auxiliary	Solution	Time	Temperature	material	Exhauster		
1	Automatic mounting								
2	HF treatment	Bubbling	HF	7 min	RT	PP	Yes		
3	DI water spraying	/	DI water	7 min	RT	PP	/		
4	DI water cleaning	Bubbling	DI water	7 min	R	PP	/		
5	Pre-dehydration	/	DI water	7 min	RT	PP	/		
6	Drying	/	Hot air	15 min	110 °C	SUS316	Yes		
7			Nitrogen			SUS316	Yes		
8	Automatic dismounting								

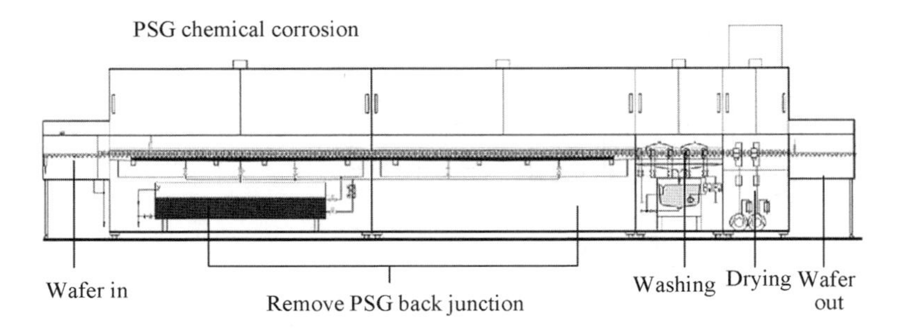

Chained silicon wafer edge PSG remover of RENA

Figure 3.31 Removing the surrounding diffusion layer while removing the PSG back junction by chemical corrosion methods (Edge Isolation Machine HYB-4000 T, Shenzhen Hongyibao Automation Equipment Co., Ltd.).

3.6.2.2 Chain-Type PSG-Removal Production System

Industrially, the PSG back junction is mainly removed by chemical corrosion methods. In recent years, RENA and Shenzhen Hongyibao Automation Equipment Co., Ltd. have developed wet chain-type production lines to remove PSG where the silicon wafer after phosphorus diffusion shall be directly put on the chain-type conveyor. It can remove the surrounding diffusion layer and simultaneously remove the PSG back junction. Accordingly, it can save the plasma edging isolation procedure (see Figure 3.31).

3.7 Preparation of Anti-Reflection Coating by PECVD and PVD Methods

3.7.1 Objectives and Principles for Anti-Reflection Coating Preparation

It generally has about 9% light reflection losses on the surface of the textured silicon wafer. Another anti-reflection coating can be deposited on the silicon wafer to further reduce light reflection and make the reflectivity down to 1–2% or so. Theoretically, the reflectivity will be zero if the refractive index of the anti-reflection coating is the geometric average of the refractive index of the materials on both sides (glass and silicon) $\left(N^2_{coating} = N_{glass}N_{silicon}\right)$. Since the refractive index has relations with wavelength, generally, the anti-reflection coating can achieve the minimum reflection at 600 nm wavelength for the solar cell. In addition, the reflectivity has relations with the thickness of the anti-reflection coating. And theoretically, the reflectivity sees the minimum when the product of the refractive index $N_{coating}$ and thickness $D_{coating}$ of the anti-reflection coating is one-fourth wavelength. If the glass refractive index $N_{glass} = 1.5$ and $N_{silicon} = 3.5$, the optimal refractive index of the anti-reflection coating is about 2.3. For the solar cell, the anti-reflection coating material must be transparent. To settle the light scattering issue at the crystal boundary, the anti-reflection coating is often deposited into an amorphous thin layer.

In the preparation technique of solar cells, PECVD is generally used to prepare the anti-reflection coating and raise the cell short-circuit current and open-circuit voltage. The hydrogen-contained silicon nitride can be used as CSSC anti-reflection coating, which has another advantage—the hydrogen can passivate the dangling bond at the crystal boundary. It has a lot of hydrogen atoms in the silicon nitride coating deposited by reaction gases SiH_4 and NH_3 on the silicon wafer surface, it can significantly improve the lifetime of the carrier in the silicon wafer.

Compared with LPCVD, since PECVD uses the plasma to generate and maintain the chemical vapour deposition reaction, its reaction temperature is much lower than that of LPCVD although their reaction system pressures are similar. For example, the silicon nitride deposited by LPCVD, the reaction temperature is usually 800 – 900 °C while the deposition temperature of PECVD is only about 400 °C.

Double MgF_2/ZnS coatings are also used for the anti-reflection coating of solar cells. Figure 3.32 shows the refractive-index of several common anti-reflection coatings and crystalline silicon as a function of wavelength.

3.7.2 Principles of Silicon Nitride Coating Prepared by PECVD

Since the quality of the silicon nitride coating deposited by PECVD has direct relationship with the efficiency of solar cells, and the PECVD equipment is also the most expensive in the CSSC production line, it is necessary to describe its deposition mechanism in details. In the PECVD technique, the reaction gas is introduced in

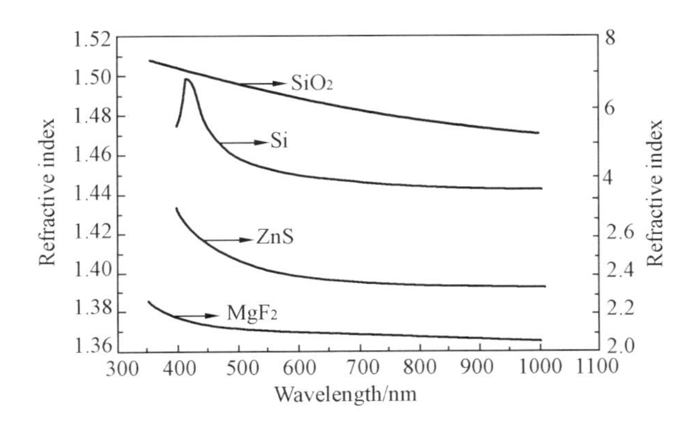

Figure 3.32 Refractive-index of several common anti-reflection coatings and crystalline silicon as a function of wavelength.

vacuum and produces the plasma between the parallel electrode plates in the manner of high-frequency discharge. The sample shall be put on the cathode with glow discharge at low voltage, and the low-temperature plasma shall be used as the energy source, and the glow discharge (or other heating equipment) shall be used to heat up the sample to the given temperature. The reaction gas shall achieve chemical deposition via a series of plasma reaction, forming a solid thin film of several hundred to thousand angstroms in thickness. The biggest difference between PECVD and other CVD lies in that the plasma contains a lot of high-energy electrons, which can offer the activation energy for CVD. The collision of electrons and vapour molecules can accelerate the vapour molecules to decompose, recombine and ionise, and generate various chemical groups with high activity, significantly reducing the deposition temperature of CVD film and making the CVD process (originally at high temperature) available at low temperature. Figure 3.33 shows the preparation process of PECVD coating. And Figure 3.34 shows the PECVD system. The PECVD system mainly consists of the process gas cabinet, the

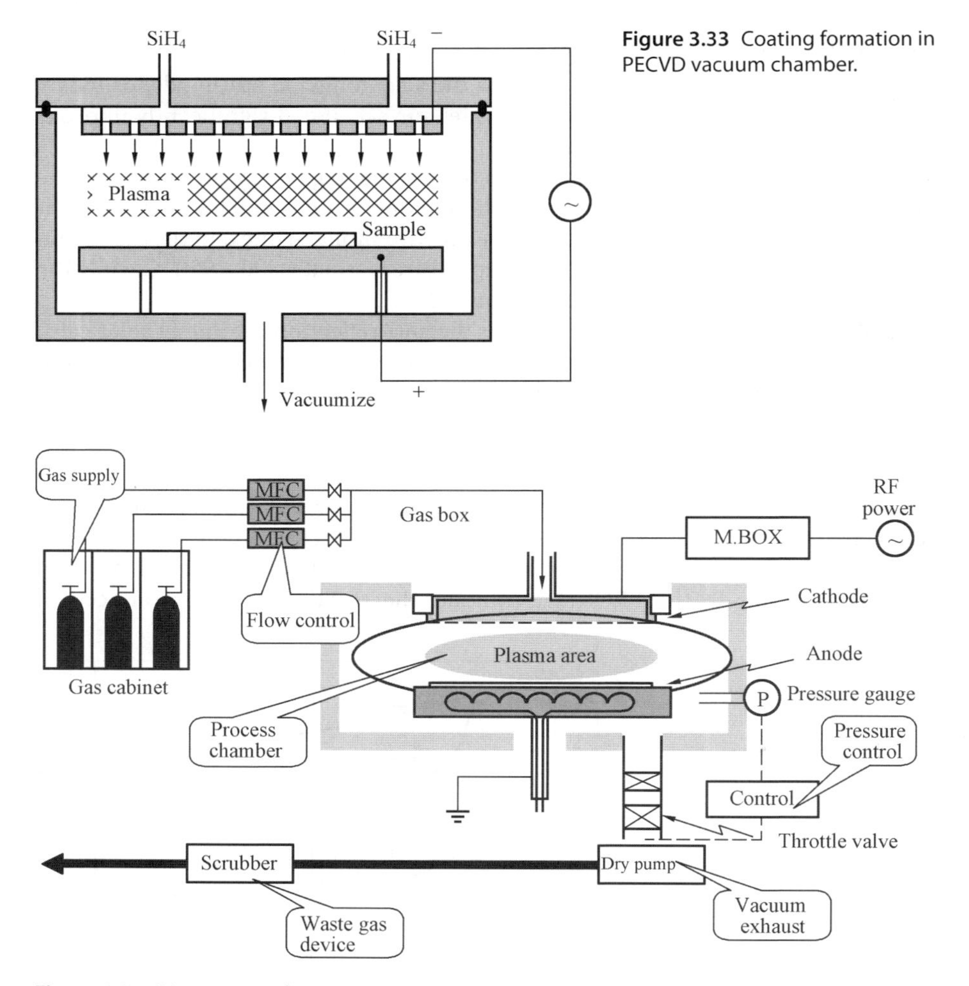

Figure 3.33 Coating formation in PECVD vacuum chamber.

Figure 3.34 Diagramme of PECVD system.

gas control system, the power and control system, the electrodes, the process chamber, the pressure regulation system, and the vacuum exhaust system and the waste gas treatment system. The RF or microwave power can be applied in the process chamber to make the gas decompose and ionise.

The RF for silicon nitride deposition is generally 13.56 MHz. Since the molecules activated are chemically active, they are apt to bond with other atoms to form the coating on the silicon wafer surface. The wafer substrate is usually required of heating to accelerate the surface reaction and regulate the hydrogen content in the silicon. The silicon nitride generated by PECVD is deviated from the traditional chemical stoichiometric ratio and it contains some hydrogen and thus it is often expressed by SiNx:H. The proper quantity of hydrogen can passivate the silicon dangling bond and reduce the coating compressing stress (Formula 3.12). The deposition reaction of silicon nitride is as follows:

$$SiH_4 + NH_3 \rightarrow SiNx\colon H + H_2 \tag{3.12}$$

PECVD is a typical cold-wall reaction where the silicon wafer is heated to a high temperature while the rest of the reactor is not. The cold-wall reaction can produce fewer particles and reduce the shutdown cleaning time. The deposition chamber has adopted the cleaning method at the original position and the gaseous by-products shall be exhausted by a vacuum pump. The relevant process parameters, however, shall be controlled to make sure that the temperature gradient will not affect the homogeneity of coating thickness.

PECVD coating formation is also dependent on many other factors, for example, the structure and spacing of the electrodes, RF power and frequency, gas composition, pressure intensity and velocity, as well as substrate temperature and so on. These factors are mainly reflected on design of the reaction chamber and system. When the silicon nitride coating is prepared by PECVD, its homogeneity and quality are generally affected by silane flow, ammonia flow, hydrogen flow, nitrogen flow, pressure in the reaction chamber, distance between top/bottom electrodes, power and frequency of the power supply and the like. Generally, when the silane flow rises, the coating will be silicified, and the coating formation speed will grow with growth of refractive index. When the ammonia flow rises, the coating will be nitrided, and the coating formation speed will reduce with reduction of refractive index. Generally, when the pressure in the chamber is appropriately increased, it will be good to raise the homogeneity of the coating. As for the influence of the distance between top/bottom electrodes, an optimal value shall be determined. Generally, when the power of the power supply rises, the coating formation speed will grow but the homogeneity will fall.

3.7.3 Direct (Tubular) PECVD and Indirect (Plate-Type) PECVD

At present, the leading PECVD equipment for SiNx deposition can be divided into direct and indirect types by structure of reactors and frequency of plasma generator. In the direct PECVD system, the sample or the support is directly a part of the electrode generated by plasma. And in the indirect PECVD system, the sample is out of the plasma zone and the plasma does not directly collide on the sample surface and the sample or the support does not form a part of the electrode.

In the direct PECVD system for SiNx deposition, the frequency of the power supply is low (e.g., 13.56 MHz, RF), and the mounting carrier is graphite boat, and the silicon wafer is a part of the circuit. Since the wafer is in the bias state, which accelerates the bombardment of ion on the wafer surface, but the plasma density is not high, the deposition speed is slow. It generally takes a dozen of minutes to deposit the silicon nitride coating. The dense silicon nitride coating can grow.

In the indirect PECVD system for SiNx deposition, the frequency of the power supply is high (e.g., 2.45 GHZ, microwave), and the mounting carrier is composite graphite rack. Neither the silicon wafer nor the rack forms circuit with the plasma generator. Accordingly, it has nothing to do with plasma. In the indirect PECVD system, the excited NH_3 plasma shall enter the reactor via a narrow quartz tube, and after reacted with silane, it will deposit silicon nitride coating on the silicon wafer. The plasma produced by this method has high density and faster deposition speed than that by direct PECVD. In addition, the robots can be used to load and unload the silicon wafer (especially the thin one). The silicon nitride coating produced by indirect PECVD, however, has lower density.

Both direct and indirect PECVD techniques can produce the silicon nitride with good homogeneity, and both have been widely applied to cell manufacture. Figure 3.35 shows the structure of the two PECVD systems. For direct or indirect PECVD systems, the proportion of silane and ammonia shall be all optimised during depositing silicon nitride coating, and the coating thickness can be monitored by an ellipsometer. The temperature of the substrate shall be controlled at 400 °C or so.

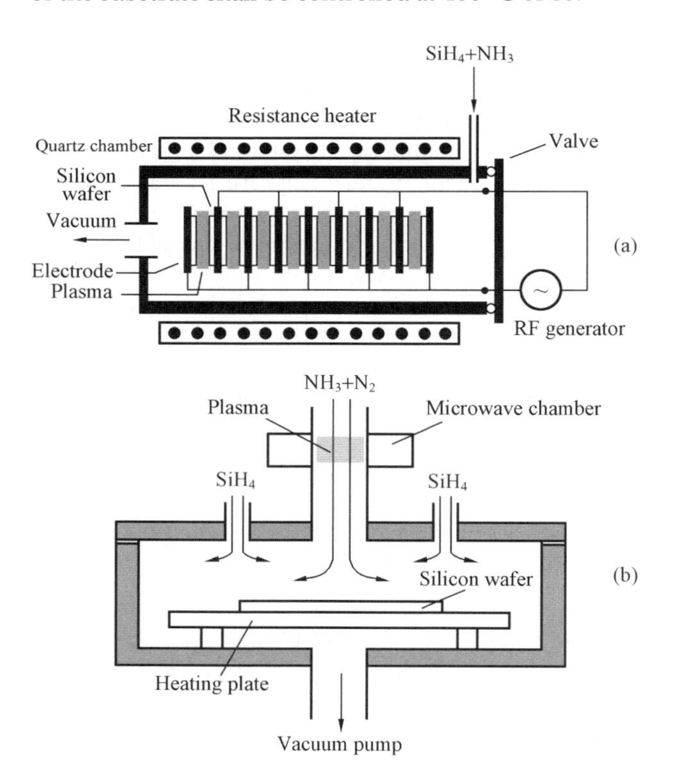

Figure 3.35 Diagramme of tubular direct PECVD system (left) and chain-type indirect PECVD system (right).

3.7.4 Typical PECVD Systems

3.7.4.1 Tubular Direct PECVD System

In the tubular PECVD system, a graphite boat that can accommodate several pieces of silicon wafer is inserted into the quartz tube for deposition and heating by resistance wire. The main manufacturers include Centrotherm, Germany, the 48th Research Institute of CETC, Beijing Sevenstar Electronics Co., Ltd. For M822200-1/UM hose-type PECVD, developed by the 48th Research Institute of CETC, each tube can produce 4000 pieces per day, which has been applied to many solar cell production lines, and the operation is smooth. The tubular direct PECVD system is shown in Figure 3.36.

Take the PECVD system of the 48th Research Institute of CETC as an example. In the system, the working frequency is RF, and the pulse plasma is used, and the graphite rack and the silicon wafer are participated in building of plasma field. It is tubular with four tubes on one line where each tube can be mounted with 168 pc (156 × 156 mm)/192 pc (125 × 125 mm). It can meet the demand of the product line of 25 MW solar cell per year. The system consists of the heating system, the gas path system, the boat pusher system, the RF system, the vacuum system, the computer control system and the quartz tube and so on. The heating system includes the heating furnace cabinet, heating furnace, the power components, and the SCR and so on. The gas path system includes the gas source cabinet, the mass flowmeter, the filter, the pneumatic valve, check valve, the reducing valve, and the manual valve and so on. The system is supplied by five gas lines—$SiH_4/NH_3/N_2/O_2/CF_4$ at 2–3 kg of pressure and purity above 5 N. The cantilever is used for soft landing, and the boat pusher system consists of the computer automation control system, the guide rail, screw rod, boat rack and so on. The fine adjustment is available for the boat pusher system in X, Y and Z directions. The boat pusher system is the core of the system since it can directly affect the field distribution, plasma generation, air flow direction, coating quality, coating formation pressure and so on, and even the gas flow has relations to its structure. The RF power components mainly include the RF power supply, the matching system and the graphite electrode and so on, where the RF is input to the reaction system in the shortest route to reduce interference and power consumption.

3.7.4.2 Plate-Type Direct PECVD System

Shimadzu, Japan has developed the plate-type direct PECVD system, which combines the virtues of direct and plate-type systems, that is, dense nitride coating,

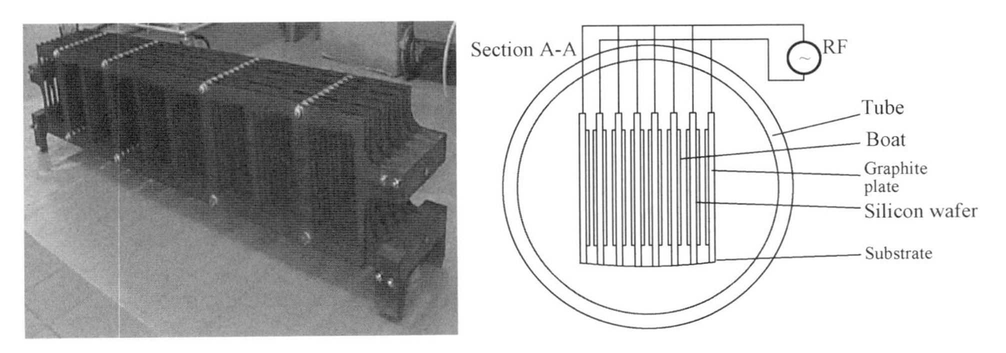

Figure 3.36 Tubular direct PECVD system and graphite rack boat.

large production capacity, easy mounting/dismounting and convenient automation integration and so on. In the plate-type direct PECVD system, several pieces of silicon wafer are mounted on a graphite or carbon fibre support, and then put into the metal deposition chamber with plate-type electrodes and then form the discharge circuit with the sample support, and the process gas will form the plasma in the alternating field between two electrode plates, as shown in Figure 3.37.

3.7.4.3 Plate-Type Indirect PECVD System

Figure 3.38 shows the indirect PECVD system, represented by ROTH-RAU.

SiNA series anti-reflection coating system of Roth-Rau consists of three main chambers: the feed chamber, the process chamber and the discharge chamber. The three chambers, independent from each other, are separated by the pneumatically controlled valves. Since the infrared heating lamps are installed at the position of the feed chamber, the silicon wafer can be rapidly heated up to 350 – 450°C. The process chamber, the core of the system, can be divided into three zones by function. The first is the heating zone where the surface temperature of the silicon wafer to be deposited will be kept at the

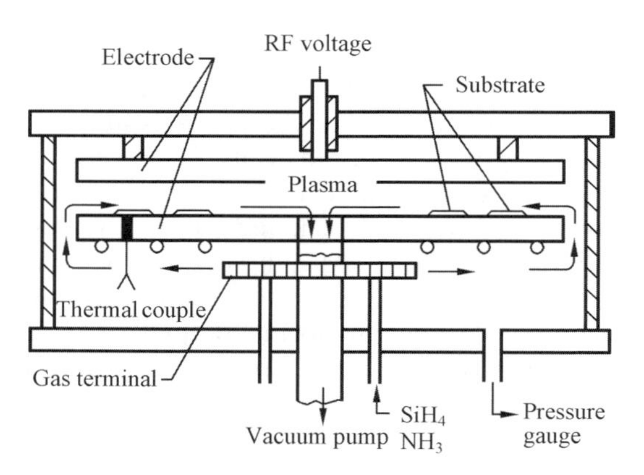

Figure 3.37 Plate-type direct PECVD system.

Figure 3.38 Tiled indirect PECVD system and graphite rack plate, ROTH-RAU.

set value. The second is the deposition zone, which consists of several rows of identical linear plasma sources at microwave frequency of 2.45 GHZ, and the process gases used are SiH_4 and NH_3, and the standard process pressure intensity is 10–30 Pa, and the carrying plate moves in the zone at a constant speed. The effective zone of the plasma is limited to the area longitudinally stretching several centimetres between the carrying plate full of silicon wafer and the plasma glow zone. Since the wafer faces down during deposition, it is inevitable to have contamination by the minor particles generated in the process. The third is the cooling zone where the load plate and the deposited wafer shall be cooled down before leaving the system.

The SiNA systems of various types are provided with matching vacuum pumps. Each set of vacuum pumps consists of one root pump and one master mechanical pump. The process pressure intensity is controlled by and independent inverter regulating pump.

All the special gases are controlled by mass flowmeters. The gas control system on the low-pressure side, for example, the mass flowmeter and the valve, are installed in an independent gas cabinet. All the gas pipelines of the access equipment are welded, and the reaction gas connections are all metal sealed, which shall be subjected to strict leak test. All reaction gas channels and pumps are provided with gas gauges.

The system is also provided with a carrying plate return system, which is installed at the bottom of the chamber. The system is compatible with many silicon wafer mounting robots.

The process course from feed to discharge is subjected to on-line control. The programme is conducted by industrial computers. All the functions are operated by the private login system. These measures can ensure safe system operation. All operation functions are included to the user-oriented WINDOWS NT software. The software can display the system running status and physical conditions via a series of on-line charts and it can help the user start or stop the process and set the working parameters and other operations.

3.7.5 Preparation of Silicon Nitride Coating by Physical Vapour Deposition (PVD)

3.7.5.1 Principles of Silicon Nitride Coating Prepared by PVD

The principle of silicon nitride physical sputtering is that argon and ammonia are applied to the sputtering equipment, and then the voltage applied to make the argon rapidly move towards the cathode and bombard the silicon target, and the sputtered silicon target atoms will gradually gather and deposit on the silicon wafer surface and react with the ionised ammonia plasma, generating the silicon nitride coating on the wafer substrate surface. The diagrammes of silicon nitride physical sputtering and the silicon target used are shown in Figure 3.39 and Figure 3.40. The sputtering mechanism is described below:

- After voltage is applied, the positive ion of argon ① will move rapidly towards the cathode ② and bombard the silicon target ③, making the silicon target atoms ④ and the secondary electrons ⑤ bombarded and transmitted;
- The target atoms gradually gather and deposit on the wafer surface ⑥;
- With NH_3 ionisation, the plasma generated ⑦ (7a, 7b, etc.), for example, Si^+, NH^+, ${NH_2}^+$ and the electrons, will generate rapidly the silicon nitride coating;

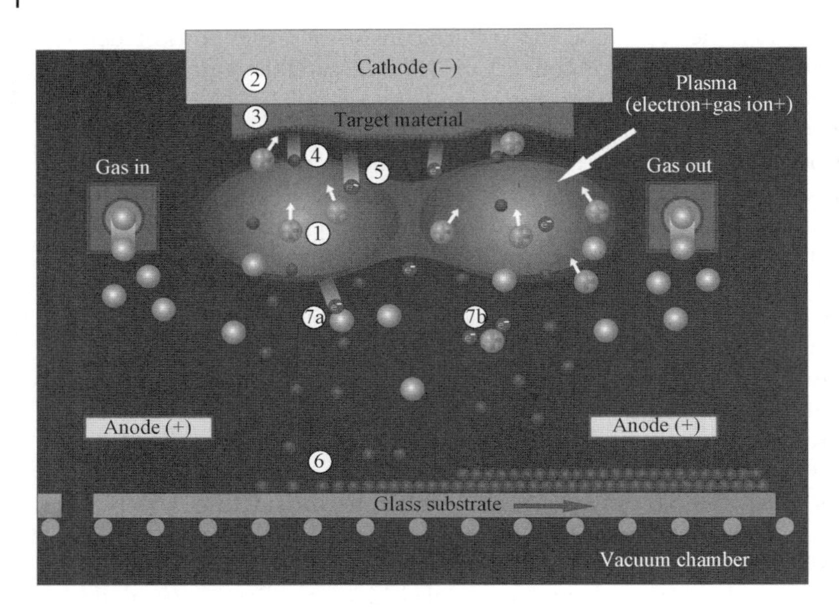

Figure 3.39 Diagramme of silicon nitride physical sputtering.

Figure 3.40 Silicon target used by physical sputtering method.

- The magnetically controlled field (not shown) can effectively improve the sputtering output;
- The distance between the silicon target and the wafer is generally several centimetres;
- The substrate is heated by the heater installed on the back of the substrate.

Sputtering is mostly a physical process. The sputtering deposition system adopts the high-energy argon particles to collide the silicon target, which is because the argon ion is heavy and its injection energy under the action of the field is large enough to strike out one or more target atoms in the target material, and a typical sputtering ion has an energy range of 500–5000 electron volt. In addition, since argon is inert gas, it can avoid reacting with the grown coating or the target material. The silicon atoms in the silicon target will be stricken out during sputtering. And these atoms will go through the vacuum zone and finally deposit on the silicon wafer. The silicon target earthing is the cathode and the wafer substrate has positive potential (the anode). The purity of the target material is required above 5 N.

During sputtering, the target will get hot. As a result, it shall be cooled down to keep low temperature. The wafer substrate is heated by infrared, and the optimised substrate temperature of PVD is about 250 °C with temperature homogeneity of ±10%. The appropriate process parameters can be used to avoid overheating. The number of cathodes sputtered can be provided as demanded by the user. The PVD process can be suited to the wafer in various sizes.

The deposition chamber of the sputtering system consists of the silicon target, the wafer substrate and the vacuum environment and so on. The system is designed with double vacuum valves, which will move the wafer from the atmospheric condition to the background vacuum in segments. The system adopts automatic mechanical transmission to make the wafer move between chambers. The automatic mechanical transmission is magnetically coupled for several chambers to reduce the foreign particles.

In PVD, the gas flow can be regulated to control the refractive index, hydrogen content and thickness of the silicon nitride coating, and it has good refractive index homogeneity (±2.5%), good anti-reflection and passivation performance (see Figure 3.41 for details).

3.7.5.2 Comparisons of Silicon Nitride Coating Deposited by PVD and PECVD

PVD adopts argon and ammonia to deposit the silicon nitride coating where the temperature of the silicon wafer substrate for the depositing silicon nitride coating is less than 300 °C, lower than that of PECVD. PECVD needs silane and other explosive gases, and thus the gas emission and monitoring needs additional production cost. PVD, however, does not use any explosive gas, much safer and more convenient than PECVD, and pollution-free for the surroundings. In addition, the PVD sputtering system covers less land than PECVD system at unit output, and its productivity is larger than PECVD system. The process chamber of PVD is cleaner than that of PECVD during coating deposition, with cleaning period larger than one week. Moreover, the damaged wafer rate is low (<0.2%), it is applicable to produce at high efficiency.

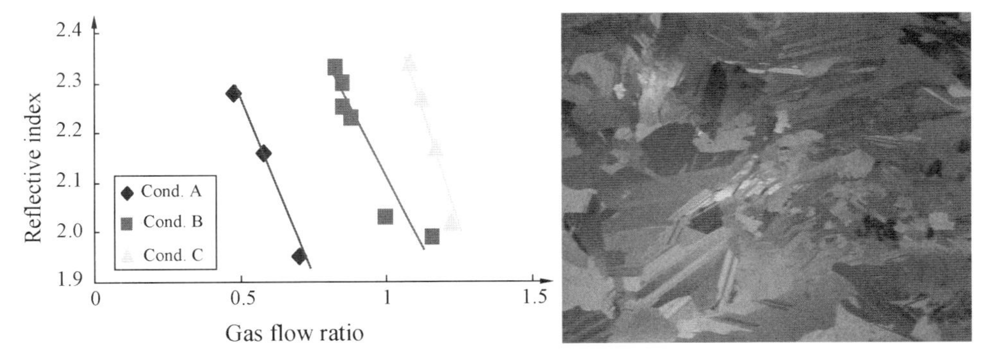

Figure 3.41 Control of silicon nitride coating thickness by regulating gas flow.

PVD can prepare the silicon nitride coating and sputter the metal film, which has been applied to polycrystalline/monocrystalline silicon solar cell and thin film solar cells.

3.7.5.3 ATON Sputtering System Produced by Applied Materials, USA

ATON sputtering system, produced by Applied Materials, USA, is provided with a RF cathode or a double-plane magnetically controlled cathode (TWIN-MAG) to conduct dynamic plating. The solid silicon target as the silicon source, and nitrogen, hydrogen and ammonia are applied. It can dynamically deposit at a high speed up to 40 nm/min with hourly output up to 1800 pc or even to 3000 pc. One PVD system can produce 20–100 MW solar cells per year. The size and shape at the break of the cathode barrier can be optimised to achieve good coating homogeneity (±2%). Figure 3.42 shows Astro-SiN50 system and PVD process chamber.

Astro-SiN50 with carrying plate in 1200 × 1500 mm, can carry 48 × 156 mm × 156 mm or 80 × 125 mm × 25 mm wafer. The material of the carrying plate is CFC, and it can automatically return. The wafer is mounted by robots. The output is 3000 pc/hr for 125 × 125 mm wafer and 2200 pc/hr for 156 × 156 mm wafer.

3.7.6 Measurement of the Thickness and Refractive Index of the Anti-Reflection Coating by Ellipsometer

The ellipsometer, based on non-contact optical thin film thickness measurement techniques, is mainly used to measure the refractive index and thickness of the transparent thin film. It can detect the thickness and refractive index of the silicon nitride anti-reflection coating grown by PECVD system. The basic principle is that when the linear polarised laser source reflects on the sample, it will turn to elliptical polarised

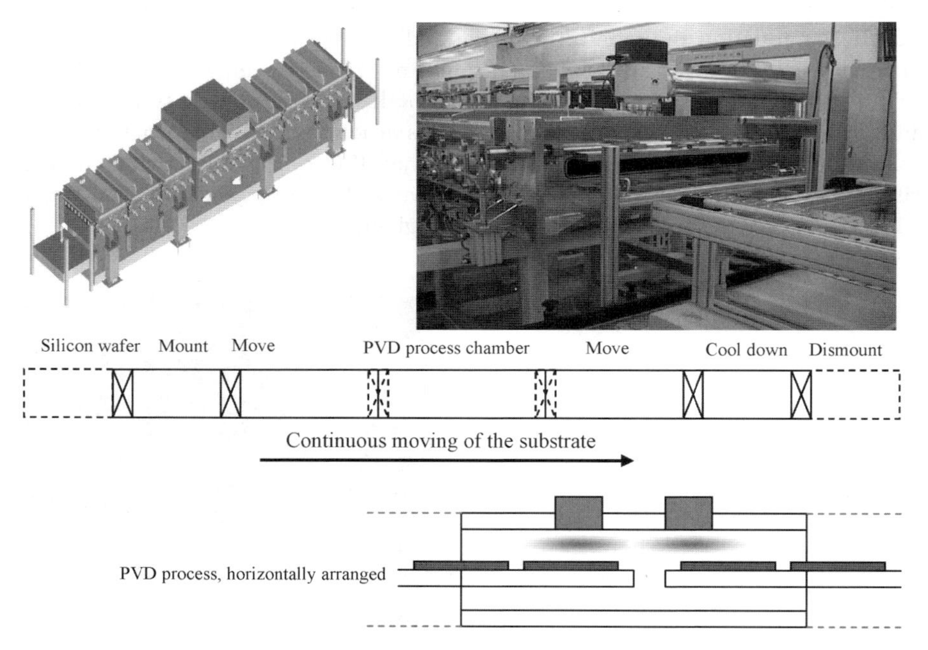

Figure 3.42 Astro-SiN50 system and PVD process chamber.

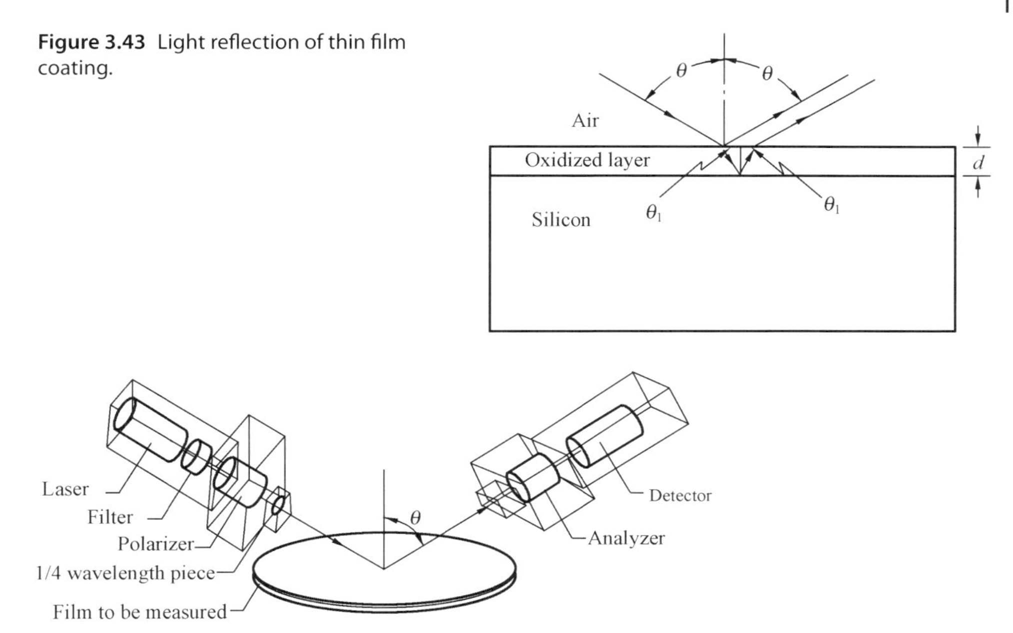

Figure 3.43 Light reflection of thin film coating.

Figure 3.44 Basic principle of the ellipsometer.

light. The elliptical polarised angle and intensity can be measured by polariscope, and then, based on the known inputs (e.g., the reflection angle), the accurate thickness and refractive index of the thin film can be worked out. The reflection of the thin film coating on the linear polarised light is shown in Figure 3.43 where the incidence light is linear polarised light and the reflection light is eliptical polarised light (see Figure 3.44 for the ellipsometer). The full automation ellipsometer can be directly applied to the production flow for on-line detection and process control.

The ellipsometer can measure the thin films of various types with thickness in dozens of angstroms, including media, metal and coated polymer. The basic requirement is that the film shall be transparent or semi-transparent.

3.8 Preparation of Top/Bottom Electrodes (Surface Metallisation)

3.8.1 Technical Requirements and Production Flow for Top/Bottom Electrode Preparation

Preparation of the top/bottom electrodes is to prepare the electrodes on the irradiating and the bottom surfaces of the solar cell to output the photo-generated current. It is generally viewed as the last procedure for production of solar cells. The following methods are mainly used to prepare the electrodes: vacuum evaporation, chemical plating, slurry printing and sintering and so on. The screen printing is viewed as the most common production process for the solar cell electrodes, that is, the metal electrodes of some patterns are printed on the silicon wafer as designed and sintered to ohm contact.

The slurry shall penetrate the meshes of the screen patterns, and a scraper shall be used to compress on the slurry position and move towards the other end. In this case, the slurry will be squeezed to the wafer via the meshes on the patterns. The silver slurry and the argentalium slurry shall be printed on the irradiating and the back surfaces of the solar cell, respectively, to form the leads of the positive and negative electrodes and then subjected to low-temperature sintering and high-temperature sintering to form ohm contact and finally produce the solar cell. It shall be stressed that the cell irradiating surface must be to the maximum extent transparent to light, and the electrodes on the irradiating surface are generally designed in a comb, screen or branch structure to minimise the shady area of the irradiating surface.

In addition, to raise the open-circuit current, a layer of aluminum slurry shall be first printed before printing the back electrodes to form the aluminum silicon alloy as the BSF. Preparation of BSF can also carry out aluminum impurity gettering for the crystal silicon. Compared with phosphorus impurity gettering in diffusion, phosphorus impurity gettering has faster speed than aluminum impurity gettering but the capacity and stability of aluminum impurity gettering are better than that of phosphorus impurity gettering. For the polycrystalline solar cells, independent phosphorus/aluminum impurity gettering or their combination is generally used.

In the preparation process of CSSCs, to reduce reflection of solar rays on the wafer, the previous procedure has etched many pyramids and conic structures on the irradiating surface of the wafer, resulting in the thickness of the silicon nitride anti-reflection coating deposited later inconsistent (about 80–200 nm). When the conductive silver slurry of the irradiating surface is printed on the uneven silicon nitride coating, sintering must be carried out to make the silver element penetrate the silicon nitride coating and become in ohm contact With the bottom N^+/P junction only several submicrons deep, and it shall not damage the N^+/P junctions. The detailed technical requirements include:

① To form ohm contact on the silicon wafer deposited with silicon nitride coating and N^+/P junctions, it must strictly control the penetration depth of silver slurry to the silicon nitride: The series resistance will be excessively large if it is too shallow; otherwise, it may damage the $N+/P$ junctions. The final series resistance shall be less than 0.25 Ω;

② To ensure print quality, the conductive silver slurry on the irradiating surface shall have good flowability, for example, the cohesion of the common conductive silver slurry is generally 150 Pa.s or so;

③ After sintering, the bonding strength with the silicon wafer shall be good, for example, >2 N (Newton). And it shall not result in excessively large stress on the silicon wafer, that is, the contractibility rate shall not be too big;

④ After sintering, the silver electrode on the irradiating surface shall have certain height/width ratio, minimising the electrode shading area of the irradiating surface ($<8\%$);

⑤ After electrode metallisation, it shall good welding performance with the assembly collector bar;

⑥ Low cost;

⑦ Less pollution;

⑧ Suited to process.

It is difficult to prepare the electrodes on the irradiating surface of the solar cell. For design of the metal gate lines of the irradiating surface electrodes, it shall follow the principles that the cell series resistance shall be minimised, and the sunshine area on the cell irradiating surface shall be maximised to achieve maximum cell output. At present, the screen printing for thick films have become mature where the line width can be reduced to 50 μm and the height can reach 10–20 μm. The metal electrodes are generally composed of two parts: the bus and the fingers. The bus is bold and it shall directly transmit the current to the external via the collector bar. The fingers are thin, which shall collect the current and transmit it to the bus. The silicon wafer as a whole can be decomposed into several single cells and their electrodes are generally symmetrically distributed. Theoretically, the maximum output power of single cell can be figured out by the area of the single cell, the current density and voltage of the maximum power point, the resistance of the fingers and the resistance of the bus thin layer, the average width of the fingers and the bus, the spacing between the fingers, the resistive power losses of the grid and the bus, the power losses caused by solar ray shading and so on.

It is relatively easy to prepare and print the back electrodes of the solar cell. The requirement is that it shall cover the back to the maximum extent. The covering area will also affect the cell filling factor.

The metallisation print, dry and sintering procedures of top/bottom electrodes (from right to left) are shown in Figure 3.45: that is, (1) Take out the silicon wafer from the box or rack and mount it to the production line; (2) first print; (3) first dry; (4) second print; (5) second dry; (6) turn over the silicon wafer; (7) third print; (8) third dry; (9) fast sintering. Finally, it shall be subjected to cell test and sorting.

At present, the homemade manual and semi-automatic screen printer has been produced in batches, and the fully automatic screen printer is already under research and development. But the solar cell screen printer made by China still has some gap with that produced by BACCINI, Italy. In this section, we will introduce the print, dry, test and sorting production line of the solar cell, BACCINI.

3.8.2 Electrode Printing, Drying, Testing and Cell sorting

The production lines of BACCINI consist of the laser edging isolation (replacing plasma edging isolation), the final testing and cell-sorting equipment in addition to printing, drying and sintering (sintering equipment provided by the user; see Figure 3.46).

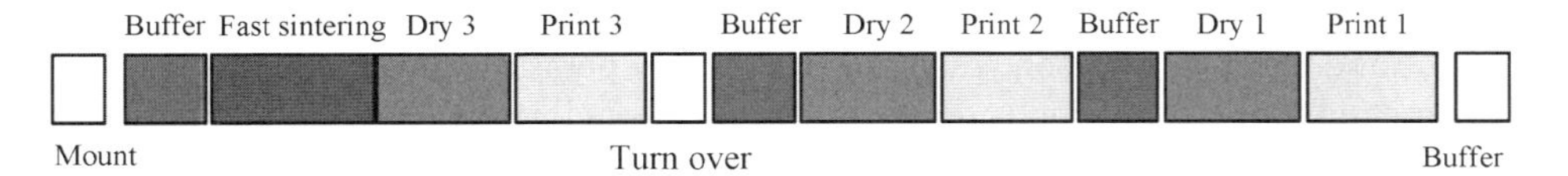

Figure 3.45 Process procedures of top/bottom electrodes (from right to left).

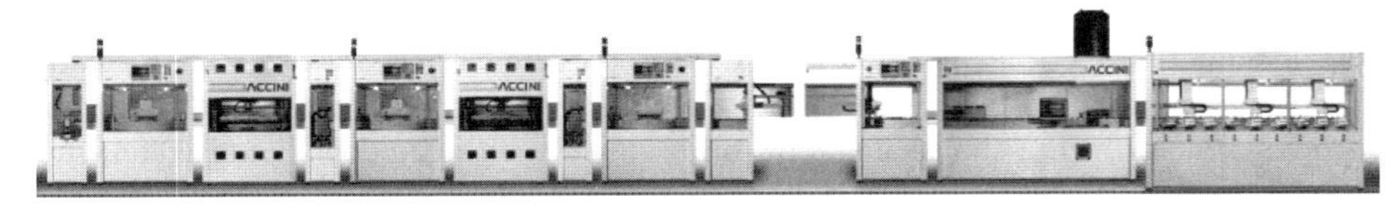

Figure 3.46 Preparation, laser edging isolation, cell testing and sorting production lines of BACCINI for top/bottom electrodes.

The fully automatic screen printer of BACCINI, drum-type, is designed with printing and drying functions. Its output is 1440 pc/hr. if the power of each cell is 3.6 W, the annual output is 36 MW.

The drum-type fully automatic screen printer consists of the printing module, the drying module, the mounting and transmission systems.

1. Features of the printing module:
   - 1440 pc/hr; damage rate < 0.5%
   - Applicable to the silicon wafer of 125 × 125 mm × 156 × 156 mm (thickness: 0.2 mm or so)
   - Maximum screen size: 380 × 400 mm
   - The print screen plate is made of nickel plate by laser grooving to ensure the durability of the screen plate and the accuracy of the grids
   - Printing accuracy: ±0.005 mm
   - Slurry driven by the automatic slurry distributor
   - Screen sockets printed by compression rollers
   - One wafer correction camera before and after printing (back lighting)
   - On-line high-speed AOI after printing
   - Inspection cameras not contaminated by fragments or slurry
   - Wafer turning device
2. Features of the drying module
   - Drying principle: gas convection (number of dryers: two or three);
   - Precision 4 temperature zone control;
   - Dense vertical design, minimal top/bottom space;
   - Temperature curve of the drying set as demanded by the user.
3. Features of mounting and transmission systems:
   It will mount the silicon wafer from the conveyor to the rotary screen printer. In total, four positions are provided. One production line can produce 1440 pc/hr at fast speed with high accuracy.

   BACCINI also produces the printers of other types and refer to 'Use Manual of BACCINI Printer' on the internet.

3.8.3 Fast Sintering Furnace System

After high-temperature sintering furnace screen printing, the slurry shall be first dried to remove the organic adhesive in the slurry. In the sintering furnace, when the silver electrode and the crystal silicon reach the eutectic temperature, the crystal silicon atom will be melted to the silver electrode at some proportion, forming ohm contact between the top and bottom electrodes.

The sintering furnace includes three stages: pre-sintering, sintering and cooling down stages. The pre-sintering stage is aimed to decompose and fire the polymer binders in the slurry and in this stage, the temperature will rise slowly. The sintering stage shall finish the eutectic reaction and other physical and chemical reactions or make the electrodes in ohm contact with the silicon wafer, and it needs high temperature. The metal electrodes after sintering shall have good ohm contact with the cell silicon wafer, the metal electrodes shall be free from bubbles with clear line and even thickness as well as fixation, but the P-N junction on the cell surface shall not be burned out. Accordingly, the furnace temperature shall be rapidly increased and decreased, that is, the peak sintering temperature curve shall be available where the high temperature shall not hold for a long time to achieve rapid sintering and cooling down. Since the slurry will evaporate in the sintering furnace, the sintering furnace shall be designed with sufficient glue removal.

Figure 3.47 shows the typical temperature curve of the slurry of the top/bottom electrodes in the sintering furnace after low-temperature dry.

The key techniques of the fast sintering furnace include even thermal field and fast cooling down in addition to peak sintering temperature curve and control techniques. The bottom guide and local tightening drive shall be added to prevent the chain-type sintering furnace mesh belt from deviation and runoff and ensure smooth transmission. To make the cell silicon wafer rapidly cooled down to below 60°C when leaving the furnace, the cooling down zone shall be provided with front air blow + rear water cooling + forced ventilation convection. To improve the cleanness of silicon wafer surface, the design of inlet + outlet + top gas entry and glow discharge at the bottom can be used. To reduce the energy consumption and ambient temperature, the new insulation design and thick insulation layer can be adopted. The mesh belt shall be made of new materials to prevent rust. For various slurries, the sintering process shall be adjusted to make the silver slurry matched with the argentalium slurry.

After the sintering process, the silver slurry shall form good ohm contact with the substrate and the slurry shall be free from skinning, bubbling or silicon wafer bending. For the 1.5 mm welding joint, the pull force of 1–3 N (Newton) in 45 ° shall be applied to ensure the welding position will not fall off.

For the ohm contact between the silicon and the interconnecting metals after electrode alloying, its performance can be identified by the V-I characteristics curve on the contact surface and the ohm law. If the V-I characteristic is in linear direction, it can be identified as ohm contact, that is, the slope is the resistance.

In the following, the fast sintering furnace, produced by DESPATCH, will be described. Figure 3.48 shows the outline, inner furnace and conveyor of the sintering furnace.

The main process parameters and characteristics of CDF7210 sintering furnace, DESPATCH, are described below:

① The conveyor of the sintering furnace can reach up to 255 inch/min, and it can meet the demand of annual output of 25–30 MW cells;

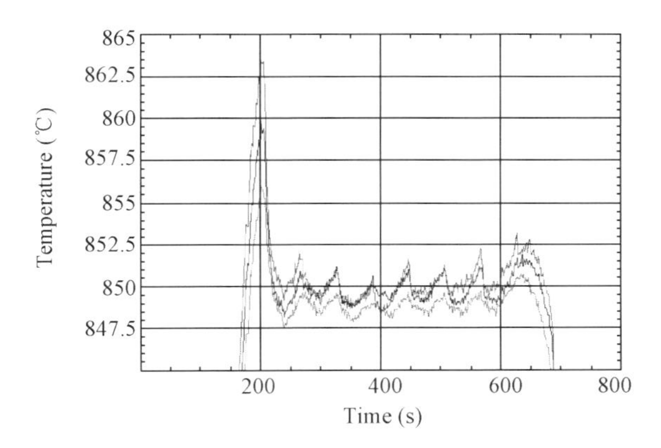

Figure 3.47 Temperature curve of a typical sintering furnace.

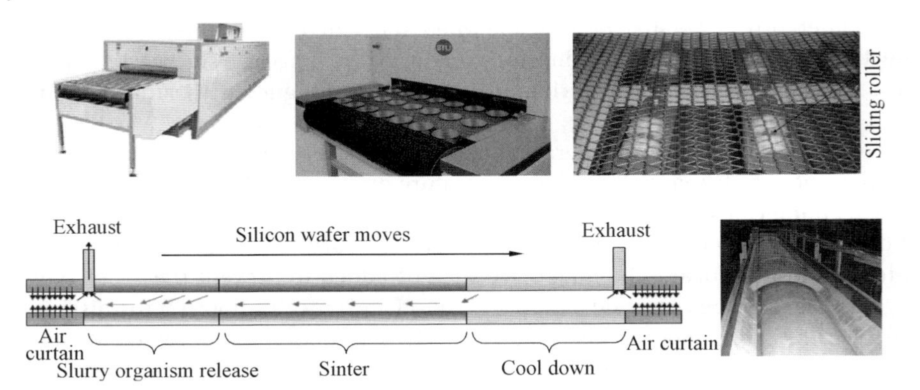

Figure 3.48 Outline, inner furnace and conveyor of the sintering furnace, DESPATCH.

② The sintering furnace is provided with thick insulation layer, and the air flow shall be preheated at low speed with big flowrate, and the hot air flow is designed in all-layer flowing manner to ensure the evenness and smoothness of the furnace temperature;
③ The conveyor drive motor of the sintering furnace will minimise the jittering of the conveyor and ensure the thin silicon wafer below 200 μm has smooth operation and low damage rate;
④ The quartz supports beneath the sintering furnace conveyor are in special design arrangement (patent technique, in special angle) to make the silicon wafer free from cold/hot zone differences;
⑤ The sintering furnace temperature can rapidly rise or fall with temperature rise rate above 80°C/sec (conveyor speed: 200–255 inch/min) from 580°C to 820°C (finally the sintering zone), and temperature fall rate above 65°C/sec from 820°C to 400°C;
⑥ In the sintering furnace, from the dry furnace to the sintering zone and the cooling down zone, the furnace is opened by automatic falling, fast and simple;
⑦ The sintering furnace can control the power of the top/bottom heating tubes in the sintering furnace in a separate manner to fully heat the aluminum slurry on the back while prevent the silver slurry beneath from overheating.

3.8.4 Electrode Slurry

Both the irradiating surface and the rear surface of the silicon solar cell shall be printed with silver slurry and argentalium slurry, which shall be sintered to conductive electrodes on the irradiating surface and the rear surface. They shall form good ohm contact with the silicon cell because the existence of contact resistance will reduce the FF of the cell. For the irradiating surface, the grid shades some area on the irradiating surface. To raise the cell conversion efficiency, the grid area shall be minimised and the height/width ratio of the gate line shall be raised. In this case, the silver thick film slurry shall be developed to achieve big height/width ratio of the gate line.

The electrode silver slurry on the irradiating surface ranks the second by cost, following the silicon wafer. It will exert great influence over the product rate of the follow-up procedure and have a direct impact on the conversion efficiency of the solar cell. The electrode silver slurry is a viscous mechanical mixture composed of the high-purity

(99%) metal silver powder particles, glass phase and organic carrier and the like. The composition of the conductive silver slurry is very strict and the content, form, particle diameter and dispersion degree of each component has close relations with the silver slurry performance. The silver slurry shall go through a very complex process in the sintering process of the solar cell, including silver powder dissolution, transport, redox reaction and nucleation growth and so on. The glass phase can corrode and penetrate the silicon nitride coating in an effective manner to make silver connected with N^+/P junctions and ohm contact. Obviously, the difficulty of low-contact resistance lies in the control of the silver slurry on the silicon nitride penetration depth, and the key of penetration control lies in the effective matching of the glass powder and the silver particle as well as the metallisation sintering technique. Theoretically, if the glass phase fuse point is low, the flowability at high temperature will be good and the silver is easy to diffuse in the glass phase during slurry sintering, and the penetration is easy to control to form good ohm contact. As a result, the size of the silver particle in the slurry, the solid content, the cohesion of the organic phase and the flow levelling are key factors.

The back electrodes have large area and it is easy to settle the contact resistance problem. Another virtue to use the argentalium slurry for the back electrodes lies in that it can form silicon-aluminum alloy BSF in appropriate sintering conditions and play a role in impurity gettering, improving the solar cell efficiency.

The sintered electrode slurry will adhere to the silicon wafer surface. But it shall be noted that the extraction of the electrode slurry during sintering will result in stress, which may result in wafer bending. The E.I. Du Pont Company has developed the slurry with extra-low bending. The sintering technique plays an important role in slurry use, and thus the matching shall be optimised. In addition, the lead-free slurry is also viewed as a key direction for slurry development.

Du Pont has built a solar cell metal electrode slurry manufacturer and a lab in Dongguan, and a R&D centre in Shanghai. FERRO has set up a lab in Suzhou. Both of the two slurry manufacturers can offer convenience to optimise the slurry sintering technique. These labs are installed with the solar cell experiment line, which is provided with printer, sintering furnace, cell analysis and test, scanner, I-V test, bonding power, bending, slurry modification, barrelling and powder mixing and the like. The associated sintering curves are provided for different types of solar cells, slurries and preparation process equipment. Since the high-temperature sintering lasts for a short time, to make sure the macromolecule organic solvent of the slurry can fully evaporate, the cross section of the slurry shall be dense to form good ohm contact. These labs can help mesh screen design and sintering technique optimisation, for example, sintering furnace temperature distribution, slurry and silicon wafer mutual action and the like.

It shall be pointed out, to reduce the shading of the irradiating surface on the light, the contact resistance between the grid and silicon shall be further reduced. In recent years, the solar cell of laser grid buried contact (LGBC) and plated metal electrodes have been developed to improve the grid height/width ratio. And the laser is used to make the grid region obtain higher doping concentration than the irradiating surface. In addition, to further reduce the production cost of the solar cell, some companies are developing to replace the costly silver slurry by plated copper and nickel as the electrodes of the irradiating surface. In this case, nickel can isolate copper and form ohm contact with silicon. The technique has achieved high cell efficiency in the lab (see Section 5.1 for details).

3.8.5 Aluminum Impurity Gettering

In addition to phosphorus impurity gettering, aluminum impurity gettering is also a common technique for the polycrystalline solar cells. The aluminum thin film deposition can be used as the back electrode, the aluminum back surface field and aluminum impurity gettering. The aluminum back surface field uses the high/low P^+/P junctions to reduce surface recombination of carriers and getter the impurities, playing a good role in raising the long wavelength response.

Based on the print, sputtering, and evaporation techniques, the aluminum impurity gettering generally prepares a thin aluminum layer on the silicon wafer surface and then carry out thermal treatment at 800–1000 °C, making the aluminum film and silicon alloying to form AlSi. At the same time, aluminum will diffuse in the crystal silicon and form a doped P-type layer with high aluminum concentration near the AlSi layer. During aluminum alloying or the follow-up treatment, the metal impurities in silicon will diffuse to the AlSi layer or the doping layer of high aluminum concentration, significantly reducing the metal impurity concentration.

The mechanism of aluminum impurity gettering is similar to phosphorous impurity gettering. The high defect density in the AlSi layer and the high solid solubility of metal impurities in in the AlSi layer are the main reason for the metal to be gettered.

3.9 Cell Testing and Sorting

3.9.1 Objectives of Solar Cell Testing and Sorting

Before the solar cell is packaged into a module, the cells must be tested one by one and sorted by current and power to ensure consistent product performance; otherwise, nonequilibrium power generation may have great impact on the solar cell modules, the array and the PV generation system.

In the silicon wafer test, the xenon lamp is often used to simulate the sunshine to conduct open-circuit current, short-circuit current, PV conversion efficiency and maximum output power tests. The check standards of cells are listed below:

GB/T6496.1-1996—Measurement of PV I-V characteristics
GB/T6496.2-1996—Specifications on standard solar cells
GB/T6496.1-1996—Procedures for temperature and irradiance corrections to measured I-V characteristics of crystal silicon PV components (IEC60891)

Figure 3.49 shows the picture of a typical on-line photoelectric production line of solar cells.

3.9.2 Cell-sorting Equipment

For a solar cell production line with 25 MW annual output, it shall produce 2500–3000 qualified crystal silicon cells in a size of 125 × 125 mm or 156 × 156 mm per hour. Obviously, it is difficult to finish it manually. In addition, the solar cells shall be sorted

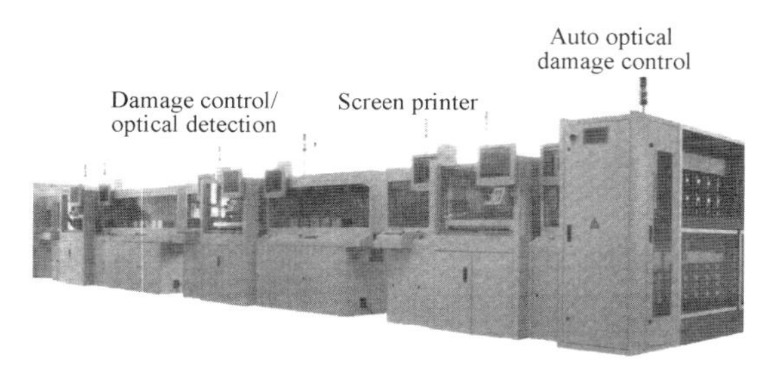

Figure 3.49 On-line monitoring equipment of solar cells.

by short-circuit current, open-circuit voltage and output power during preparation of PV modules to achieve consistent performance for series/parallel cells. As a result, the equipment manufacturer both at home and abroad have developed various cell parameter sorting equipment to conduct only detection and sorting by open-circuit current short-circuit current and PV conversion efficiency on the production line. Several cell-sorting equipment will be introduced below.

1. **Solar cell-sorting machine, Beijing Delicacy Laser Optoelectronics Co., Ltd.**
   LSK-FXJ5 solarcell-sorting machine, produced by Beijing Delicacy Laser Optoelectronics Co., Ltd., uses the pulse xenon lamp to simulate the sunshine (width of flash pulse: 5–35 ms; measurement range of light intensity: $50-120$ mW/cm^2 continuously adjusted).
   - Test parameters: Isc, Voc, Pmax, Vm, Im, FF, E_{ff}, Rs and light intensity;
   - Test range: voltage up to 40 V, current up to 10 A;
   - Test resolution: voltage: 10 mV, current: 20 mA;
   - Measurement time of each cell: 3 sec (excluding loading/unloading time);
   - Test area: 200 × 200 mm;
   - Outline: 800 mm(L) × 800 mm(W) × 1950 mm.
2. **SCHMID cell-sorting machine**
   The main features of C-SORT3000 cell-sorting machine, SCHMID, are described below: The output can be up to 3000 pc/hr, the silicon wafer is $150-300\mu$m in thickness and 32 classes are available. It covers 15 m^2. The cells shall be loaded on a conveyor composed of four channels, and the robot shall put the cells accurately into the rotary disc. Each cell shall go through four measurement stations and then the tested cells will be put on a conveyor of one channel, and two robots will sort them by the associated class (parameters of each class selected by the user).
3. **BACCINI test and sorting system**
   BACCINI cell test and sorting system is matched with the sintering furnace. It will put the printed top/bottom electrode silicon wafer to the sintering furnace conveyor, and after sintering, it will unload them from the sintering furnace conveyor for cell sorting and test. The cells on the production line can be sorted into 24 sets, or even 72 sets by short-circuit current, open-circuit voltage and conversion efficiency.

Table 3.4 shows TSA and TSFA cell-sorting machines of BACCINI.

Table 3.4 Performance of the solar cell-sorting machine, BACCINI.

Type	TSA- High-speed, put into operation in 2001	TSFA Extra high-speed, put into operation in 2007
Silicon wafer	100×100 mm^2–156×156mm^2 Thinnest: 160 μm	125×125 mm^2 ~ 156×156 mm^2 Thinnest: 160 μm
Mount, transport, dismount	Automatic (on-line, or independent)	Automatic (on-line, or independent)
Mount	Furnace, wafer box, rack	Furnace, wafer box, rack
Silicon wafer dismount and load to the sorting box	Sorting box, wafer box	BACCINI sorting box (c/w transition connector)
Auto load/unload the sorting box	NO	YES
Optical control	AOI system Colour—bending control—stability test—micro-crack, integrated to laser edging isolation	AOI system Colour—bending control—stability test—micro-crack, integrated to laser edging isolation
Sun simulation (light source)	Xenon lamp, AM1.5 Maximum irradiation area, diameter 500 mm Two-step flash Lifetime > 500000 times Intensity of light: 80–110, 40–550 mW/cm^2	Xenon lamp, AM1.5 Maximum irradiation area, diameter 500 mm Two-step flash Lifetime > 500000 times Intensity of light: 80–110, 40–550 mW/cm^2
Electric load of solar source	1–15 V; 8,12,16 A Reference cell rage: 80 mA	1–15 V; 8,12,16 A Reference cell rage: 80 mA
Cell sorting	Minimal 24 types; maximal 72 types Output: 1440 pc/hr Damage rate < 0.5%	Minimal 24 types; maximal 72 types Output: 1500–2400 pc/hr Damage rate < 0.5%
Compression air	600 kPa, 20 m^3/h, the whole line	600 kPa, 20 m^3/h, the whole line
Power	Maximum 8 kW/the whole line	Maximum 8 kW/the whole line
System size	(maximum: 9.0 m; minimum: 3.0 m) × 2.15 m × 1.9 m	(maximum: 9.0 m; minimum: 3.0 m) × 2.15 m × 1.9 m

3.10 Automation of CSSC Production Techniques

As the output of solar cells rapidly rise, automation has caught the attention in the cell production lines at large scale because it plays a crucial role in production efficiency and finished product rate. The section will introduce the application of the chain-type production line and the robot.

3.10.1 Promotion of Cascading/Chain-Type Production Lines

The traditional solar cell production lines are characterised by silicon wafer loaded in the carrier boat and the quartz tube, independent equipment and intermittent production. As the wafer becomes thinner and thinner and the output of the production line becomes larger and larger, the cascading-type production lines have fully developed in recent years to reduce wafer damage rate and improve the cell quality and productivity. The production flow charts and workshop arrangement for the two typical systems—tubular/ cascading-type are shown in Figures 3.50, 3.51, 3.52 and Figure 3.53.

We will compare the cascading-type system and the basket chemical corrosion system based on the PSG procedure. The cascading-type system has the following advantages:

- The horizontal chain-type transmission system is easy to mount the wafer, which can reduce the silicon wafer damages and is applicable to the silicon wafer in various sizes (while the wafer in various sizes shall be put into the baskets of various sizes);
- It can avoid chemical reagent from entry and exit and further prevent the chemical reagent from spraying when the basket enters and leaves the pickling bath;
- It is easy to achieve stable process control via continuous chemical reagent make-up;
- It is easy to achieve dirt-free drying via the dry injector;
- The system is extendable, which is good to integrate the silicon wafer automatic separation, mounting and sorting.

3.10.2 Mounting/Dismounting the Silicon Wafer by Robots Instead of Manual Operation

The production line of solar cells has seen larger and larger scale, and the silicon wafer thickness has become thinner and thinner in recent years. The production line of annual output in a range from 30 MW to 50 MW shall produce thousands of quality cells per hour. To reduce the damage rate and improve the cell efficiency and performance consistency, many cell manufacturers have begun to adopt the robots and integrate the chain-type production line with on-line monitoring to achieve automatic transport, automatic monitoring and automatic control in the production process of solar cells.

MANZ and SCHMID, have developed the robots (see Figure 3.54) with output can reach 2400–3500 pc/hr. It is very suited to transport the thin wafer. The robots are designed with pull down and drawing techniques (patented), for example, 'Bururi gripper', and can pick up without contact from the silicon wafer surface. The gripping point is reliable and it will never touch the edge of the wafer.

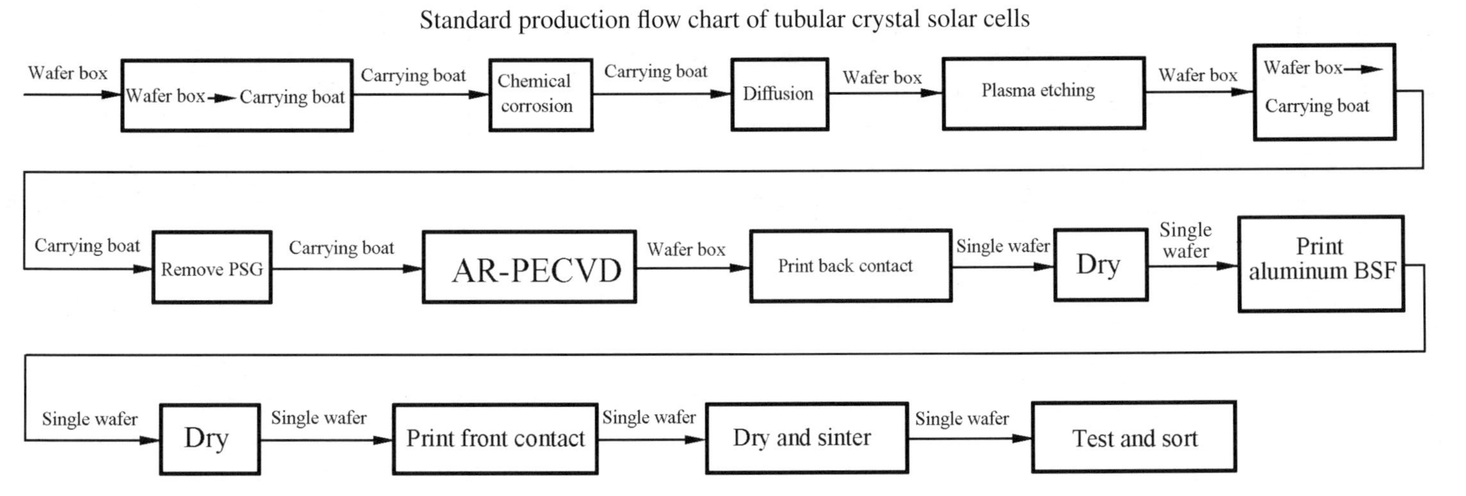

Figure 3.50 Production flow chart of typical tubular solar cells.

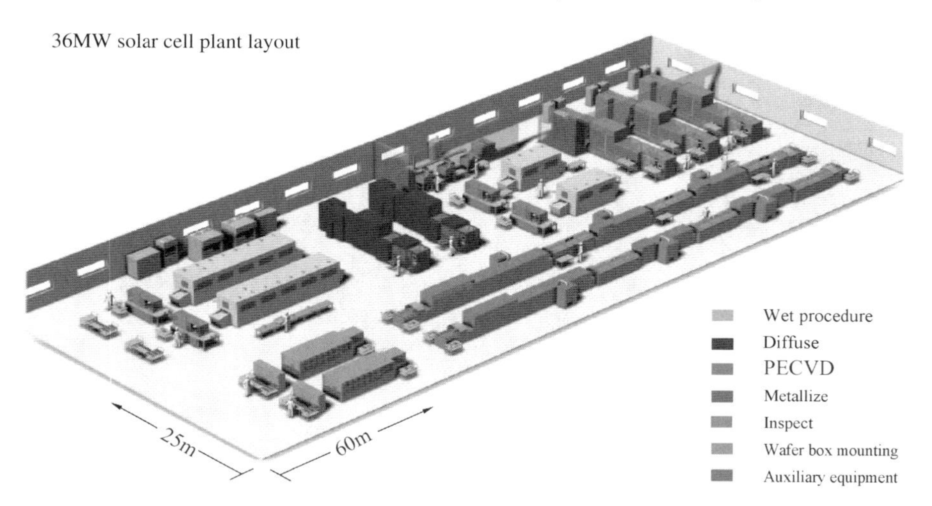

Figure 3.51 Layout of typical tubular solar cell workshop and equipment.

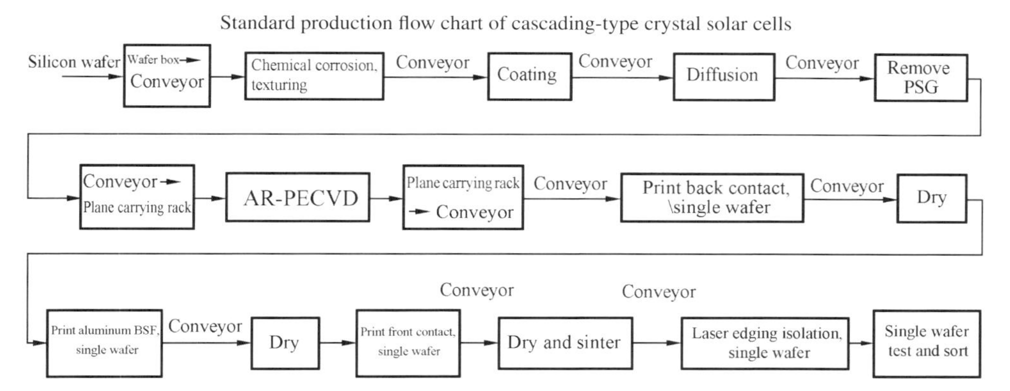

Figure 3.52 Production flow chart of typical cascading-type) solar cells.

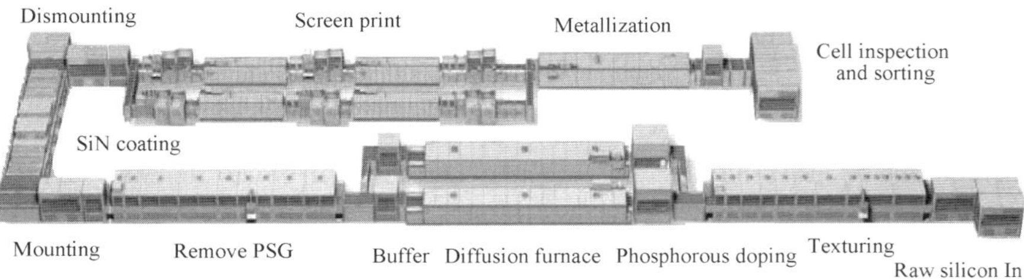

Figure 3.53 Production flow chart of typical cascading-type CSSC production lines.

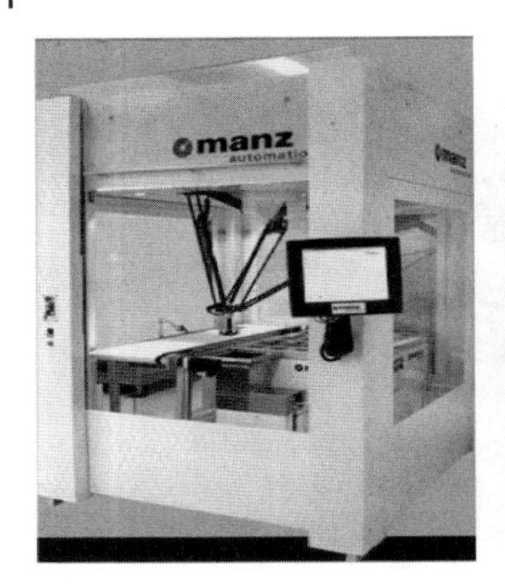

Moving robots

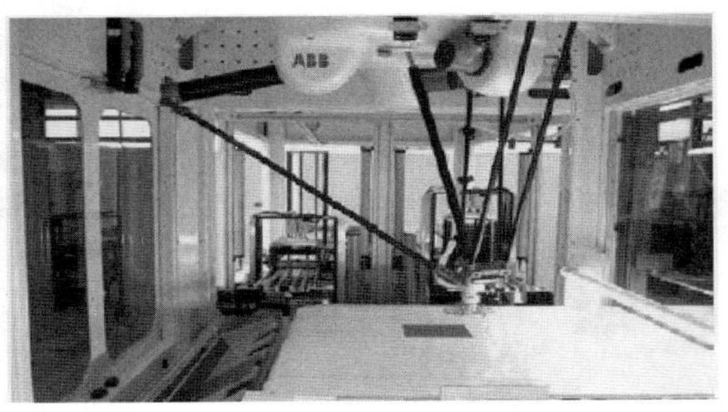

Figure 3.54 Robots of the solar cell production line, MANZ and SCHMID.

The CCD and laser positioning system are provided for the robots to load/unload the silicon wafer. The robots can pick up the cells from the moving conveyor in an accurate manner. In case the on-line detector is installed, it can inspect all the processes, significantly improving the production efficiency and wafer availability. For example, after the wafer is cleaned, the automatic on-line inspection and sorting system (including wafer resistivity, carrier lifetime, thickness and micro-cracks) can be installed. After the silicon nitride anti-reflection layer is coated, the inspection and sorting system for coating thickness and refractive index can be provided. After the screen is printed, the gate line graphic inspection can be provided. After the cell is finally produced, the cell performance can be also inspected and sorted.

3.11 Parameter Measurement in CSSC Production Process

3.11.1 Solar Simulator

The photoelectric performance such as V-I characteristics and the like shall be measured for the solar cell in case of solar simulation. The solar simulator, the private tester for solar cell testing, generally consists of the light source, constant temperature testing rig, the electronic load and the signal amplifier, computers, data processing software and the like. The light source adopts the pulse xenon lamp because its spectrum is close to the sunlight (see Figure 3.55). The light of the pulse xenon lamp shall first pass the collimator objective, optical filter (with certain spectrum transmission characteristics) and reflector (with certain spectrum reflection characteristics), which shall make the spectrum mismatching $\leq \pm 5\%$, that is, meeting the general measurement demand of solar cells. The international standard testing condition for solar cells are AM1.5, 1000 W/m^2, 25°C. The basic structure of the solar simulator and the measurement system are shown in Figure 3.56.

3.11.2 Measurement of V-I Characteristics and PV Conversion Efficiency for Solar Cells

The load V-I characteristics curve is a key reference to analyse the working characteristics of the solar cell. When the solar rays radiate on the solar cell and the load is

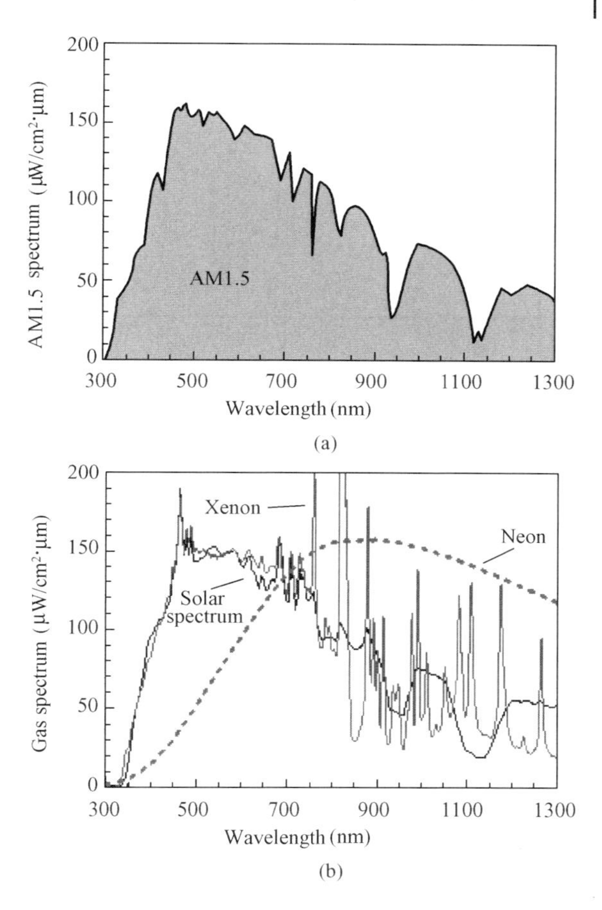

Figure 3.55 Spectrum of the pulse xenon lamp, close to the sunlight (a) AM1.5 solar spectrum; (b) Spectrum of the pulse xenon lamp, close to the AM1.5 sunlight.

connected, the PV generation can be measured. Figure 3.57 shows the V-I characteristics curve at various solar ray intensities. The dark line shows the V-I characteristics in the dark conditions, and it is similar to that of the ordinary diode. As the solar intensity rises, the V-I characteristics curve will gradually move downwards. The dotted line shows the characteristic curve in the solar simulator, that is, the typical V-I characteristics of the solar cell. The short-circuit current I_{SC} is the measured current when the two terminals of the solar cell are short-circuited. The open-circuit voltage U_{OC} is the measured voltage when the load is disconnected. The output power of the solar cell has relations with the load. The product of current with voltage at each point on the curve is the output power at the given load. The conversion efficiency of the solar cell is the ratio of the maximum output power of the solar cell to the accepted solar power.

To measure the V-I characteristics and PV conversion efficiency of the solar cell in an accurate manner, the circuit of the measurement system must be well designed because the voltage drop produced by the contact resistance of the connecting cable will result in measurement error if the output voltage of the solar cell is very low. Four-end connections are generally used as shown in Figure 3.58 where r is the contact resistance (including the cable resistance). The voltmeter shall be directly connected on the two

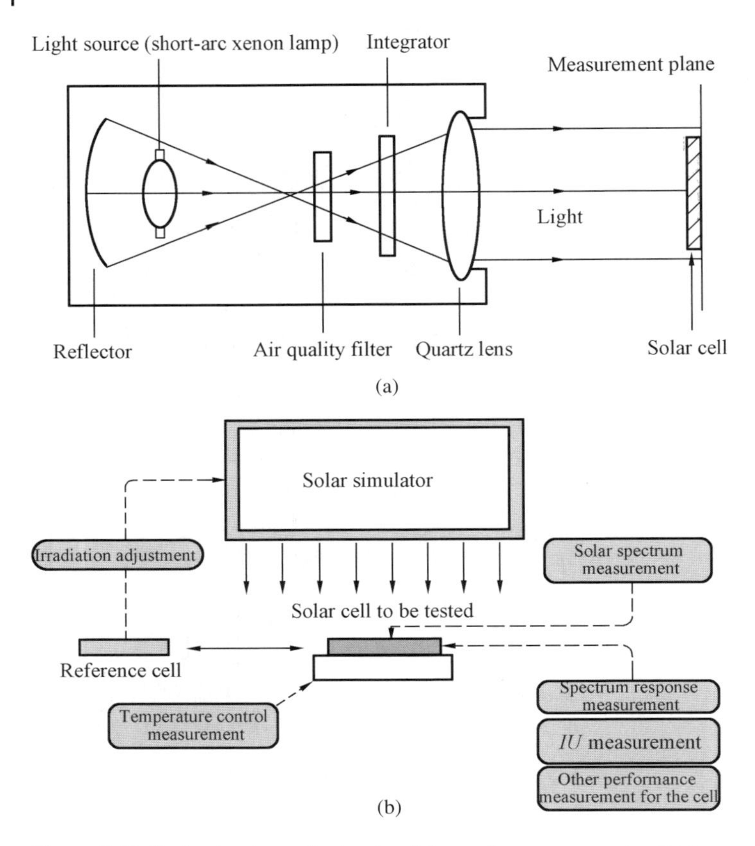

Figure 3.56 Basic structure of solar simulator and measurement system.

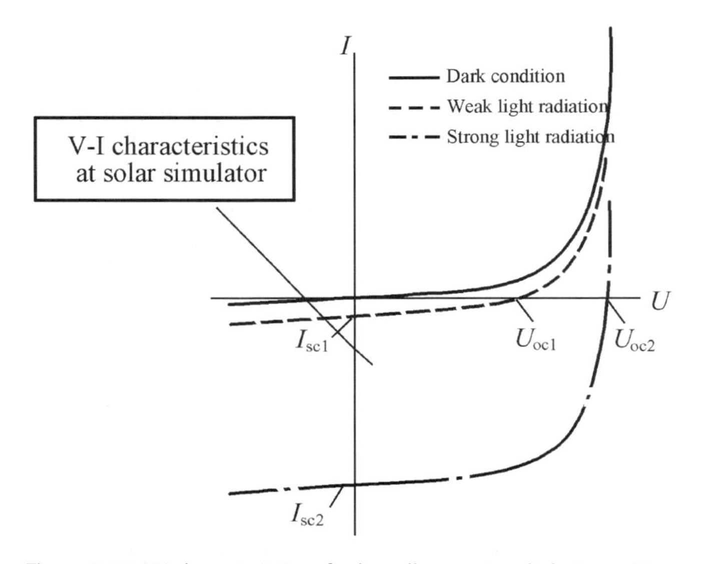

Figure 3.57 V-I characteristics of solar cells at various light intensities.

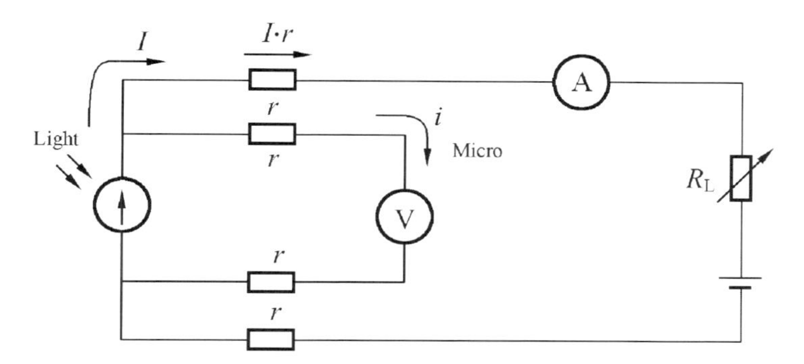

Figure 3.58 Four-end measurement method for V-I characteristics of solar cells.

ends of the solar cell. The circuit, known as 'four-end measurement method', is designed to measure the V-I characteristics of the solar cell.

3.11.3 Measurement of Spectral Response for Solar Cells

The spectral response of the solar cell refers to the relationship between the light current generated at unit irradiation and the incidence light wavelength. It reflects the capacity of the electron-hole pairs, generated by the incidence photon absorbed by the cell, to form the light current. The spectral response characteristics include a lot of information of the solar cell, which can reflect not only the quality of the cell materials but also the quality of the anti-reflection coating and the interface layers. A series of monochromatic light with various wavelength shall shine on the solar cell, and then the short-circuit current density and irradiation can be measured to work out the spectral response of the solar cell. Figure 3.59 shows the diagramme of spectral response measured by a monochromator.

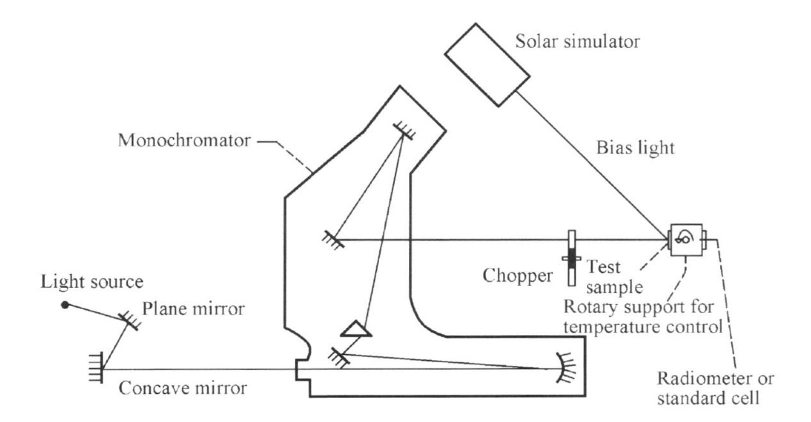

Figure 3.59 Diagramme of cell spectral response measured by a monochromator.

3.12 Product Quality Control and Cost Analysis for Solar Cell Production Lines

3.12.1 On-line Inspection of Solar Cell Production

The cell manufacturers have accumulated a lot of experience in practise to improve the qualified rate of CSSCs, especially the on-line inspection during solar cell production, which mainly includes the followings:

1. Silicon wafer quality sorting and inspection in the raw material procedure;
2. Silicon wafer corrosion thickness measurement in the chemical cleaning procedure;
3. Surface texturing micro-observation in the silicon wafer surface texturing procedure;
4. Inspection of phosphorous diffusion layer thickness, square resistance and minority carrier lifetime in the junction preparation procedure;
5. Micro-observation of the edging isolation width and inspection of cell front/rear insulation after the edging isolation procedure;
6. PSG removal effect check in the PSG removal procedure;
7. Check of the anti-reflection coating thickness and refractive index, and solar reflectivity in the silicon nitride coating procedure;
8. Printed pattern check in the electrode print procedure;
9. Measurement of electrode ohm contact in the electrode alloying procedure;
10. Performance inspection and sorting of the solar cell.

3.12.2 Traditional Process Quality Control on the Solar Cell Production Line

3.12.2.1 Working Environment

The purification grade in the cleaning and diffusion rooms shall be 1000–10,000; the ambient temperature shall be 18–28°C with humidity $\leqq$ 60%. The process gas purity is generally 4–5 N. The chemical reagent is usually of electronic grade. The tools and molds shall minimise metal ion pollution. The operator shall wear masks, gloves and clean clothes.

3.12.2.2 Quality Control of the Cleaning and Texturing Process

The cutting damage layer of the silicon wafer is about 10–20 μm deep on the wafer surface. For the monocrystalline silicon wafer, the sodium hydroxide is generally used for the corrosive agent due to its anisotropic corrosion characteristics. The isopropanol can be added to the corrosive liquid, which can help the hydrogen bubble, generated in the reaction process, release from the wafer surface and form good pyramids. For polycrystalline, as the corrosion time prolongs, steps will form between the crystal boundaries on the wafer surface, and the steps will increase with long time. The step increase will accelerate recombination of the minority carriers, exerting adverse influence over the follow-up diffusion and screen print processes and finally affecting the cell conversion efficiency. As a result, focus shall come to control the thin film thickness, adjust the concentration of the corrosive liquid and the corrosion time and make up the liquid in time.

In the cleaning procedure, the resistivity of the DI water shall be no less than 18 MΩcm, and the resistivity of the cleaned water shall be larger than 16 MΩcm. It shall observe the silicon wafer damages and surface contamination, and take actions in time during texturing and cleaning.

3.12.2.3 Quality Control of Diffusion Process

During diffusion, it shall pay attention to diffusion evenness, and the square resistance is usually 50–70 Ω/cm, and the unevenness shall be controlled within ±2 Ω/cm. It shall monitor the different wafers of one batch and the different positions of one wafer to ensure that the longitudinal and axial diffusion evenness of the quartz tube can meet the requirements. When the axial evenness of the quartz tube reduces, it can be improved by temperature distribution regulation at the furnace inlet, centre and tail. When the longitudinal evenness of the quartz tube reduces, the gas flow shall be increased to improve the diffusion evenness.

In addition, the diffused wafer shall be slowed transported outwards to avoid thermal stress in the wafer, which may result in thermal defects. During diffusion, special attention shall be paid to quartz cleanness and avoid secondary pollution.

The purification degree of the power house in the diffusion procedure shall be no less than 10,000.

3.12.2.4 Quality Control in Wafer Edging Isolation and PSG Removal Procedures

In the phosphorus diffusion procedure, a PSG layer of high phosphorus concentration will be formed on the wafer edge and surface. This layer has loose structure and many defects, and it is apt to absorb humidity, reducing the solar cell performance and having adverse impact on the follow-up procedures. As a result, it must be removed. In addition, the PSG layer on the edge may make the electrodes before/after the cell short-circuit. The fluorinion will react with silicon during glow discharge to generate volatile products and etch the PSG at the wafer edge. During the process, the multimeter can be used to measure the edge resistance to judge the etching status. The PSG on the wafer surface is generally removed by soaked in hydrofluoric acid. The hydrophobic nature of the wafer surface shall be observed to judge whether the PSG has been completely removed.

3.12.2.5 Quality Control in PECVD Procedure

The anti-reflection effect of silicon nitride coating is dependent on the refractive index and thickness of the coating. The optimum anti-reflection coating thickness is about 80 nm, and the refractive index falls in a range of 2.3–2.5. When the process gets stable, the colour of the coating can be observed by naked eyes to judge the stability of the coating thickness and process. When the colour has obvious difference, it indicates the silicon nitride coating has problem in evenness, and the equipment shall be checked, for example, the RF or the microwave frequency shall be adjusted. The ellipsometer can be used to monitor the refractive index and thickness of the silicon nitride coating.

3.12.2.6 Quality Control in Print and Sintering Procedures

The two procedures—screen print and sintering—supplement each other. The sintering conditions (sintering temperature, time and atmosphere) are keys to screen printed electrodes. They may affect the electrode contact resistance, electrical performance output and electrode firmness. The metal electrodes printed on the wafer surface shall form good ohm contact between silver silicon and sliver-aluminum-silicon and not damage the cell P-N junctions. It may deteriorate the cell if the sintering temperature is excessively high and the time is too long; and the electrode firmness may be poor, and the series resistance will rise with reduced filling factor if the sintering temperature is excessively low and the time is too short.

During screen print, attention shall be paid to the evenness of electrodes, symmetry of patterns, and consistency of BSF printed thickness. The electrodes on the front shall be free from any broken lines or dirty, and the printed thickness shall be even.

It shall be noted during sintering that the slurries of various brands and types have different sintering curves. The organic solvent in the slurry shall be fully volatilised during sintering, and the slurry shall be free from bubble or skinning after sintering. The slurry shall be continuous and compact and free from cellular structure. The gate lines shall have even colour, smooth surface and free from any particles. The series resistance shall fall in the specified range.

It is still very difficult to fully control the solar cell process in the existing conditions, and it is also very difficult to predict the cell conversion efficiency based on the electrical parameters of the silicon wafer.

3.12.3 Cost Analysis for CSSCs

Since 2010, the price of polycrystalline raw materials has greatly reduced and the price of the silicon solar cell has also fallen from €2.2/Wp (in 2009) to €0.35/Wp (in 2014) in the world. In 2009, a consulting company worked out the cost composition of solar cells on the basis that the unit cost of monocrystalline silicon wafer (156 × 156 mm) is €4/pc where the silicon wafer cost accounts for 75%. Here the cost composition of the solar cell is calculated suppose that the unit price of the silicon wafer in the same specifications was €1/pc (in 2014). In this case, the silicon wafer cost accounts for 43%. The comparisons between them are shown in Table 3.5.

The above table shows, although the silicon material has been greatly reduced, it still accounts for over 40% of the total cost. The silicon cost for the semiconductor, for example, the integrated circuit and the like, only accounts for 3–4% the total cost. As a result, it shall pay more attention to the cost of silicon materials during solar cell production. The second focus shall come to the silver (argentalium) electrode slurry, equipment depreciation and labour, whose cost is all 6% or so. It shall be noted that the cost compositions only reflect the situation in Europe where the production equipment has high automation level and the labour salary is also high.

Table 3.5 Estimates of CSSC cost composition.

Item	Silicon wafer	Silver slurry	Consumables	Power	Equipment depreciation	Salary	Power house and facility
Ratio, 2009	75%	6%	3%	1%	6%	6%	3%
Ratio, 2014	43%	13.7%	7%	2%	13.7%	13.7%	7%

4

Preparation Methods of Thin Film Silicon Solar Cells

4.1 Advantages and Prospects of TFSSCs

4.1.1 Advantages of TFSSCs

Compared with crystalline silicon solar cells (CSSCs), the thin film silicon solar cells (TFSSCs) have the following advantages:

1. Low consumption of raw materials, and many substrates available
   ① In the TFSSCs, the silicon material is several microns thick, only 1/100 of the CSSCs;
   ② The low-cost materials, for example, glass, stainless steel and plastic, can be used as the substrate;
   ③ The main raw materials for growing thin film (SiH_4 and H_2) are noncorrosive and suited to sustainable development.
2. Low production cost
   ① The thin film materials and the components can be produced simultaneously in a simple and short process flow;
   ② Since the low-temperature production process (about 200 °C) is adopted, its energy consumption is much less than that of CSSCs;
   ③ Easy to realise large-scaled continuous production.
3. Good appearance, high reliability, large size, and suited to building-integrated photovoltaics (BIPV)
   ① The thin film shall be directly deposited on the irradiating surface, and the sealing material, for example, EVA, will be set on the back so that it will not block the irradiation of the solar rays on the cell. Even though EVA becomes yellow, it will not reduce the power of the TFSSC module;
   ② The TFSSCs are more reliable and endurable than CSSCs since their modules are designed with integrated internal connections. Since the electrodes on dozens of CSSCs are welded together, it is apt to have such faults as short circuit or open circuit between cells and the like;
   ③ Nowadays, The TFSSCs' module is in the range of 1.4–5.7 m^2 of dimension and rich with colours. And it is easy to make them transparent. They can be integrated to the buildings, especially used as the glass curtain wall.

Technology, Manufacturing and Grid Connection of Photovoltaic Solar Cells, First Edition. Guangyu Wang.

4. Small high-temperature attenuation, and good weak light response
5. Compared with CSSCs, the a-Si cells have lower efficiency and temperature coefficient, about −0.1%/K, that is, the cell efficiency will reduce only 0.1% when the outdoor temperature of the module rises by 1 °C i.e., the rise of ambient temperature will not result in significant reduction of cell efficiency. The temperature coefficient of CSSCs, however, is −0.4%, and as the ambient temperature rises, the cell efficiency will decrease dramatically;
6. They have a relatively good performance in a dimly lit environment. In the range of visible light (1.7–3.0 eV), the absorption coefficient of a-Si materials is larger than crystalline silicon by one order of magnitude. Thus, the TFSSCs can still generate power in dim light.

Main existing problems of TFSSCs:

1. At present, the average cost of TFSSCs is still larger than CSSCs but it may be regarded that the TFSSCs and modules have a large space for cost reduction with technology improvement and application expansion since the consumption of raw materials is few, and the module preparation process is simple;
2. The a-Si solar cells will see unstable performance after irradiation, reducing the conversion efficiency. The conversion efficiency decreases down exponentially, which mainly falls in the period from three months to one year since it is put into service. The reduction, however, has been deducted when the solar cells are sold and the price is based on the stable power. In recent years, the a-Si/μC-Si stacked solar cell has been developed, which has low light attenuation and is viewed as a promising PV technique.

4.1.2 History and Prospects of TFSSCs

The first-generation technology of TFSSCs is a-Si TFSC, whose disadvantages include low efficiency, poor stability and small cost advantage. Since the disordered structure of the a-Si material has metastable characteristics, the a-Si TFSC has low efficiency. Since the a-Si material has big poor gap width, many defects and severe carrier recombination and narrow response range to solar spectrum.

In the recent years, the second-generation technology of TFSSCs has been developed for the PV industry—amorphous silicon/microcrystalline silicon (a-Si:H/μC-Si) stacked TFSCs. Since μC-Si can greatly extend the response range of the solar long wave spectrum (i.e., it can absorb the energy of the incidence light with long wavelength), it is benefical to the improvement of the PV conversion efficiency. On the other hand, it can significantly improve performance stability, and the light attenuation of the μC-Si cell is low since μC-Si has better order than a-Si. Because μC-Si can grow in the same equipment with a-Si:H, the preparation process is simple. It has become a key development trend to combine them together—to use a-Si:H (width of forbidden band: 1.7 eV) as the top cell and μC-Si (width of forbidden band: 1.1 eV) as the bottom cell.

μC-Si, a semiconductor material with indirect band gap, has low light absorption coefficient. To fully absorb the sunlight, the absorption layer shall be thick (1–2 μm). The μC-Si thin film grows at a very low speed (<0.3 A/s) in the traditional PECVD methods, it often takes several hours to grow the μC-Si absorption layer. To reduce the production cost, the deposition speed must be raised. Many researchers are working to explore the methods to improve the deposition speed so as to minimise the growth time of the μC-Si absorption layer. At present, the followings are available: very high frequency (VHF) PECVD, hot wire (HW) CVD, radio frequency high pressure depletion (RF-HPD) PECVD and so on. Some results show, the μC-Si thin film can deposit at a speed above 10 A/s when VHF-HP-PECVD is used; and it can deposit at a speed up to 45 A/s when the external bias is applied to VHF-HP-PECVD and the cell stable efficiency can reach 7% and it only takes 5min to grow the absorption layer. These results are ready for industrialisation.

Another key issue affecting TFFSC industrialisation is expensive equipment, which is generally several times that of CSSCs. Since its substrate is in large size, high automation is required to achieve massive production. In addition, the production equipment must be based on material and process technique study. As a result, most of the TFSSC equipment is still nonstandard and needs improvement. Since equipment production has close relations with process techniques, the equipment manufacturers need the return for their R&D input.

The followings are being taken for TFSSC improvement:

① The multi-junction stacked structure of various band gaps are adopted to improve efficiency and stability;
② Surface texture and bottom reflection techniques are adopted to improve the light trapping effect;
③ Thinner i-layer is used (to raise the built-in field and reduce light-induced attenuation);
④ The cell structure has been improved.

So far, the TFSSCs have achieved great results and the stable efficiency of the single-junction, double-junction and triple-junction cells has reached 8–10%, 13%, 15%, respectively. As the multi-junction stacking rises, the attenuation of conversion efficiency will fall down where the attenuation of the single-junction, double-junction and triple-junction cells is 24%, 15%, and 13%, respectively. It is under industrialisation and massive production.

More materials, in addition to glass, can be used as the substrate of TFSSCs, including porcelain, graphite, plastics and stainless steel and the like. It is generally considered that the following six principles shall be followed for substrate selection:

(1) Low price;
(2) Certain mechanical strength;
(3) Capable of matching with the silicon expansion coefficient;
(4) Low porosity rate to ensure continuous thin film;
(5) Relatively low impurity content;
(6) Flat and smooth surface with high reflectivity;
(7) Not react with silicon at high temperature.

For the insulated substrate material, the emitter and the base electrodes shall be all installed on the substrate surface, and several cells shall be simultaneously integrated. It is reported that the maximum efficiency of the solar cell prepared on silicon oxide porcelain, mullite porcelain and graphite substrates is 9.4%, 8.2% and 11.0%, respectively. The stainless steel substrate mainly faces the following problems: contamination of metal on the thin film and the thermal deformation of thin substrate. A U.S.-based company announces when the polycrystalline silicon (microcrystalline silicon) thin film is deposited on the porcelain substrate, the grains are small and even and the efficiency can reach 16.6%. The porcelain substrate has a virtue that its impurities are generally present in the form of oxides and thus they are difficult to diffuse to the silicon film at high temperature. As a result, it is acceptable to contain more impurities than the silicon substrate material.

In addition to the silicon-based thin film, some other promising non-Si-based thin film materials are also available for TFSCs, including CdTe, CIGS, III-V group, organic thin film, dye photosensitive thin film and perovskite thin film. They have the virtues similar to TFSSCs. At present, the following three are undergoing industrialisation: the GaAs-based TFSC, CdTe TFSC and CIGS TFSC. And the organic TFSC, dye photosensitive TFSC and perovskite TFSC are still at the initial stage for research and development in a short time. In this chapter, we will introduce the silicon-based TFSC to obtain the basics on TFSCs. And for several non-Si-based TFSCs, their generation principles and preparation processes will be further introduced as the new next-generation products in Chapter 5.

4.2 Structures and Power Generation Principles of TFSSCs

4.2.1 Structures of a-Si:H and μC-Si THSCs

The typical a-Si:H and μC-Si solar cells are all based on the P-I-N structure (Figure 4.1), that is, the very thin P and N layers ride on both sides of the intrinsic layer, generating

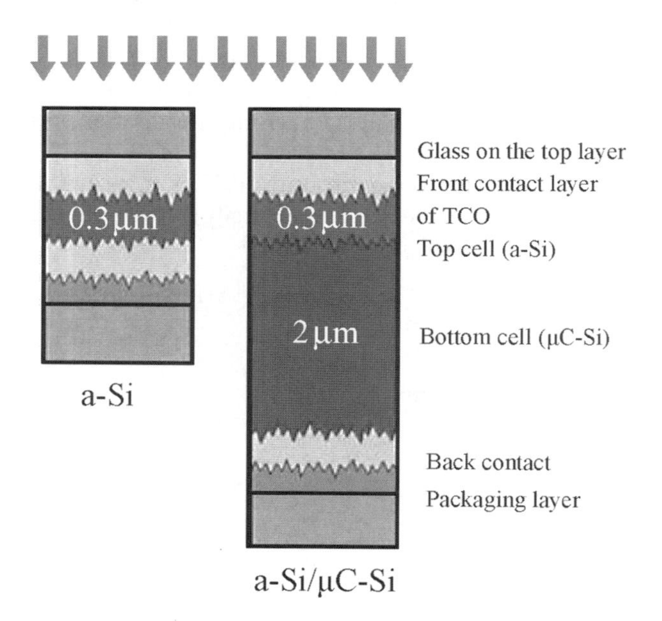

Figure 4.1 Structural diagramme of a-Si:H and a-Si:H/μC-Si stacked solar cells.

the built-in field. When the carriers generated after the I layer absorbs light are separated by the built-in field, it will generate the PV effect. The a-Si:H/μC-Si cell stacked by a-Si:H and μC-Si cells can promote light absorption because the a-Si band gap of the top cell is 1.7 eV (a-Si), which is good for absorption of the short wavelength light, and the μC-Si band gap of the bottom cell is 1.1 eV (μC-Si), which is good for absorption of the long wavelength light.

In the structural diagramme of the a-Si:H single-junction cells and the a-Si:H/μC-Si stacked cells, the transparent conductive film forms ohm contact with the a-Si:H P-type zone, the metal film on the rear of the μC-Si cell can be used for back contact and light reflection.

Figure 4.2 shows the approximate thickness and growing method of the layers in the a-Si:H/μC-Si stacked TFSCs. The transparent conductor oxide (TCO) is deposited on the 4 mm-thick glass plate by the PVD method, and then the PECVD method is used to deposit the a-Si top cell about 300 nm in thickness and the μC-Si bottom cell about 1500 nm in thickness, and finally, the PVD method is used to deposit the back contact metal. The top/bottom electrode wires shall be connected from the TCO and the back metal layers and then connected with the junction box on the right.

Figure 4.3 shows the production process of a-Si:H/μC-Si stacked TFSCs.

4.2.2 Power Generation Principle of TFSSCs

Although a-Si has high absorption coefficient, the carriers have short diffusion length and it is difficult to realise effective PV conversion by P-N junctions. A thick intrinsic

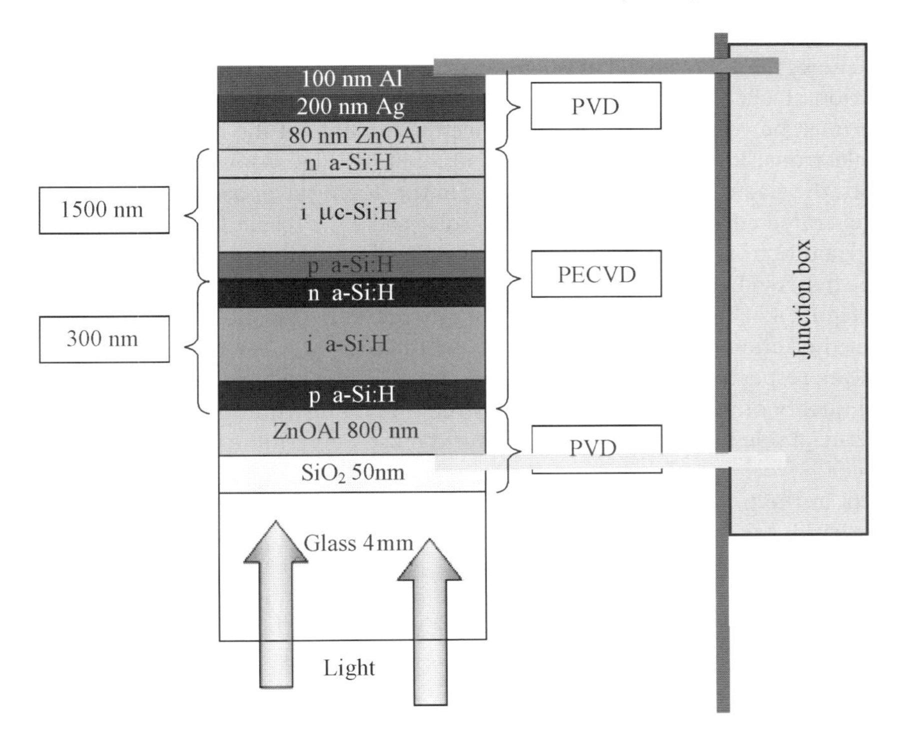

Figure 4.2 Typical structure of a-Si:H/μC-Si stacked TFSCs.

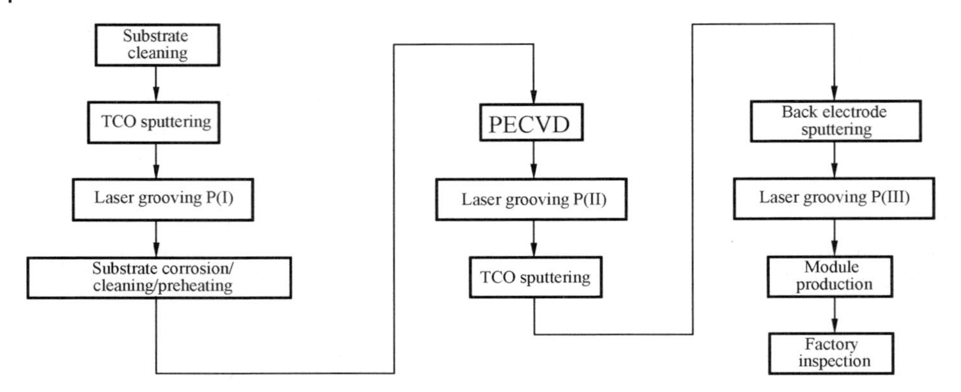

Figure 4.3 Production process of a-Si:H/μC-Si stacked TFSCs.

zone shall be inserted between the P and N zones to form the P-I-N junction. Similar to the P-N junctions, the built-in field is dependent on the P and N doping layer in the P-I-N structure, and it will be extended in the intrinsic zone. This is because the doping concentration in the intrinsic zone is low, and the width of the space charge layer (the depletion zone) mainly falls in the intrinsic zone. When the solar rays radiate on the cell with the P-I-N structure, the photons will be effectively absorbed by the intrinsic zone in sufficient thickness, and the photo-generated carriers generated in the intrinsic zone will be separated the intrinsic zone under the action of the built-in field. Since the P and N layers are very thin, the carriers are easy to pass through them and be collected. This is the basic power generation principle of the cells with P-I-N structure.

Figure 4.4 shows the energy band of P-I-N single-junction solar cells and P-I-N/P-I-N double-junction stacked cells. When the solar rays enter from the P-type layer, the P-type layer must be very thin (10–30 nm) since the mobility of the holes is less than that of the electrons. In addition, the carriers in a-Si:H of high doping concentration are very short, the electrons and holes generated in the doping layer cannot generate all the photo-generated current demanded by the solar cell. As a result, the TFSSCs have a different structure from the crystalline silicon solar cells. And a very thick I (intrinsic) layer must be inserted between the a-Si P layer and the N layer to build the built-in field as shown in Figure 4.4. The I (intrinsic) layer will generate sufficient electrons and holes, which, under the action of the built-in field, will diffuse towards the N-type layer and the P-type layer, respectively, to generate the photo-generated voltage. In consequence, the material quality of the intrinsic zone, the strength and distribution of the built-in field play a crucial role in collection of carriers and mostly determine the performance of the solar cell.

The defects in the intrinsic zone can exert influence over carrier collection in two ways: on one hand, its charge status can exert direct influence over the field distribution in the intrinsic zone; on the other hand, it can serve as the carrier recombination centre. According to the Fermi level position, the defect states near the P-type layer (i.e., the front half of the I layer) are in positive charge while those near the N-type layer (i.e., the rear half of the I layer) are in negative charge. These charged defects will intensify the field on the P/I and N/I interface. If the I layer has too small thickness, the light absorption will be insufficient and it is adverse to generate the photo-generated carriers, reducing the short-circuit current of the solar cell. If the I layer is too thick,

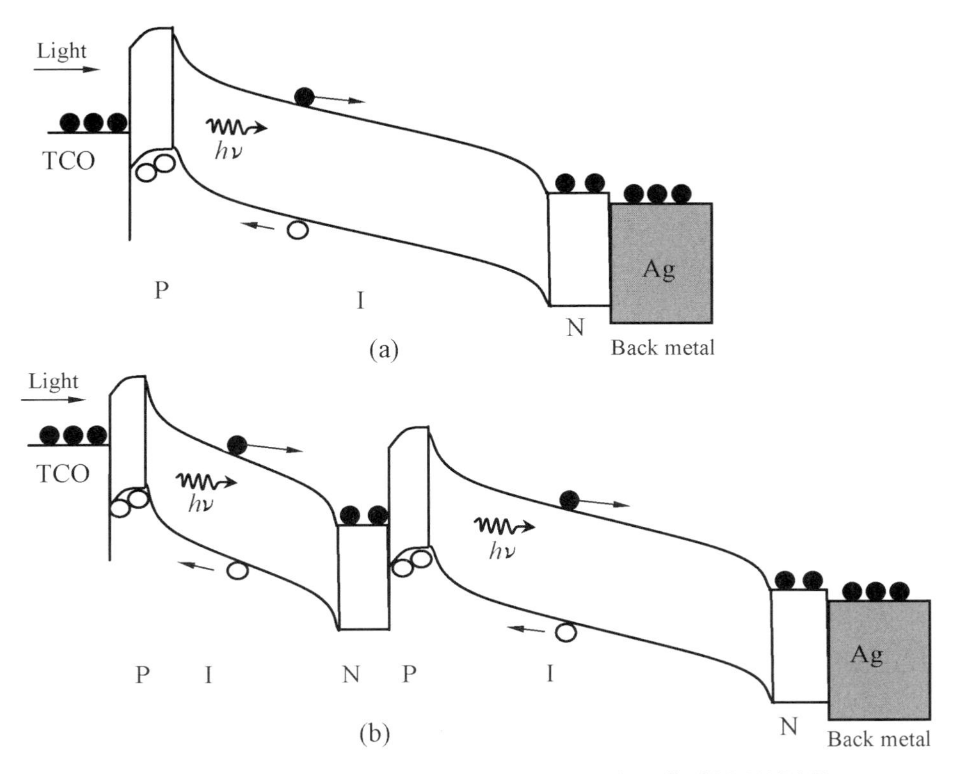

Figure 4.4 Energy band diagramme: (a) P-I-N single-junction solar cells; (b) P-I-N/P-I-N double-junction stacked cells.

it will weaken the built-in field intensity in the intrinsic zone. Since the carriers in the intrinsic zone move mainly via drifting, and the result will cause more recombination chances for carriers and reduce collection of the photo-generated carriers. In addition, the low conductive intrinsic zone will raise the series resistance in the cell, which is adverse for FF improvement. Accordingly, to tackle the conflict between carrier generation and collection, the thickness of the intrinsic zone shall be optimised so that the photo-generated carriers can reach the doping layer by moving a short distance.

The existence of interface state shall be noted in the a-Si cell structure. The holes in the P-type layer will recombine with the electrons from the N-type TCO conductive film at the TCO/P interface, resulting in depletion of the P-type charges (depending on the TCO material performance) and the reduction of the contact barrier of TCO/P interfaces. The interface state in the P/I interface region plays a crucial role in the cell performance. The minor charge at the P/I interface will exert great influence over the solar cell efficiency and the stability after solar ray radiation because the charge at the P/I interface will result in re-distribution of the built-in field and degrade the cell performance. The improvement of the interface characteristics on TCO/P-type layer and P/I-type layers will help raise the open-circuit voltage. In addition, the back surface field (BSF) also plays an important role in performance improvement for TFSSCs.

In the PIN structure of a-Si TFSCs, the S-W effect will exert great influence when the intrinsic zone is thicker and it is more likely to result in light attenuation. This is mainly because diffusion and movement of hydrogen in the a-Si film will result in the rise of space charge density near the P-I interface, increase of photo-generated carrier recombination, and attenuation of a-Si performance. The establishment and nature of these metastable defect centres are closely related to the content, distribution and bonding form of hydrogen in a-Si.

Hydrogenation can improve the short-range order of a-Si, and reduce the bond angle deviation of a-Si:H, and the middle-range order will be present in the network. Since the defect density of a-Si:H is low, doping of N and P types is available (although doping efficiency is low). Since the defect-state density of the doped a-Si:H will rise, resulting in massive recombination of photo-generated carriers, the doping layer can be only used to build the built-in field and ohm contact and it cannot be used as the light absorption layer in the amorphous TFSSCs. In a-Si, the mobility of holes is far less than that of electrons and the photo-generated current is mainly attributed to the electrons. The a-Si containing hydrogen has a little nanocrystalline structure with minimum size of 1–2 nm and good stability. For the single-junction a-Si solar cells, the a-Si TFSCs with a little nanocrystalline structure have high PV conversion efficiency and low light attenuation.

To design the TFSSCs of high efficiency, the theoretical simulation shall be compared with the actual application over and over to discuss the key issues in the thin film solar cell, for example, transport of carriers, recombination losses and field distribution.

4.2.3 Light Absorption of a-SiC:H/μC-Si and a-Si:H/a-SiGe:H Stacked Solar Cells

The textured ZnO:Al film can be used as the TCO material and ZnO/Ag as the back contact to realise 'light trapping' in an effective manner. The solar rays can be repeatedly reflected in the multi-layer thin film cells to pass through the intrinsic zone (the main zone to generate the photo-generated current) several times. In this way, even though the I layer is thin, it can still realise good light absorption and high efficiency of solar cells.

Figure 4.5 shows the comparisons between the a-Si:H/μC-Si stacked solar cell and the a-Si single-junction TFSC on absorption spectrum. The stacked layer can absorb the short/long-wavelength, and it has more advantages than a-Si TFSCs.

For the stacked multi-junction TFSCs, the junctions can be relatively thin to improve the built-in field. Ideally, the photo-generated voltage of double-junction cells approximates to the voltage sum of two sub-cells while the photo-generated current is dependent on the smaller current of the two sub-cells. The two sub-cells are connected

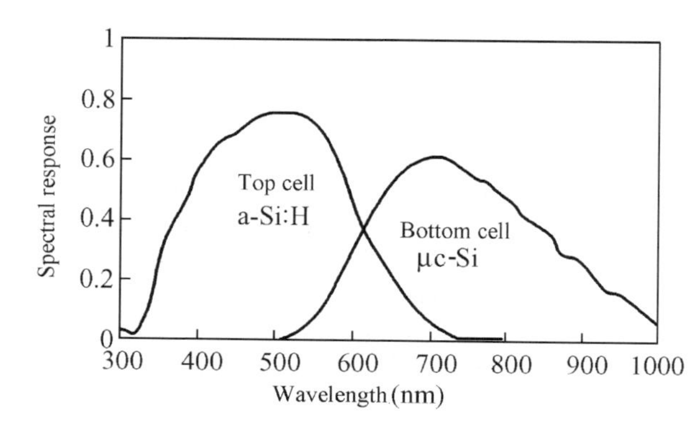

Figure 4.5 Comparisons between the a-Si:H/μC-Si stacked solar cell and the a-Si single-junction TFSC on absorption spectrum.

by the N-type layer of the top cell and the P layer of the bottom cell, which forms a reverse PN junction, and the photo-generated current flows in the form of tunnel recombination. The stacked silicon-based TFSC has higher conversion efficiency and better stability than the single-junction cells.

To improve light absorption, a-SiC:H—the amorphous silicon carbide with wide band gap—shall be adopted as the P-type window of the a-Si top cell to absorb the blue and green light. Since most infrared light cannot be absorbed by the a-Si top cell, the material with small width of forbidden band, for example, μC-Si or a-SiGe, shall be used as the bottom cell to absorb more red and infrared light penetrating the a-Si top cell. In this way, the a-Si and μC-Si stacked solar cells can be built.

The band gap of a-SiGe:H is smaller than that of a-Si:H, and it can be also used as the bottom cell to form the a-Si:H/a-SiGe:H stacked solar cell. Compared with the a-Si:H/a-SiGe:H stacked solar cell, the a-Si:H/μC-Si stacked one has larger short-circuit current because the long wave response of the μC-Si bottom cell is better and the μC-Si bottom cell has better stability than a-SiGe. But the band gap of a-SiGe:H will reduce with increase of Ge content in the band gap, and it can form the triple-junction or multi-junction stacked cells to enhance absorption of solar spectrum.

4.3 Preparation Techniques of TFSSCs

4.3.1 TCO Sputtered on Glass Substrate

The TCO is used as the front contact electrode of TFSSCs. Since it plays an important role in TFSSCs, it shall have small resistance, high light transmissivity and reflective index matched with the incidence light. In addition, the TCO at the bottom can enhance light reflectivity in the cell with the back metal thin film.

TFSSCs have the following requirements on TCO:

① High transmissivity;
② Low square resistance;
③ Good evenness;
④ Matched reflectivity;
⑤ Good chemical stability.

At present, the zinc oxide doped with aluminum (ZAO) is often used as the TCO of TFSSCs. Since the zinc oxide is stable at the hydrogen plasma condition, its transmissivity is also high. The intrinsic zinc oxide, however, has a problem—the conductivity is not high. The aluminum or gallium is often doped to raise its conductivity. To improve the scattering effect, the chemical etching method can be used to raise the roughness. Before preparing TCO, the neutral detergent shall be first used to clean the glass substrate and the deionised water used for washing, and then the medium-frequency pulse magnetically controlled sputtering method shall be used to deposit ZAO. After soaked in the 0.5% diluted hydrochloric acid for some time, the textured structure will be finally formed on the surface. At present, the coated TCO glass is available on the market.

The experiments show, if the substrate temperature sputtered by magnetic control is low (e.g., 220 °C), the thin film prepared will be not so dense, the cohesion to the substrate will be weak, and the anti-corrosion performance in the hydrochloric acid will be also poor. If the substrate temperature sputtered is raised to 250 °C or so, the thin film will be dense, the cohesion will be big, and the anti-corrosion performance in the hydrochloric acid will be also good (corrosion time of hydrochloric acid: about 20 sec), and the surface shape will be very good.

It shall be pointed out, LPCVD, MOCVD and other methods can be also used to deposit TCO on the glass substrate in addition to magnetically-controlled sputtering, but the deposit effects are different. Table 4.1 shows the comparisons of TCO by these deposition techniques.

4.3.2 P-Type (a-SiC:H) Film Deposited by PECVD Method

The a-SiC:H with wide band bap can be used as the P-type window of TFSCs, and its band gap will rise with growth of C content. Its defect density, however, is **too** high, so it is not suited for the intrinsic absorption layer. When glow discharge is used to decompose the mixed gas of SiH_4 and CH_4, and B_2H_6 is used as the dopant to produce the P-type window layer of the a-SiC:H solar cell, the band gap can reach above 2.0 eV in the width, and it can significantly improve the open-circuit voltage, the short-circuit current and the conversion efficiency.

Before depositing the amorphous silicon carbide thin film, the textured conductive glass shall be first washed by the detergent and the deionised water, and then dried and put into the PECVD discharge chamber. The gas sources for depositing p(a-SiC:H) are the mixed gas of SiH_4, CH_4, B_2H_6 and H_2. B_2H_6 is used as the P-type dopant, and H_2 as the diluent gas. To improve the a-SiC:H performance, SiH_4 and CH_4 shall be diluted by hydrogen of high content. CH_4 doping is aimed to improve the optical performance of the window layer of a-Si:H solar cells. The concentration of CH_4 can be changed to produce the P-type (a-SiC:H) thin film with various C contents and different optical performances, for example, various optical band gaps, light/dark conductive ratio and the like. The P layer with wide band gap can improve the built-in field of the I layer and further raise the open-circuit voltage.

The typical process conditions are shown below for preparation of a-SiC:H P-type thin film:

Temperature T: 200–250 °C
Pressure P: 400–500 mTorr
RF current: 5 A
Time t: 40 s
Thickness d: 80 Å (angstrom)

4.3.3 I (a-Si:H) Intrinsic Zone Deposited by PECVD Method

To reduce the interface state and provide adequate photo-generated carriers, TFSSCs are often designed in the PIN structure, that is, a thick intrinsic zone is deposited between the N and P layers. The intrinsic zone is the main generation zone of the photo-generated carriers, and the a-Si:H solar cell has two basic requirements on the intrinsic zone: first, the intrinsic zone shall have certain thickness to enhance light absorption in the intrinsic zone and raise the number of carriers; second, the space charge density of the intrinsic zone shall be minimised to improve the lifetime and mobility of the photo-generated carriers as well the built-in field so as to raise the

Table 4.1 Comparisons of TCO by different deposition techniques.

Preparation technique	Temp. of substrate	Deposition rate	Evenness	Repeatability	Cost	Conductivity	Transmittance
Thin film	High	High	Poor	Intermediate	Low	Intermediate	Intermediate
Evaporation deposition	Low	High	Intermediate	Intermediate	Intermediate	Intermediate	Intermediate
CVD	High	High	Good	High	Intermediate	Intermediate	Intermediate
Sputtering	Low	Low	Very good	Very good	Intermediate	Very good	Very good
Ion plating	Ambient temperature	Low	Very good	Very good	High	Very good	Very good

collection efficiency of the photo-generated carriers and the cell stability. As a result, the intrinsic zone thickness must be optimised.

The gas source for depositing the intrinsic zone (a-Si:H) is the mixed gas of SiH_4 and H_2. Before depositing the intrinsic zone, the deposition chamber and the relevant gas pipeline shall be washed by a lot of electronic-grade SiH_4 to minimise the pollution caused by various impurity atoms. The performance of the a-Si:H intrinsic zone is mainly dependent on the RF frequency, substrate temperature, reaction pressure and gas flow during preparation. In addition, H_2 can be used to dilute SiH_4, which is good to grow the orderly a-Si:H thin film and improve the conductivity.

The typical process conditions are shown below for the intrinsic zone of a-Si:H thin film:

Temperature T: 200–250 °C;
Pressure P: 600–700 mTorr;
RF current I: 3.5 A;
Time t: 2500 s;
Thickness d: 5000 Å

4.3.4 N-Type (a-Si:H) Layer Thin Film Deposited by PECVD Method

The gas source for depositing the N-type (a-Si:H) thin film is the mixed gas of SiH_4, PH_3, H_2 and He where PH_3 is used for material doping. The structure of the N-type (a-Si:H) thin film is dependent on the PV performance, the substrate temperature, gas source ratio, reaction pressure, RF discharge power, gas flow and so on.

The typical process conditions are shown below for the N-type thin film of a-Si:H solar cells:

Temperature T: 200–300 °C;
Pressure P: 800–1000 mTorr;
RF current I: 6 A;
Time t: 90–100 s;
Thickness d: 300 Å.

4.3.5 Al(Ag) Back Electrodes Sputtered by PVD Method

The back electrodes are usually produced by magnetically controlled sputtering PVD method. The Al(Ag) back electrode plays the following roles in the a-Si:H solar cell:

a) Negative pole of the sub-cell;
b) Series circuit between sub-cells;
c) Reflection of the long-wavelength photons unabsorbed by the a-Si layer to raise the optical utilisation rate of the solar cell.

To prevent diffusion of Al(Ag) back electrodes to the N-type layer (a-Si:H), a thin ZAO layer shall be coated on the sputtered back electrode.

4.3.6 Integration of TFSCs and Modules

The TFSCs have a virtue—it can simply connect the internals of the monolithic circuit in series, as shown in Figure 4.6. The laser beam shall be used to scribe the absorption layer of thin film silicon. In this way, the TFSC of large size can be scribed into sub-cells so that

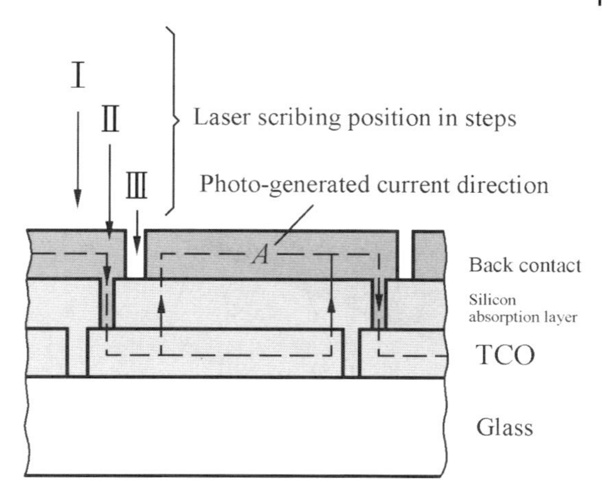

Figure 4.6 Series connections of TFSC monolithic circuit by three laser scribings.

they can have good contact with the Al(Ag) electrodes and the TCO. In the preparation process, the red Nd-YAG laser shall first scribe TCO into several strips of equal distance, and the PECVD method shall be used to deposit silicon, and the green Nd-YAG laser will be used for the second scribing (penetrating the silicon layer). For series connection, a slight transverse deviation shall be present in the second scribing. After the back metal is deposited, the third laser scribing shall be carried out. In this way, the back metal can be connected with the front of the neighbouring cell, realising series connections of cells.

The a-Si/μC-Si stacked cells have similar process procedures with a-Si TFSCs. The thin film cell structure can be produced as monolithic circuit. Similarly, TCO film shall be first scribed by the red YAG laser into strips and then the PECVD method shall be used to deposit the thin film silicon. Finally, the green YAG laser shall be used to scribe the thin film silicon layer. To achieve series connections, the second scribing shall have slight transversal deviation. The third step is to produce the back contact where the metal shall be connected to the front of the neighbouring cells. Figure 4.7 shows both the single-junction and the multi-junction TFSCs can achieve series connections between cells by means of the monolithic circuit.

Since the TFSSCs are designed in the monolithic circuit structure, it is very easy to produce the modules. For the module of the CSSCs, the cells shall be sorted by the open-circuit voltage and the short-circuit current and then welded into cell sets and finally connected in series or parallel to the modules of some area. Different from the CSSCs, since the cells of TFSSCs are connected in series or parallel during the thin film cell production process, the modules are only subjected to cell performance inspection in the condition of solar simulation and then EVA can be installed for packaging, as shown in Figure 4.8.

Figure 4.9 shows the picture of TFSC modules.

4.4 Main Production Equipment for TFSSCs

4.4.1 Production System of TFSSCs

The production system of TFSSCs consists of the magnetically controlled sputtering system, the laser grooving system, PECVD and inspection equipment etc. Figure 4.10 shows the production line of TFSSCs.

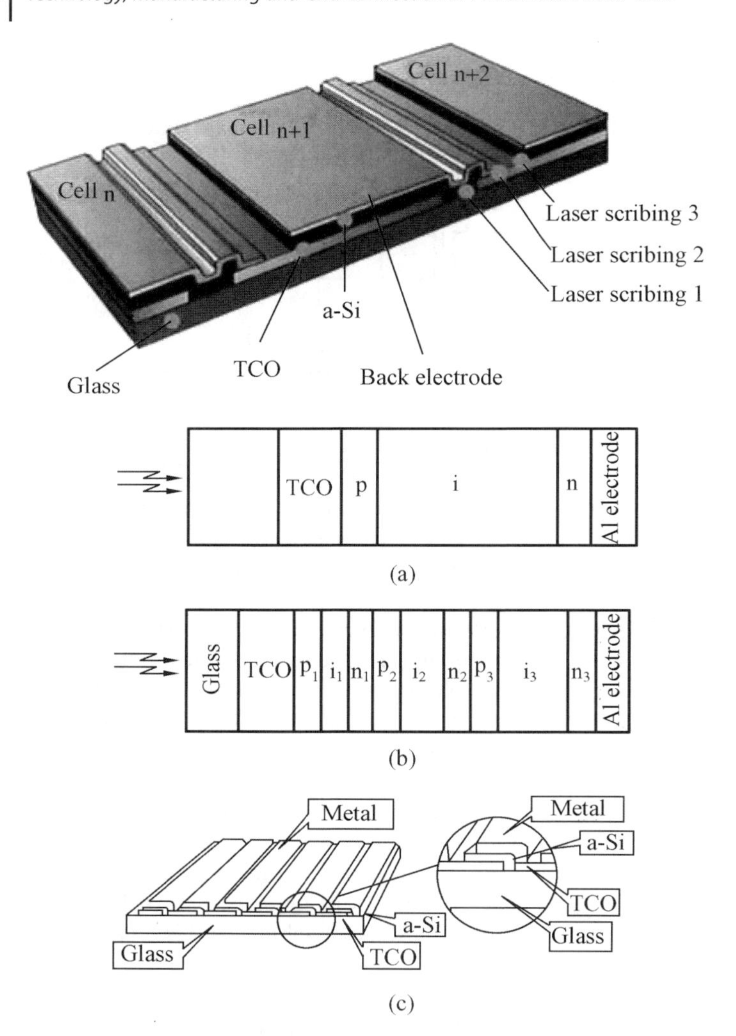

Figure 4.7 Series connections of monolithic circuit for TFSSCs: (a). single-junction cell; (b). multi-junction cell; (c). cell integration.

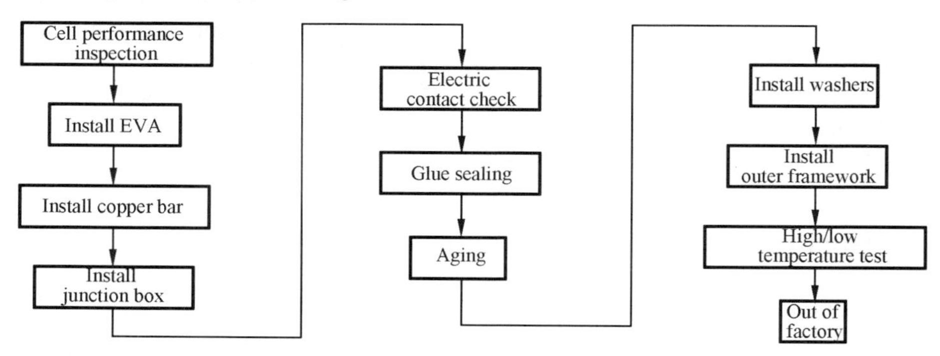

Figure 4.8 Packaging process chart of TFSC modules.

Figure 4.9 Picture of TFSSCs.

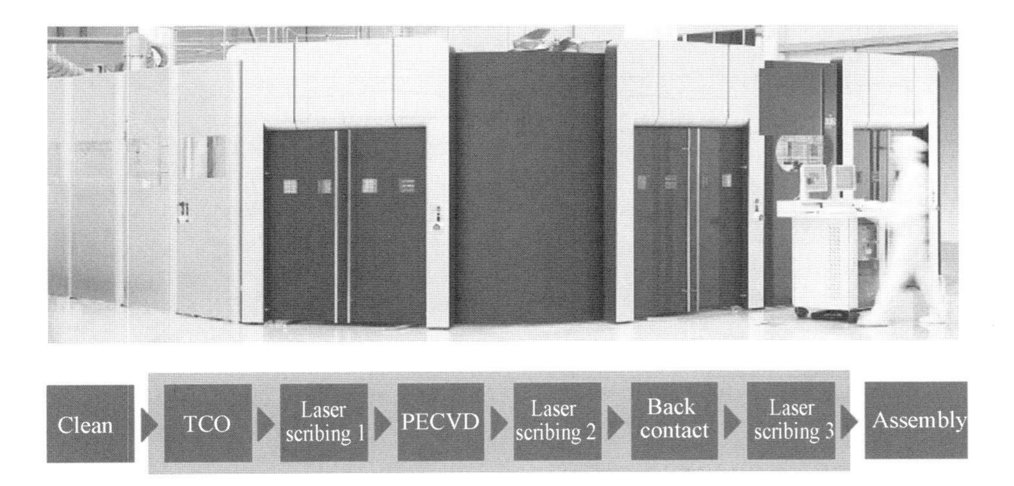

Figure 4.10 Production flow diagramme of TFSSCs.

In the early twenty-first century, many world-known PV equipment manufacturers have entered the manufacturing sector of TFSSC equipment, for example, Applied Material (USA), Oerlikon (Germany), Leybold Optical (Germany) and Ulvac (Japan).

As a typical case, the production line of Leybold Optical is designed in the following size:

1401 mm(L) × 1101 mm(W) × 4–3.2 mm(T) for the glass substrate;
1410 mm(L) × 1110 mm(W) × 35 mm(H) for the module outline
TACT time of the production line as a whole: 77 s.

4.4.2 Glass Cleaning and Surface Texturing Equipment

For the glass substrate of TFSCs, the surface ripple shall be less than 1 mm/1 m, and the glass substrate edges shall be ground into arc. Figure 4.11 shows the typical glass transitivity.

Before TCO coating, it shall be washed by the deionised water, and after coating, it shall be corroded by 0.5% hydrochloric acid to make the TCO surface rough. Figure 4.12 shows the cleaning and wet texturing equipment.

4.4.3 TCO Sputtering Equipment and ZAO Target

Below shows the basic description on TCO sputtering equipment and ZAO target of Leybold Optical, Germany:

- The DC magnetically controlled sputtering method is used to sputter the ZAO (ZnO: Al) layer on the glass substrate in the atmosphere Ar/O_2, with thin film thickness about 1000 nm. In the sputtering equipment, the glass substrate can be continuously transported.

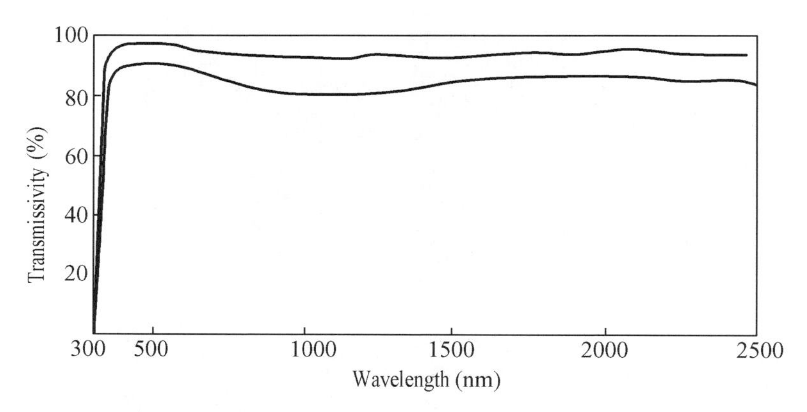

Figure 4.11 Glass transmissivity of Typical TFSSCs.

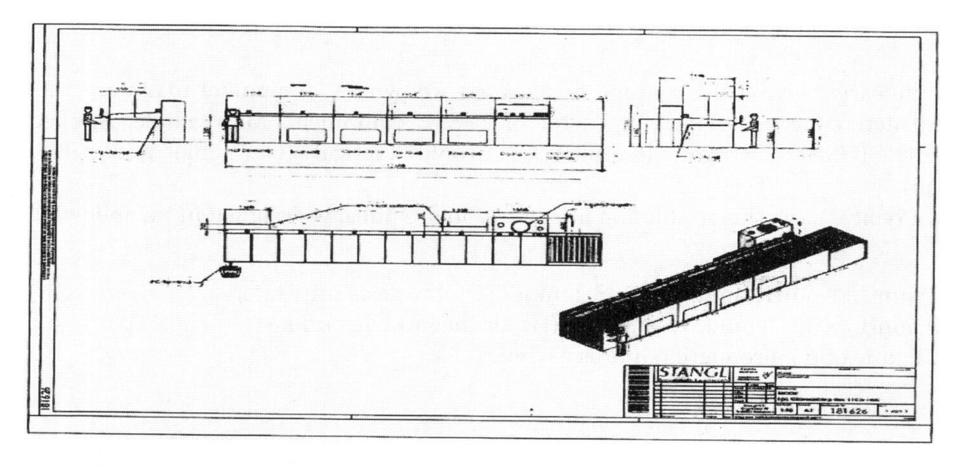

Figure 4.12 Cleaning and wet texturing production line.

- TCO sputtering system consists of two types of units: Type A—the loading/unloading chambers and the buffer chambers, and Type B—the DC sputtering chambers. The system is designed with 11 chambers where:
  The loading/unloading chambers—M2 and M10, are all provided with the vacuum pump system. In M2, the substrate shall be heated to 100 °C, and M10 is for cooling down.
  The buffer chambers—M3, M4, M8 and M9, all designed with the vacuum pump system.
  The sputtering chambers—M5, M6 and M7, all designed with the vacuum pump system.
- The tubular magnetically controlled tube with ZAO porcelain target is 5″ in diameter. The ZAO porcelain target is 1500 mm in length, 14 mm in thickness, and 125 mm in inner diameter. The magnetically controlled tube is driven by a rotary motor, and the target horizontal position can be properly adjusted for coating. The availability of the porcelain target is larger than 70%.
- The distance between the target and the substrate is adjustable in a range of 60–140 mm.
- The manipulator will put the substrate (1401(L) mm × 1101(W) mm) into the carrying vehicle (1600 × 1300 mm). During the loading/unloading process, the manipulator only touches the back of the substrate.
- In the loading/unloading chamber and the buffer room are provided with equiflux type heater, and the sputtering zone is provided with single-sided heater. The TCO sputtering system and the ZAO target of Leybold Optical are shown in Figure 4.13 and Figure 4.14.

The thin layer resistivity of the ZAO coating can be measured by the four-probe method where the square resistance shall be less than 10 Ω/□.

Figure 4.13 TCO sputtering system (double ZAO targets), Leybold Optical.

Figure 4.14 ZAO rotary target, Leybold Optical.

4.4.4 PECVD System for Thin Film Silicon Deposition

In the TFSSCs production equipment, the PECVD system is the most crucial one to produce the absorption layer of silicon thin film. The absorption layer, composed of P, I, N, can be also divided into amorphous and microcrystalline. Since their doping contents and deposition time are different, they cannot be conducted in the same process conditions. The equipment manufacturers have designed and manufactured different process production lines. Generally, PECVD consists of one chamber with several coatings and several chambers with one coating. The one chamber with several coatings means that the coatings of various types can be deposited in the same vacuum reaction chamber where it is difficult to control the cell performance due to gas pollution but the system is cheap and simple. Several chambers with one coating mean that the different coatings will be deposited in different vacuum reaction chambers where the cells have good performance but the system is expensive and complex. In the production line of several chambers with one coating, the horizontal clustering, vertical folding, linear series and linear parallel types are available, and each has its own advantages and disadvantages.

Figures 4.15–4.20 show the PECVD system of the one chamber with several coatings, the horizontal clustering, vertical folding, linear series and linear parallel types, respectively.

Although the PECVD system of one chamber with several coatings has a big virtue—low cost, it has a big problem—cross pollution of reaction gases. The system manufacturers of several chambers include UlVAC (Japan), and Applied Material (USA). These systems can avoid cross-pollution of reaction gases in an effective manner and reduce the impurity content in the intrinsic zone, improving the efficiency of solar cells. In addition, it can be deposited simultaneously in several chambers, raising the productivity. But the several-chamber system has the following shortcomings: high cost and more components to be maintained. The following manufacturers can produce the linear parallel type of several-chamber system: Oerlikon (Germany), and Leybold Optical (Germany). The system has a virtue that the whole production line does not need to stop in case one chamber fails and independent maintenance is available. But the system has high cost.

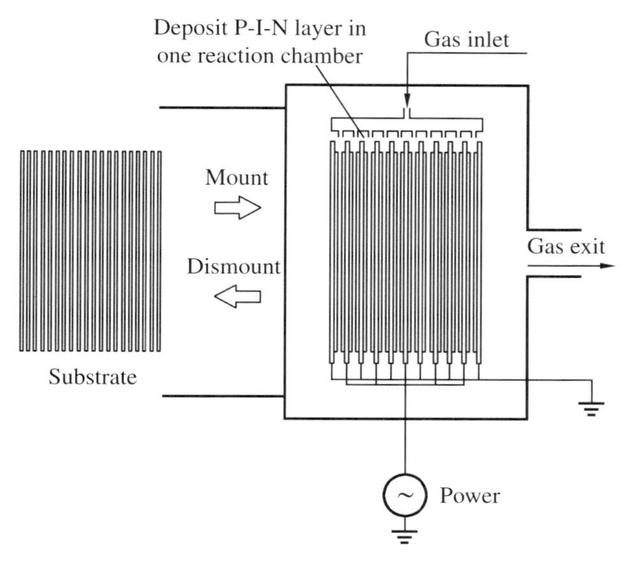

Figure 4.15 Deposition chamber of PECVD production line in one chamber with several coatings.

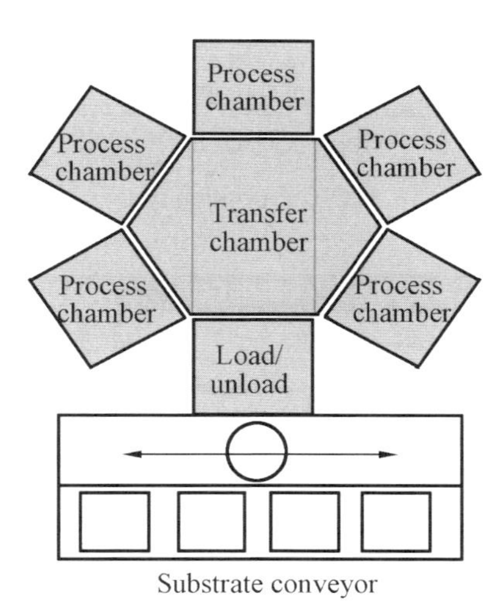

Figure 4.16 PECVD technique, horizontal clustering.

Figure 4.19 shows the PECVD diagramme of linear parallel type, Leybold Optical., Germany. The figure on the right shows the support to deliver the glass substrate to the process chamber. The support can move on the track in the system and when the gate valve of a chamber is opened, the glass substrate can enter vertically into the process chamber. Below shows the main description on the PECVD system, Leybold Optical, Germany:

- The PECVD system can be provided with 19 chambers, and driven by 13.56 MHz RF. On the back of each chamber is connected to the bypass via high vacuum valve. The NF_3 plasma dry corrosion method can be used to clean the chamber on a regular

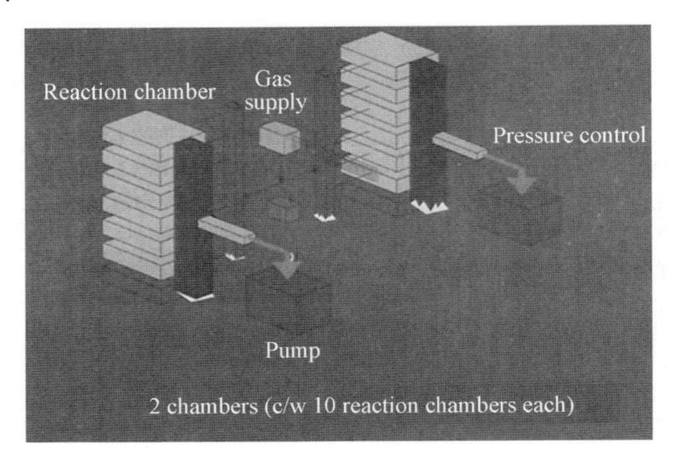

Figure 4.17 PECVD technique, vertical folding.

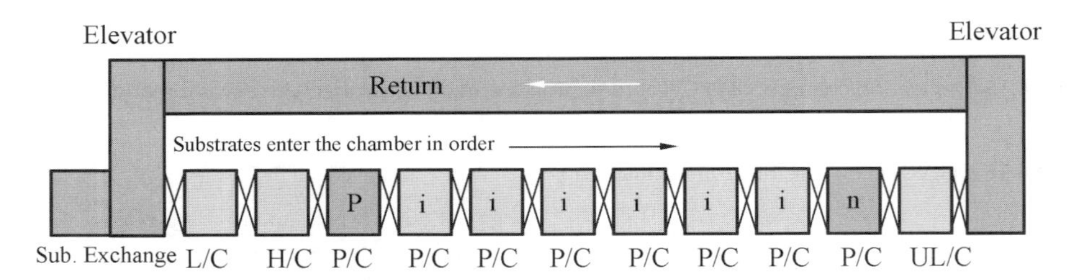

Figure 4.18 PECVD depositing system, linear series.

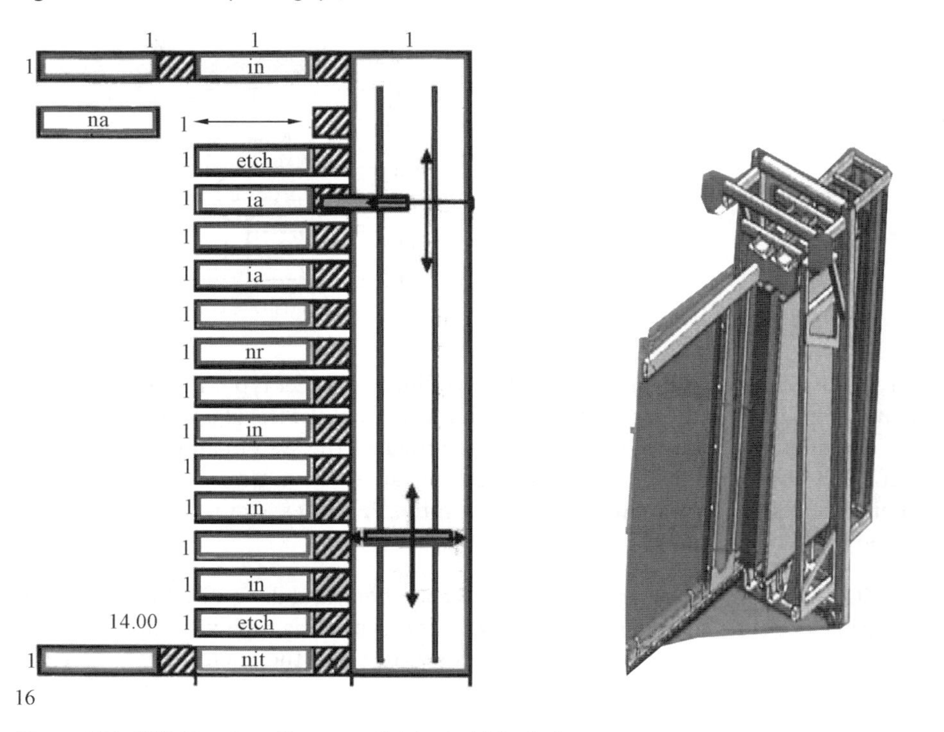

Figure 4.19 PECVD system, linear parallel, Leybold Optical.

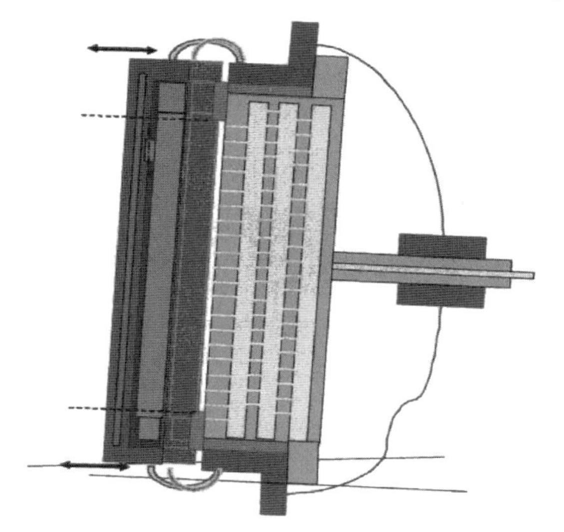

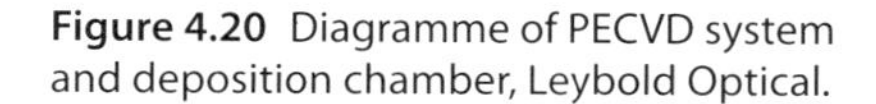

Figure 4.20 Diagramme of PECVD system and deposition chamber, Leybold Optical.

basis to prevent cross pollution. One or two manipulators shall move in the channel to transport the substrate.

- Four types of chambers are available: 'ia' (intrinsic amorphous layer), 'iμ' (intrinsic microcrystalline layer), 'n' (N-type layer) and 'p' (P-type layer). More chambers can be extended as demand. And the gas systems between the independent reactors can be separated from each other.
- Each chamber can be maintained and repaired independently with ease, and in this case, the other chambers can be in the vacuum state.
- The manipulator can execute two actions: One is to open the valve when the substrate is put into the chamber and the other is to move the earth electrode to a place about 10cm from the live electrode and then push it between two electrodes.
- One manipulator can carry two pieces of substrate at one time.
- The heater can heat two pieces of substrate to 200 °C or so.

Figure 4.20 shows the PECVD system of Leybold Optical.

4.4.5 Back Contact Sputtering Equipment

The back contact (BC) sputtering system of Leybold Optical is shown in Figure 4.21, which is described as follows:

- The BC system is designed to deposit the back contact of a-Si or a-Si/μC-Si stacked cells. The BC totally consists of three layers: zinc oxide doped with aluminum (AZO 80 nm), silver (200 nm) and aluminum (100 nm). The substrate shall be transported by a carrying vehicle.
- The BC system has two types of chambers: Type A—the loading/unloading chambers, the buffer rooms and the extension rooms; Type B—including the chambers of DC sputtering for 3 alloys. There are a total of 10 chambers, where, The loading/unloading chambers—M2, M8, and M9, all designed with the vacuum pump system. M2 shall

Figure 4.21 BC sputtering system, Leybold Optical.

be heated to 100 °C, and M8 and M9 serve as the cooling chambers. The buffer chamber—M3 and M7, all designed with the vacuum pump system. M3 and M7 shall be heated to 200 °C. The extension chamber—M4 and M6, all designed with the vacuum pump system. The sputtering chamber—M5, including the cathode gates of three targets—ZAO, silver and aluminum, all designed with the vacuum pump system.

- The porcelain ZAO, silver and aluminum target is 1500 mm in length, the ZAO and aluminum target is 14 mm in thickness, the silver target is 10 mm in thickness, and their inter diameter is all 125 mm. The magnetically controlled tube is driven by a rotary motor, and the horizontal position of the target can be adjusted to carry out coating. The availability of the target is larger than 70%.
- The distance between the target and the substrate is adjustable in a range of 60–140 mm.
- The manipulator will put the substrate (1401 mm(L) × 1101 mm(W)) into the carrying vehicle (1600 × 1300 mm). During the loading/unloading process, the manipulator shall only touch the back of the substrate.

4.4.6 Laser Scriber

The laser scriber of Leybold Optical is shown in Figure 4.22, which can be described as follows:

- The laser scriber system is designed with two laser scribing and etching systems S1 and S2. The laser wave of S1 is 1064 nm in length, and it is used for the first P(I) laser scribing; and the wave of S2 is 532 nm in length, and it is used for the second P(II) and third P(III) laser scribing. The differences of the two lie in the laser components and sensors. Both of them shall be designed with air conditioners.
- Size of the scribed glass substrate: 1400 × 1120 × 4 mm (width of useless edge: 5 mm); grooving width: 30–40 μm.
- Y-axis movement range: 1500 mm; maximum movement speed: 2000 mm s^{-1}; repeatability: 1 μm.

  X-axis movement range: 1300 mm; maximum movement speed: 2000 mm/sec; repeatability: 1 μm.

  Z-axis movement range: 11 mm.

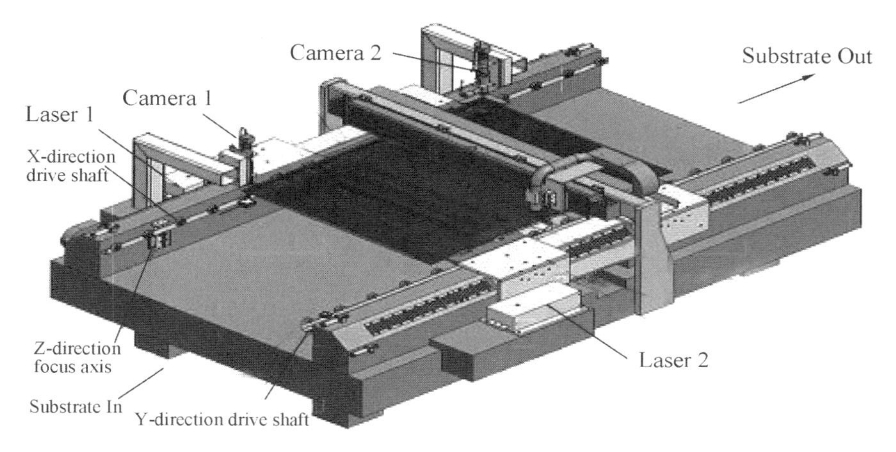

Figure 4.22 Laser scriber diagramme.

- To make the circulating time less than 70 sec, two laser beams in parallel can be obtained by means of optical principles, and the distance between them can be manually adjusted to a range of 5–15 cm.
- For the 1120 mm × 1400 mm substrate, the distance between the 113 lines is 10mm.

4.4.7 Testing Equipment

1. **Crystallisation rate test**
   RAMAN spectrometer is used. The helium-neon laser of 632.8 nm wavelength serves as the light source, and three peaks—480 cm^{-1}, 510 cm^{-1} and 520 cm^{-1}—are fitted to work out the crystallisation rate of silicon thin film $X_C \cdot X_C = (I_{510} + I_{520}) / (I_{480} + I_{510} + I_{520})$, where, I_{520} is the scattering peak area of the crystal silicon structure near 520 cm^{-1}; and I_{480} is the scattering peak area of the a-Si structure near 480 cm^{-1}. The peak near 510 cm^{-1} is the grain boundary, and I_{510} shows the integrated intensity.
2. **Testing of conductivity: dark conductivity σ_d and light conductivity σ_{Ph}**
   After the thin film is deposited, a layer of silver shall be thermally evaporated to form the coplanar electrode structure. Then 3 V DC bias shall be applied and based on Ohm law, the current passing the electrostatic testing circuit can be worked out in light and dark conditions.
3. **Testing of thin film thickness**
   First, the transmission and reflection spectrum can be measured by an optical multi-channel analyser and then the profiler shall be used for correction. Finally, the thickness of the sample can be worked out. Based on this, the deposition rate of the thin film can be worked out.
4. **Measurement of resistivity**
   The four-probe method is used to measure the square resistance for TFSSCs, as shown in Figure 4.23.

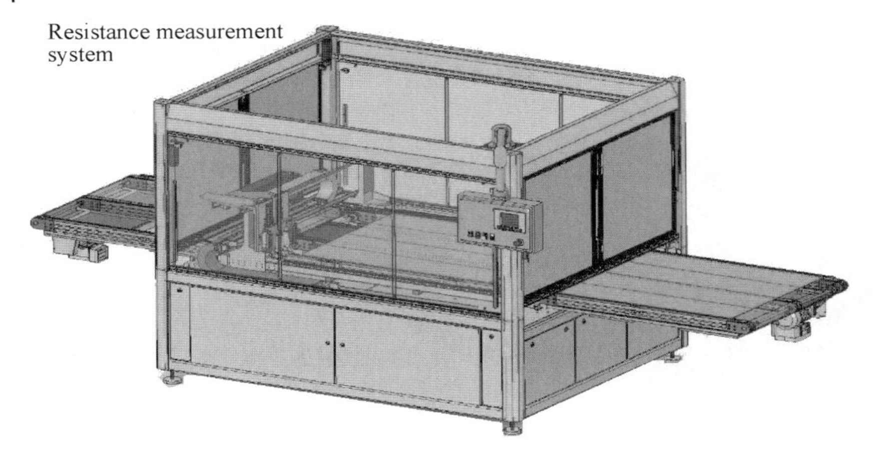

Figure 4.23 Square resistance of TFSSCs measured by four-probe method.

4.5 Discussion on Some Issues Concerning TFSSC Preparation

4.5.1 Performance, Preparation and Testing of TCO

The TCO has been widely applied to TFSSCs, for example, those made of indium-tin oxide (ITO), tin oxide SnO_2:F (FTO) and ZnO:Al (ZAO).

TCO plays two roles in TFSSCs: On one hand, it can make the solar rays penetrate the substrate and enter the solar cell. In this case it shall have high optical transmissivity and its surface shall be textured to minimise reflection. On the other hand, it can provide the electrode to collect current. And in this case, it shall have high conductivity.

The resistivity of ITO can be very low, down to 10^{-4} Ω cm, and its transmittance rate in the visible spectrum range can reach above 90%. Obviously, it is a quality TCO. It is also a key material in liquid crystal display. The indium (In) resource, however, is in shortage and thus expensive, which limits its application to TFSCs. Follow that, TFSSCs is provided with textured SnO_2:F(FTO), which has been realised industrialised production. Although SnO_2 thin film also has good electrical and optical performance, its optical characteristics will deteriorate in the hydrogen plasma atmosphere and high growing temperature (e.g., 500 °C or so), which limits its application to TFSSCs as the TCO. In addition, FTO is toxic and expensive. As a result, ZnO:Al(ZAO) has witnessed more applications in the industry in recent years. ZAO has the good PV characteristics comparable to that of FTO (low resistivity, textured structure, and high transmissivity), and it has such virtues as low price, abundant raw materials, and non-toxicity. It can have good stability in the hydrogen plasma atmosphere and grow at low temperature, and thus it has become the competitive TCO of silicon-based TFSCs. Table 4.2 lists the characteristics of some TCOs at the ambient temperature.

In the ZAO preparation processes, the magnetically controlled sputtering method is the one that has gone through the most studies and researches and achieved the widest applications. The target is the metal or porcelain one of 4N–5N pure zinc doped with about 2% aluminum. Because the sputtered ions have high energy, and the sputtered

Table 4.2 Characteristics of some TCOs at the ambient temperature.

Material	Resistivity (Ωcm)	Width of forbidden band (eV)	Reflective index
SnO_2	10^{-2}–10^{-4}	3.7–4.6	1.0–2.2
ITO	10^{-3}–10^{-4}	3.5–4.6	1.8–2.1
ZnO_2	10^{-3}–10^{-4}	3.1–3.6	1.86–1.9

film and the substrate have good cohesion, the film thickness is controllable with good repeatability. The sputtering technology, however, cannot be directly applied to the ZAO textured thin film. The glass substrate shall be coated with ZAO and then put into the 0.5% × hydrochloric acid for some time so that the zone with loose bonding or poor cohesion on the film will be corroded and the originally smooth surface can be corroded into uneven textured structure.

In the preparation technique of TFSCs, the four-probe tester shall be used to measure the square resistance of the film, and the ellipsometer to measure the thickness of the film, and the UV-visible spectrophotometer to measure the transmissivity, and the x-ray diffractometer XRD to show the structure orientation of ZnO, and the scanning electron microscope or atomic force microscopy (AFM) to show the surface shape of the film, and the Van der Waals method to carry out Hall measurement for carrier concentration, mobility and resistivity.

4.5.2 Influence of PECVD Process Parameters on Deposition and Crystallisation Rates of Silicon Thin Film

When the PECVD method is used to grow a-Si and μC-Si on the glass substrate, the deposition rate of silicon thin film is low and it is difficult to control the crystallisation rate. Many literatures show, the deposition rate and the crystallisation rate have relations to many factors, for example, the hydrogen dilution, pressure, deposition temperature, distance between the electrode and substrate and so on.

4.5.2.1 Hydrogen Dilution

Profound research has been conducted to the influence of the hydrogen dilution on the a-Si material. Many experiments show, the dilution condition of high-concentration hydrogen is good to prepare the quality a-Si and a-SiGe thin film. As the hydrogen is diluted to some degree, micro-grain content will grow and form longitudinal continuous structure in the material. In this case, it is also observed that the dark conductivity σ_d and the photosensitiveness will see sharp change, indicating that the material structure is transiting from amorphous to microcrystalline.

4.5.2.2 Gas Pressure

When the RF PECVD is used at low pressure, the deposition rate will rise with pressure growth, which is because the possibility of the neutral particles and the live ions collided with the silane molecule increases. The optimal pressure of the glow plasma deposition chamber shall take into account the influence of the pressure on the plasma stability. Generally, higher pressure is required when the distance between the cathode and the substrate is small, and vice versa.

Excessively large gas flow is adverse to microcrystalline thin film because the μC-Si thin film needs some gas dwelling time to grow.

4.5.2.3 Deposition Temperature

In the PECVD technique, the substrate temperature is a key factor to a-Si film quality. On one hand, the rise of substrate temperature is good to raise the diffusion coefficient of the reaction particles on the substrate surface, and reduce the defect state and hole density. On the other hand, excessively high substrate temperature will reduce the hydrogen content in a-Si, raising the defect density. As a result, there is an optimal substrate temperature, which is generally in a range of 200–300 °C, to deposit a-Si thin film. Figure 4.24 shows the relationship between the defect density and the substrate temperature in a-Si.

The optimal substrate temperature shall also take into account the deposition rate. The high deposition rate needs high substrate temperature. In addition, the optimal deposition temperature has also relations to the hydrogen dilution. When a lot of hydrogen atoms cover the growth surface, it can effectively raise the surface diffusion coefficient of the reaction particles. In this case, the optimal substrate temperature may be a little lower.

4.5.2.4 Distance Between the Electrode and the Substrate

The deposition rate of the film will rise with growth of electrode distance, which is because the gas is more likely to dwell in the reaction chamber. The crystallisation rate, however, will rise first and then fall down with growth of electrode distance, which is because the silane decomposition rate and the atomic hydrogen density will rise as the distance between electrodes increase and it is good for crystallisation. If the distance between electrodes is excessively large, the gas in the reaction chamber cannot be fully decomposed and the atomic hydrogen density will fall down, reducing the crystallisation degree. To achieve the component-grade microcrystalline silicon of low defect state, the distance between electrodes shall be optimised.

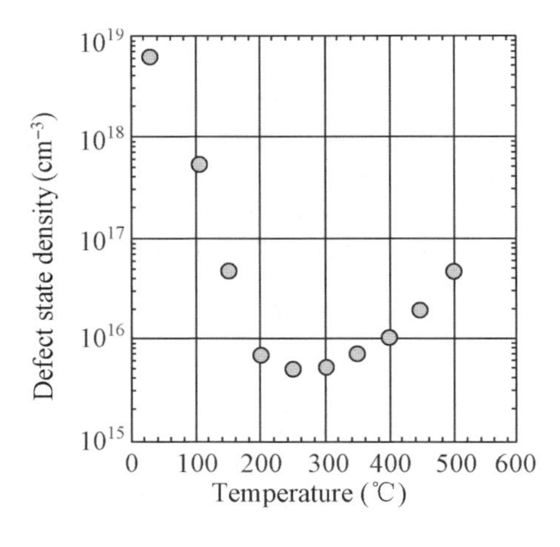

Figure 4.24 Relationship between defect density and the substrate temperature in a-Si.

4.5.2.5 Power Excited by Plasma

The power excited by plasma is a prerequisite to keep the plasma stable and it is also an important means to raise the deposition rate. The excessively high power, however, will produce a side effect adverse to the material quality because the concentration of the electrons in the plasma will rise and the associated electron of high energy will generate SiH_2 and SiH groups as the power grows. In addition, the high-power plasma will produce dust, bombarding the deposition surface and raising the defect state. As a result, the low excitation power is generally selected when the deposition conditions are satisfied.

4.5.2.6 Frequency Excited by Plasma

At present, the mature method is RF-PECVD, which can be used for a-Si TFSCs, but the method has a low deposition rate to deposit μC-Si film. The very high frequency (VHF) PECVD technique can accelerate the deposition rate of silicon film, and the micro-wave PECVD technique is under development. When the frequency rises, however, the evenness will become poor and it is also difficult to realise high frequency shielding.

4.5.3 VHF-PECVD Method to Deposit Silicon Film

It has a low deposition rate to produce μC-Si:H film by the traditional RF PECVD method. And the VHF PECVD technique can significantly raise the deposition rate of silicon film, especially for the thick intrinsic a-Si layer and the μC-Si layer. Compared with the traditional RF PECVD method, the VHF PECVD technique has a more complicated process to excite silane to form plasma. Profound and effective researches have been conducted on VHF-PECVD method to prepare μC-Si:H film both at home and abroad.

The experiment results of Yang Huidong and so on show, when the frequency rises from 80 MHz to 140 MHz, the deposition rate of μC-Si: H film will rise from 1.1 nm s^{-1} to 1.6 nm s^{-1} and the deposition rate of the traditional RF-PECVD (frequency: 13.56 MHz) is only 0.1 nm s^{-1} in the same conditions. As a result, the VHF-PECVD method is an effective method to grow the μC-Si: H film at a high speed. The main reason lies in that the excitation voltage is often only dozens of volts in the VHF-PECVD method and it shall reach several hundred volts in the RF-PECVD method. The field in the plasma reduces, decreasing the average energy of the electrons in the plasma. In consequence, the electron density in the VHF-PECVD technique will rise and the concentration of atomic H and SiH in the silane plasma will grow in the same plasma power conditions, effectively improving the deposition rate of μC-Si: H film. In addition, when the average energy of electrons falls down, the high-energy electron-molecule collision will be weakened, and the low-energy electron-molecule collision will be enhanced. Accordingly, it is easy to achieve crystallised growth of μC-Si: H by the VHF-PECVD technique.

Figure 4.25 shows the variation of the deposition rate with plasma excitation frequency. Figure 4.26 shows the variation of crystallisation volume fraction X_C and the average grain size of the silicon film with plasma excitation frequency by the VHF-PECVD technique. Figure 4.27 shows the relationship of the deposition rate and the crystalline phase ratio with the deposition pressure in the reaction chamber.

It is proposed to use the RF-PECVD (13.56 MHz) technique for producing the P-layer a-Si and the N-layer a-Si, and the VHF-PECVD (60 MHz) technique for producing the

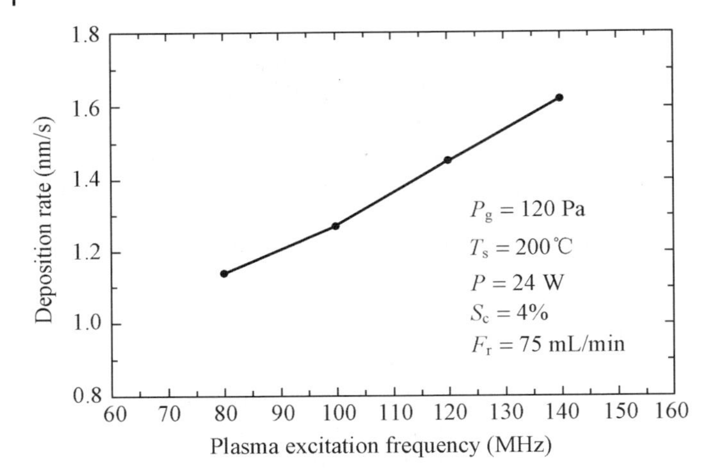

Figure 4.25 Variation of the deposition rate with plasma excitation frequency.

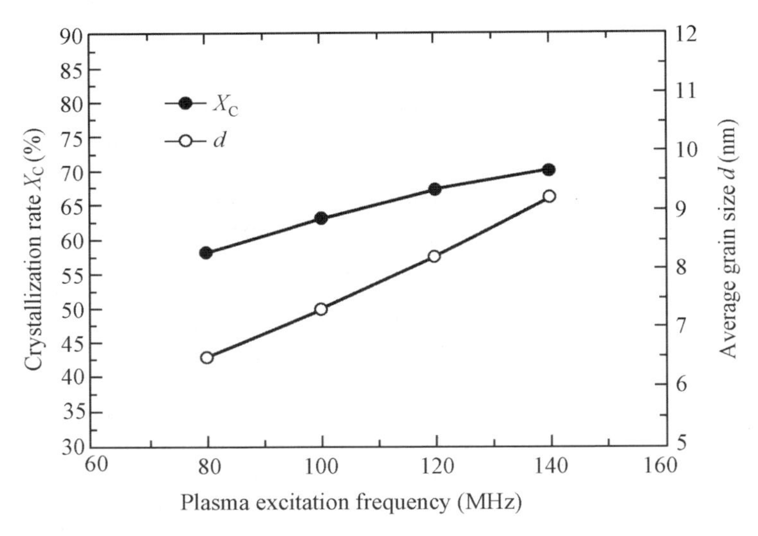

Figure 4.26 Variation of crystallisation volume fraction XC and the average grain size of the silicon film with plasma excitation frequency.

intrinsic zone so as to reduce the mismatching of lattices at the I/P transition layer and the heterogeneous interface and the interface state. At the same time, it can intensify the built-in field in the I layer near the interface and reduce recombination of the photo-generated carriers at the interface, raising the open-circuit voltage and improving the filling factor.

4.5.4 HWCVD Method to Deposit Silicon Film

When the hot wire is used to decompose silane to deposit a-Si:H film, the H content in a-Si can be reduced below 1% and the ordered silicon network structure can still be obtained, resulting in a-Si microcrystallisation.

The HWCVD technique has the following virtues: fast deposition at low temperature, simple equipment, and easy to realise massive production. The tungsten wire with

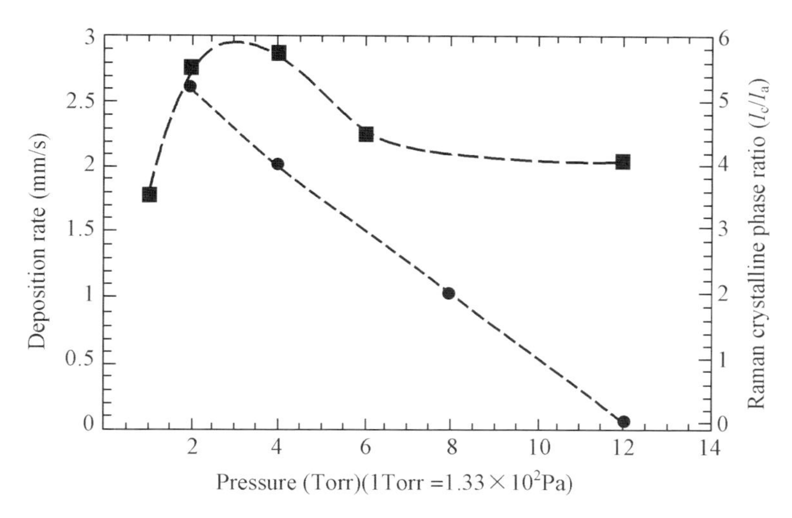

Figure 4.27 Relationship between the deposition rate (■), the crystalline phase ratio (●) of the silicon film and the deposition pressure in the VHF-PECVD technique.

0.5 mm in diameter is powered and heated, which can be also used as the reaction catalyst, and the distance between the hot wire and the substrate shall be controlled in a range of 2–4 cm. The tungsten wire is about 2000 °C. The substrate is about 250 °C. As the reaction time lasts, the crystallisation peak will show up in the Raman spectrum while the a-Si peak will gradually fall down, indicating that in the HWCVD technique, the a-Si film will first deposit on the substrate surface. Then the growth rate of the film will slow down and the crystallisation degree will gradually increase since the hydrogen etching function rises. After the film grows, the hydrogen heat treatment can significantly improve the crystallisation rate. To prevent the tungsten wire from polluting the substrate surface, the hydrogen in the proper quantity can be connected before the experiment is conducted.

The following conclusions are drawn from experiments and theoretical simulations and calculations: The silicon films produced by HWCVD and PECVD techniques are different from each other in such aspects as shape and surface roughness, indicating they have different growth dynamics processes. AFM and x-ray reflection techniques have been used to study the evolution process of film surface, and the results show the HWCVD technique corresponds to random deposition growth and the PECVD technique to the finite diffusion growth.

Compared with the PECVD technique, the key in the HWCVD technique lies in the equipment to produce even silicon film on a large area. In addition, it is also a problem that the heating wire will pollute the material.

4.6 Adjustment of TFSSC Energy Band Structure

4.6.1 Methods to Adjust the Band Gap of Thin Film Silicon

4.6.1.1 Significance of Energy Band Structure Adjustment for TFSSCs

The a-Si TFSCs are usually designed in the PIN structure. To improve the PV conversion efficiency, the P-type layer is generally made of the material with wide band gap as the window layer to enhance absorption of the short-wavelength light while the N-type layer is often made of the material with narrow band gap to enhance absorption of the long-wavelength light in the cell design. To make better use of the solar spectrum, the stacked solar cells, which are composed of the materials with different gaps, have been developed on the basis of the characteristics of the a-Si film that the band gap can be adjusted by changing the process conditions.

4.6.1.2 Gap Adjustment by a-Si Hydrogen Content and Deposition Temperature

The a-Si film preparation has a key characteristic—the preparation conditions can be changed to adjust the structure and composition of the film and further adjust the band gap of a-Si. The simplest way is to change the hydrogen content in the vapour phase, which can make the a-Si:H gap vary in a range of 1.5–1.8 eV. In addition, the substrate temperature can also change the absorption coefficient and the band gap. The experiments show, in the PECVD process, the band gap will be 1.8 eV when the substrate is 200 °C and the absorption coefficient of 1.2 eV is 0.25 cm^{-1}; and the gap will be 1.7 eV when the substrate is 280 °C and the absorption coefficient of 1.2 eV is 1.20 cm^{-1}.

It is insufficient to adjust the band gap only by changing hydrogen content and substrate temperature. And the a-Si chemical composition is often changed to produce the a-Si alloy film with various gaps.

4.6.1.3 a-SiC Carbon Material to Widen the Gap

The research shows the alloying of a-Si:H with C, N or O can widen the gap. Amongst these alloy films, a-SiC:H plays a crucial role. Its optical gap can reach to the maximum 3.0 eV. Accordingly, it is often used as the window material of TFSSCs.

4.6.1.4 a-Si Ge Material to Narrow Down the Gap

The research shows the alloying of a-Si:H with Sn, Ge can narrow down the gap. The most important narrow-gap alloy film is a-SiGe:H, and the gap can be adjusted in a range of 1.1–1.7 eV when the Ge content is changed in the alloy, and the optical gap will fall down with rise of Ge content where the minimum can reach 1.1 eV. Because the defect state of the material will rise with growth of Ge content, the a-SiGe:H gap width of the component grade is generally no less than 1.4 eV.

4.6.2 a-SiGe TFSCs

The application of a-SiGe to a-Si:H/a-SiGe:H stacked solar cells has become mature. The reaction of Si_2H_6 and GeH_4 can reduce the temperature to prepare microcrystalline SiGe into films. As the substrate temperature and power rise, the material may change from amorphous to microcrystalline state. The optimal temperature range is 230–280 °C. The growth of hydrogen dilution in the reaction gas can also help improve the material quality. In the condition of high hydrogen dilution and low temperature (180 °C), the a-SiGe:H film of high quality and narrow gap can be produced.

The mixed gases of Si_2H_6 and GeH_4, instead of SiH_4, are often used. This is because the decomposition rate of GeH_4 is 3–4 times faster than SiH_4, which may result in uneven Ge content in the film. The decomposition rate of Si_2H_6 and GeH_4 is similar to each other.

Both a-SiGe:H and μc-Si:H can be used as the material of the bottom cell with the narrow forbidden band. Figure 4.28 shows the relationship between the absoprtion

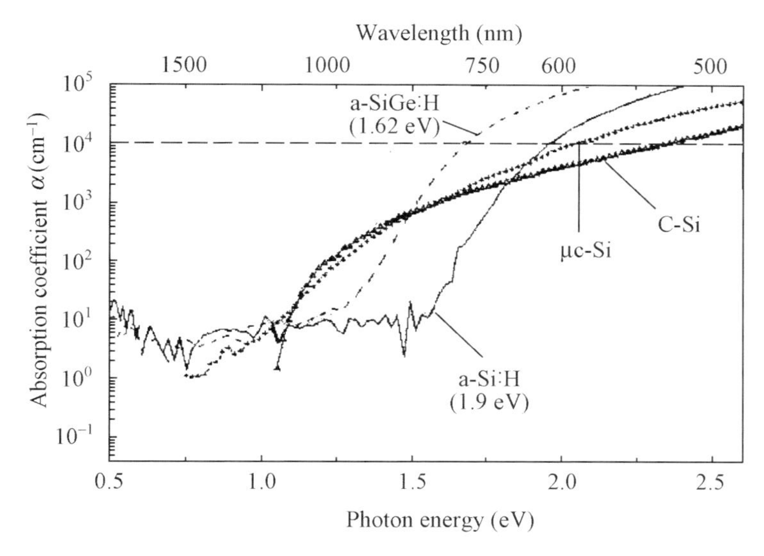

Figure 4.28 Relationship between the absorption coefficient of c-Si, a-Si:H, a-SiGe:H and μc-Si:H and the photon energy.

coefficient of c-Si, a-Si:H, a-SiGe:H, μc-Si:H and the photon energy. In the figure, the band gap is 1.9 eV for a-Si:H and 1.62 eV for a-SiGe:H. The quality μc-Si:H has larger absorption coefficient than c-Si, and its absorption coefficient is even larger than a-Si:H in the near infrared wave (>750 nm). Compared with a-SiGe:H, however, the absoprtion coefficient of μc-Si:H is less than a-SiGe:H in some wavelength range, especially in the visible light range where the absorption coefficient of a-SiGe:H is larger by one order of magnitude than μc-Si:H. As a result, for the same cell output, if a-SiGe:H is used as the bottom cell, its thickness can be less by one order of magnitude than μc-Si:H.

Since the a-SiGe:H film can be very thin, it will consume less special gases, and its production period is also shorter. Accordingly, a-Si:H/a-SiGe:H components have become a powerful rivalry of a-Si/μC-Si TFSSCs. The price of GeH_4, however, is high, and the a-SiGe:H TFSCs will have the light attenuation inherent to the amorphous silicon. As a result, the μC-Si and a-SiGe:H are still the hottest ones for double-junction cells.

For single-junction cells, a-Si is often used. To prepare the intrinsic zone of the bottom cell of the double-junction cells, the process is generally adjusted to grow microcrystalline silicon so as to minimise the band gap of the bottom cell. During design of the intrinsic zone for triple-junction cells, the SiGe of various components are generally doped to change the gap of the intrinsic zone. Table 4.3 shows the material, light band gap, and thickness of the intrinsic zone of the triple-junction cells.

Obviously, the three intrinsic zones of the triple-junction solar cells are different, which have good response characteristics to the blue, green and red light ranges, respectively. As a result, the triple-junction and multi-gap cells can improve the PV conversion efficiency. Their initial efficiency has reached 15.1%. Figure 4.29 shows the spectral response of double/triple-junction solar cells.

Table 4.3 Basic data of top/middle/bottom cells of triple-junction solar cells.

	Top cell	Middle cell	Bottom cell
Band gap Eg	1.8 eV	1.6 eV	1.4 eV
Material	a-Si	a-SiGe	a-SiGe
Ge	0	10–20%	−40%
Thickness (nm)	80–100	150–200	150–200

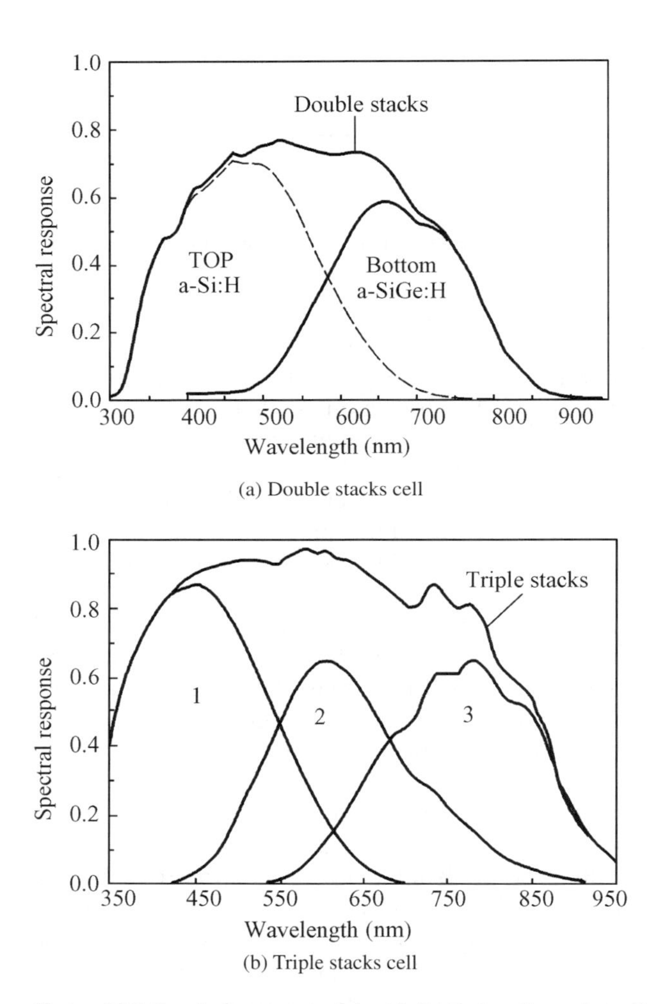

Figure 4.29 Spectral response of double/triple-junction solar cells.

4.6.3 Boron, Phosphorous and Hydrogen in a-Si Film

To prepare N-type a-Si film, phosphorane shall be used as the doping gas, and to prepare P-type a-Si film, borane shall be used as the doping gas. Different from the crystal silicon, the doping of a-Si film is not in the form of diffusion, and the doping gas is decomposed together with the silane during film growth, that is, the impurity atoms are doped while the a-Si film is forming.

When a-Si is growing, borane and phosphorane can be alternatively connected to produce the a-Si TFSCs with PIN structure. But it is better to carry out the doping process in different chambers, which may raise the cost but can improve both the cell efficiency and the repeatability.

In the a-Si thin film produced by the PECVD method, it generally contains 10–15% hydrogen. On one hand, the hydrogen can saturate the silicon dangling bond; and on the other hand, the high hydrogen content far exceeds the density of the dangling bond and the excess hydrogen will have different states in the a-Si and take several positions of low excitation energy, forming SiH_2, $(SiH_2)_n$, SiH_3 and other groups as well as micro-holes and other effects. In this case, the amorphous TFSSCs will have an efficiency degradation in the solar rays for a long time. Obviously, hydrogen plays an important role in the atomic structure of a-Si, and it will have direct impact on the performance of a-Si TFSCs.

4.7 Physical Principle of PECVD and Deposition of Silicon Thin Film

It is known that the epitaxial silicon of hydrogen must be above 1000 °C. Obviously, this is inappropriate for TFSSC production techniques because the thin film silicon shall be deposited on glass or other substrates, which cannot bear such a high temperature. As a result, the active layer of TFSSCs is generally made by the PECVD method. Below will introduce the generation mechanism of plasma, which can grow silicon thin film at about 200 °C.

Similar to CSSCs, the PECVD system is the most important one in the production process of TFSSCs. Since the plasma generation mechanism and its distribution in the PECVD system play a crucial role in understanding the early ageing of a-Si solar cells and parameter adjustment of thin film silicon process, it is necessary to describe the principle of plasma generated by glow discharge.

4.7.1 Glow Discharge and Plasma Generation

Glow discharge can generate plasma. The so-called glow discharge, in fact, is to apply the thin gas to the vacuum system to discharge between two electrodes. The I–V characteristic curve of glow discharge, as shown in Figure 4.30, consists of several stages: Townsend discharge, early discharge, normal discharge, abnormal discharge, transition zone and arc discharge and so on. For plasma generation, the most important is the abnormal glow discharge stage with saturated current. In practise, the abnormal glow discharge stage is generally selected.

During abnormal glow discharge, a glow zone will be formed between the two electrodes, which can be divided into the following zones from the cathode to the anode: Aston dark space, cathode glow space, Crookes dark space, negative glow space, Faraday

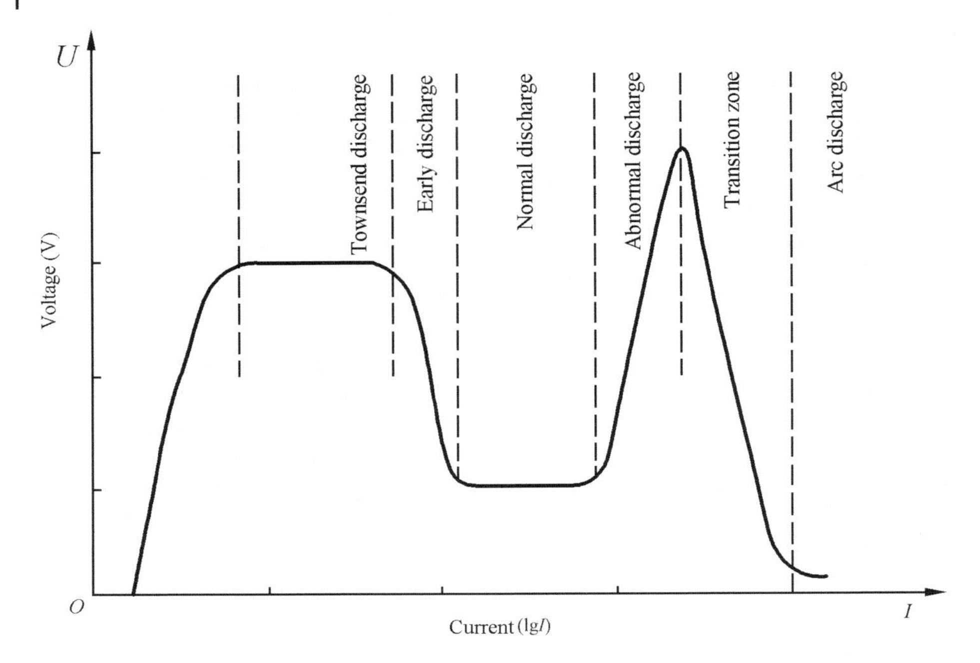

Figure 4.30 I-V characteristic curve of glow discharge system.

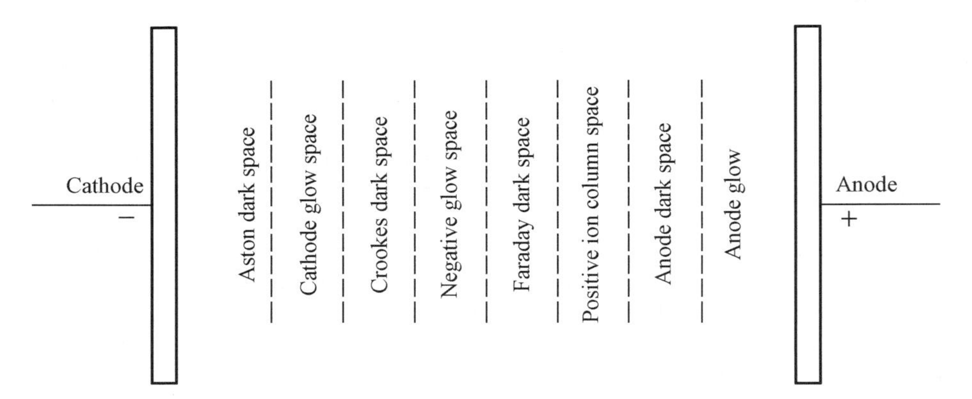

Figure 4.31 Glow space diagramme of glow discharge.

dark space, positive ion column space, anode dark space and anode glow, as shown in Figure 4.31. In these zones, only the positive ion column space shines, and the electrons and positive ions in this zone can basically meet the neutral conditions and they are in the plasma state. If the electrode spacing is properly adjusted, the plasma zone (i.e., the positive ion column space) can account for most part of the spacing between electrodes.

During glow discharge, the plasma temperature, the electron temperature and concentration are key factors, especially the electron temperature. The process of plasma generation by glow discharge is an unbalanced state. Although the reaction gas is only 100–500 °C, the electron in the plasma can reach 10^4–10^5 K after field acceleration, and the electron energy is about 1–10 eV, and the electron concentration can reach 10^9–10^{12} cm^{-3}.

The electron temperature, the most important physical quantity in the glow discharge, is mainly dependent on gas pressure and the power consumed, which can be expressed as (Formula 4.1):

$$Te = CE/K^{1/2}p \tag{4.1}$$

Where, C is a constant, E is the field, p is the gas pressure, K is the loss coefficient of the loss energy due to collision and it is the function of E/p.

In practise, it can be divided into DC glow discharge, low-frequency glow discharge, radio frequency (RF) glow discharge, very high frequency (VHF) glow discharge, and microwave glow discharge by power and frequency. Based on the electrode types, the glow discharge system can be divided into external coupling inductance, external coupling capacitance, internal coupling plate capacitance and external magnetic field and so on. And the common equipment is RF capacitive type. In the process of TFSSC production, the RF glow discharge and VHF glow discharge are generally used. In the process of CSSC production, the RF glow discharge and microwave glow discharge are often used.

4.7.2 Mechanism on a-Si Thin Film Grown by PECVD Method

4.7.2.1 Basic Principles of PECVD

The technology that uses glow discharge to generate plasma and that the thin film is deposited by chemical reaction is called as 'plasma enhance chemical vapour deposition (PECVD)'. Figure 4.32 shows the structural diagramme of RF PECVD system where the sample is directly participated in building of plasma field.

The RF PECVD reaction chamber is installed with cathode and anode. The reaction gas (silane) and the carrying gas (hydrogen) shall enter from one end of the chamber and generate plasma between two electrodes, which will enhance the chemical reaction, and the silicon atom generated will deposit on the heated substrate, forming the a-Si film. And the by-product gas generated will flow out of the reaction chamber with the carrying gas.

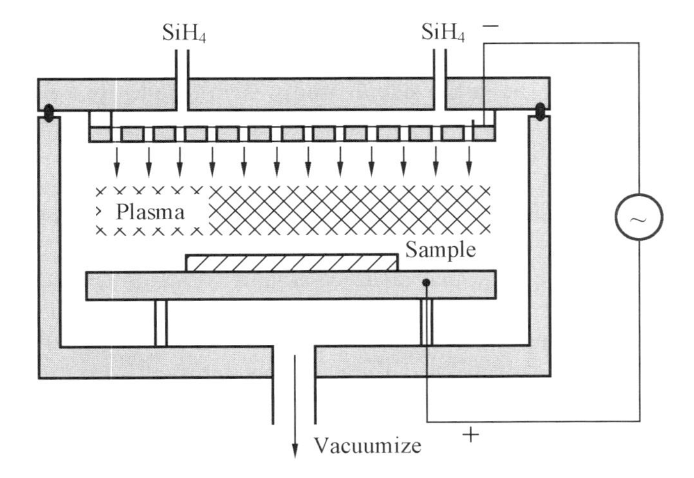

Figure 4.32 Structural diagramme of RF PECVD system.

The PECVD film generation process has the following main parameters: gas flow, gas component, gas pressure, power frequency, power of the power supply, spacing between electrodes, substrate temperature and electrode structure etc. When the substrate temperature is high, the reaction rate is generally high, and the deposition rate is mainly controlled by gas flow; the a-Si film generation temperature is much lower than that of the monocrystalline silicon integrated circuit, often below 300 °C. As a result, the variation of the substrate temperature plays a significant role in the film generation rate. For the TFSC with large substrate area, the plasma distribution evenness and gas flow distribution evenness between electrodes in the reaction chamber will, similar to the substrate temperature evenness, exert direct influence over the thickness evenness and quality of the film generated.

Since some of the process gases used by the PECVD method are toxic, or even hypertoxic, and flammable and explosive, the complete gas monitoring and alarm system must be provided. Generally, the gas cabinet shall be far away from the working area and the waste gas treatment system must conform to the requirements of the national environmental protection.

4.7.2.2 a-Si:H (a-Si-Containing Hydrogen) Deposited by PECVD Method

The a-Si thin film prepared by physical vapour deposition (PVD) method (e.g., sputtering and heat evaporation) contains excessively high 'hole' density and dangling bond density, it cannot meet the requirements on solar materials. The silicon dangling bond defects will result in pinning of Fermi level and it is difficult to form P/N types via doping. Accordingly, it is generally not used to prepare the silicon thin film by the PVD method in the solar cell sector.

The a-Si thin film prepared by the silane decomposed by glow discharge contains a certain amount of hydrogen, which can, to some extent, make up for the dangling bonds. As a result, it has become the main deposition technique for a-Sithin film of solar cells.

The profound research shows, the a-Si deposition process by the PECVD method is not a simple heat decomposition of silane where the silane is decomposed under the action of the field. It may generate Si, SiH, SiH_2, SiH_3, H, H_2 groups as well as a few $Si_mH_n{}^+$ (n, m > 1) ion groups. The groups have large differences concerning their influence over the a-Si thin film growth. Generally, SiH_2, SiH_3 are considered as the most important reaction groups. While silane is decomposed, some hydrogen atoms will also enter the a-Si thin film in addition to the silicon atoms. As a result, the a-Si thin film prepared by silane is generally the a-Si containing hydrogen, which is called as a-Si:H for short. SiH, SiH_2, $(SiH_2)n$, SiH_3 and other groups can detect the vibration absorption mode of the Si-H bonds and the infrared waves by means of the infrared spectrum, and work out the hydrogen concentration in a-Si by means of integration of infrared spectrum absorption peak.

In practise, the reaction chamber will be first pre-vacuumised and the silane diluted by hydrogen or argon shall be injected. The gas flow can be adjusted to make the pressure of the reaction chamber in a range of 13.3–1333.3 Pa. And then, voltage shall be applied between the positive/negative electrodes. The electrons, emitted from the cathode, will gain energy in the field and collide with the gas molecules or atoms in the reaction chamber to make them decompose, excite or ionised. In this case, the electron concentration can reach 10^9–10^{12} cm^{-3}, and the number of positive/negative charges will be equal, forming the plasma. The atoms finally decomposed will deposit on the substrate to form a-Si thin film.

It is the various possible chemical reactions that make the a-Si thin film very sensitive to the preparation conditions. And the equipment of different types shall have their own optimal process to produce quality a-Si thin film. The experiments prove, there are many factors influencing the thin film growth rate, thickness and quality where the most important are silane concentration (i.e., proportion with hydrogen), gas flow, gas pressure, substrate temperature, heating power and the temperature field in the reaction chamber and so on. Generally, the following conditions are suited to prepare a-Si thin film: silane concentration: about 4–10%; flow: 50–200 ml min^{-1}; substrate temperature: 200–300 °C; power: 300–500 W m^{-2}. If possible, the a-Si deposition temperature shall be minimised to save energy, reduce cost and achieve good quality. If it is larger than 500 °C, the hydrogen will escape from a-Si, losing the hydrogen passivation capacity. At the same time, temperature is the key factor for a-Si band tail structure and defect density. The experiment shows, the bond tail state and the defect density are minimal when the a-Si thin film is prepared at about 250 °C.

4.7.2.3 Growth Mechanism

Let's take a further look at the growth mechanism of a-Si thin film. Generally, the growth consists of three steps: First, the silane in the nonequilibrium plasma will decompose to generate active groups; second, the active groups will diffuse and react on the substrate surface, and the reaction layer will finally turn to a-Si thin film. The study points out, when the a-Si thin film is deposited at the temperature below 300 °C, the a-Si surface will be always covered with hydrogen atoms, which can passivate the silicon dangling bonds; during reaction, the SiH_3 group will be absorbed on the a-Si:H surface and then diffuse to the inside; finally, the hydrogen in the hydrogen-rich layer shall be removed to form the a-Si network structure. Obviously, in case of low-temperature deposition, the removal of hydrogen is the key factor to form the Si-Si weak bonds and the a-Si disordered structure. Figure 4.33 shows the diagramme of a-Si thin film grown via active group SiH_3.

Generally, the a-Si:H surface is passivated by hydrogen atoms, and the surface hydrogen concentration is 50–60% while the hydrogen concentration in the a-Si thin film is low, only 5–20%. As a result, it must remove (resolve) the surplus hydrogen atoms on the surface during growth of a-Si thin film so as to form the a-Si of the Si-Si bond.

Due to disorder, there are dangling bonds, stress and micro-hole and other structural defects in a-Si, and the band tail is present in the energy band gap. The band tail and structural defects behave as recombination centres and affect carrier transport, reducing the efficiency of the a-Si cells. When the a-Si cells are in the solar rays for a long time, it may result in photo-generated metastable state, which may raise the space charge density near the P-I interface, shrink the depletion zone, widen the low field or field-free zone in the intrinsic zone, and further increase recombination of the photo-generated carriers, resulting in 'early deterioration of a-Si cells'.

4.8 Physical Sputtering Principles and TCO and Back Metal Preparation System

4.8.1 Overview on TCO and Back Metal Deposited by Physical Sputtering

Because no hydrogen is present in the physical sputtering process, it cannot passivate the dangling bond defect in a-Si, and thus it cannot prepare the active layer of silicon thin

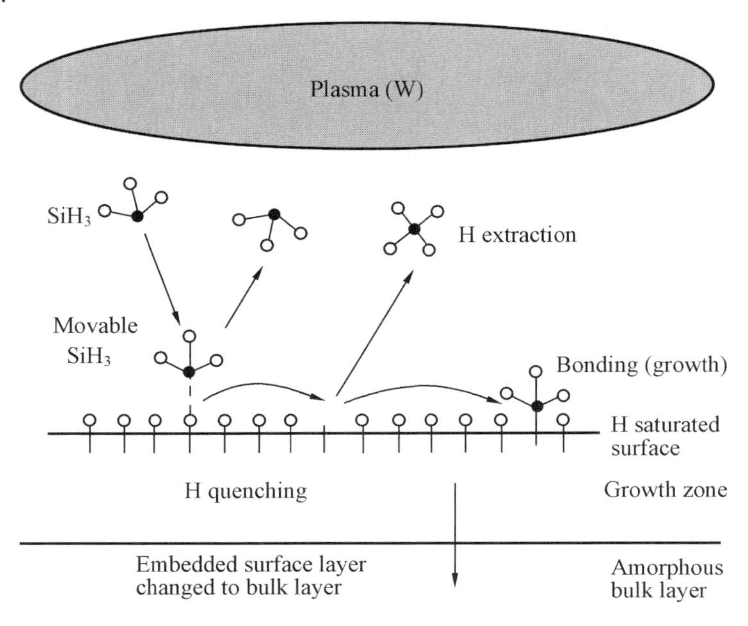

Figure 4.33 Diagramme of a-Si thin film grown via active group SiH_3.

film of the component quality. The physical sputtering, however, similar to the PECVD equipment, plays an important role in the preparation process of TFSSCs because both the surface TCO and the back metal contacts of TFSSCs are produced by the physical sputtering method. In addition, some compound TFSCs, which will be introduced in Chapter 5, are also prepared by the physical sputtering method. In consequence, it is necessary to describe the basic principles and equipment of physical sputtering.

The TCO and back contact of TFSCs are prepared by the physical sputtering and deposition process, that is, the high-energy argon ions are used to strike out the atoms in the solid target, and these atoms will penetrate the vacuum chamber and finally deposit on the substrate.

Physical sputtering has the following advantages:

① It can deposit the complex components of the original target on the substrate;
② It can deposit the metal of high fusion point and difficult to be fused;
③ It can deposit even thin film on the large area;
④ It can adopt multi-chamber integration equipment;
⑤ The pollutants on the substrate surface can be removed at the in situ before sputtering and deposition (in situ sputtering and etching).

Physical sputtering consists of six basic steps (as shown in Figure 4.34):

① It can generate positive argon ions in the plasma of the high vacuum chamber, and accelerate to the target with negative potential;
② During acceleration, the ion will gain energy and bombard the target;
③ During the physical process, the ions will strike out (sputter) the atoms from the target. The target has the components required by the thin film material;
④ The atoms bombarded will transfer to the substrate surface;

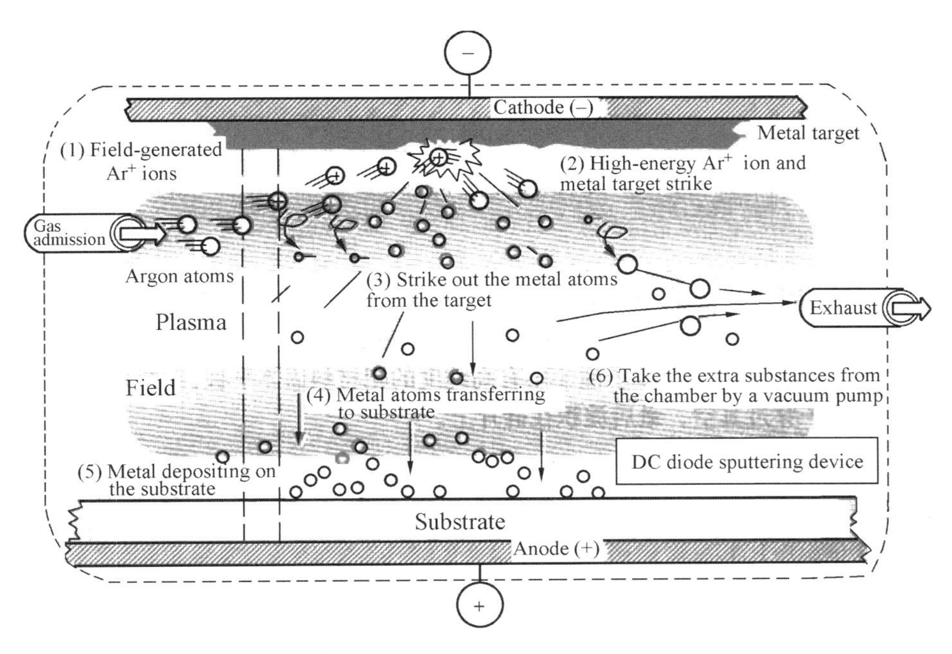

Figure 4.34 Six basic steps of physical sputtering.

⑤ The atoms sputtered will condensate on the substrate surface and form the thin film, and the thin film has the material components similar to the target;
⑥ The extra materials will be taken away by the vacuum pump.

Sputtering generally consists of DC sputtering, RF sputtering, magnetically controlled (MC) sputtering and reaction sputtering.

1. In DC sputtering, the DC voltage shall be applied between the cathode and the substrate. The DC sputtering equipment is simple, and it has high sputtering rate and fewer damages to the substrate. Since the DC sputtering voltage is up to several hundred voltages, the substrate shall have good thermal performance. And the DC cannot sputter on the insulation film.
2. The private frequency of RF sputtering is often 13.56 MHz. In the action of the alternating field, the electrons in the gas will swing with field and the gas will be ionised to plasma. During RF sputtering, the electrons move at a higher speed than the ions and more electrons will be sputtered on the target than the ions do in one cycle. As a result, after some time, the bias will be formed on the target due to accumulation of electron charges. The equilibrium built by the bias bombarding on the target is called as self bias. And sputtering is generated at the equilibrium voltage. In the two electrodes of RF sputtering, the one with substrate will be connected with the enclosure and its potential is close to the plasma and it is almost free from ion bombardment; and the other electrode is provided with the target, and it is the anode with respect to the ion potential (–), and it is under bombardment of ions. The RF can sputter the insulation film.
3. In MC sputtering, one magnetic field will be built in the anode target, which can make the electrons move on the anode surface in a spiral manner. It prolongs the

time of the electrons staying near the target and thus it can ensure the number of collisions between the electrons and the gas and fully resolve the gas into plasma. It can still maintain continuous discharge even in case of low gas molecule density at high vacuum. As a result, MC sputtering can achieve high sputtering rate at low process gas pressure, reduce the impact of gas on thin film quality. Moreover, the thin film has good evenness and high reproductivity. As a result, it has been widely used to the production sector. The target utilisation, however, is low.

4. Reaction sputtering shall add reaction gases to obtain compound thin film via sputtering, that is, the reaction gas added during thin film forming will react with the target. Here no detailed introduction will be provided.

For the gas used for sputtering thin film, first, it shall not react with the target, and in this sense, the most appropriate shall be the inert gases; second, it shall have high sputtering rate. Since the sputtering rate is related to the molecular weight of the inert gases and the helium atom is too light and generally not used. Xenon and krypton have the best sputtering rate but they are too expensive and difficult to ionise and thus argon is generally used as the gas for sputtering thin film.

The back supporting plate of the sputtering target is generally made of oxygen-free copper. The target shall be securely welded with the oxygen-free copper so that it will not fall off during sputtering. After welding, the target shall have small warpage, and the welding flux cannot penetrate to the target surface. The distance from the target surface to the sputtered substrate surface (called as the sputtering distance) is usually designed by the manufacturer, and the process technicians cannot deliberately modify it because it is related to the thin film evenness and quality. The sputtering distance set by the manufacturer generally falls in a range of 110–120 mm.

4.8.2 A Simple Parallel Metal DC Diode Sputtering System

Figure 4.35 shows a simple parallel metal DC diode sputtering system. The deposition chamber consists of the solid target, the substrate (silicon wafer) and the vacuum environment. The target earth is the cathode, and the substrate has the positive potential and thus it serves as the anode. The requirements on target production include even components, appropriate particle size and crystallisation orientation, and all of them are aimed to achieve even thin film deposition rate on the whole substrate. The purity of the target is generally larger than 5N.

The sputtering system can be provided with multiple chambers, amongst which, double vacuum locks shall be provided to move the substrate by steps from the atmosphere to the background vacuum with minimum pressure. The multi-chamber automatic mechanical transport system with special designs can move the substrate amongst the process chambers. The automatic mechanical transport of the multi-chamber integration system is realised via magnetic coupling so that the drive motor can be installed outside the process chamber to reduce the foreign particles from the chamber.

The vacuum in the sputtering chamber is of great importance. Generally, the initial vacuum is 10^{-7} torr and then the argon shall be admitted to the process chamber and sputtering will start when the pressure in the chamber rises up to 10^{-3} torr.

The argon ion, relatively heavy, is an inert gas. The live argon ions will speed up and achieve adequate dynamic energy when passing the glow discharge dark zone with voltage drop. The energy range of the typical sputtering argon ion is 500–5000 electron Volt.

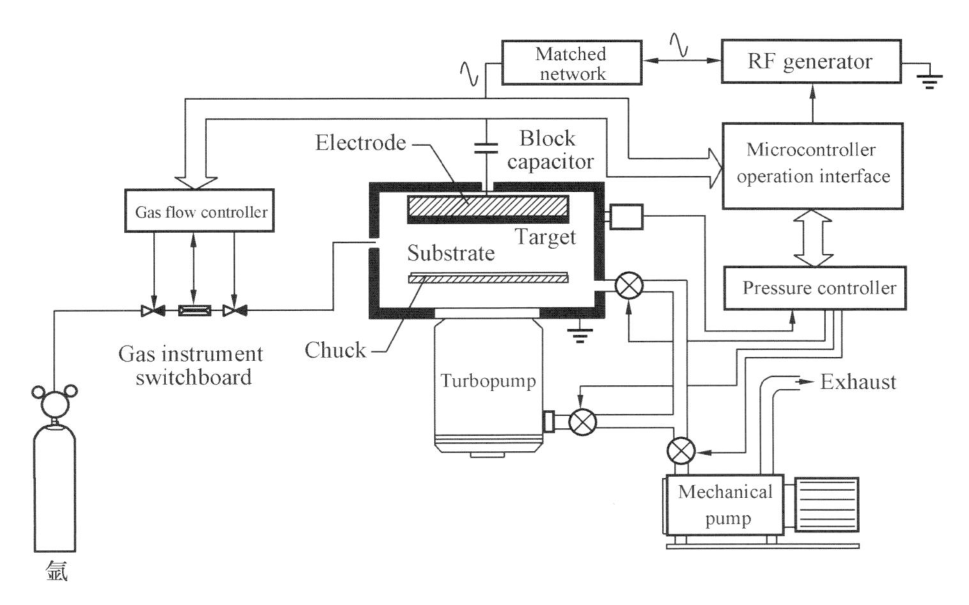

Figure 4.35 A simple parallel metal DC diode sputtering system.

After argon ions strike on the target, one or more target atoms can be stricken out from the target due to dynamic energy transfer. The sputtering yield, which is defined as the number of target atoms stricken by one argon ion, generally falls in a range of 0.5–1.5. The sputtering yield determines to a great extent the rate of the sputtering deposition. The sputtering yield has something to do with the following factors: ① incidence angle of the bombardment ion; ② target component and geometric factors; ③ quality of the bombardment ion; and ④ energy of the bombardment ion.

Since the target will be heated during sputtering, it shall be cooled down to keep low target temperature. When the target is consumed by 50–70% or so, it shall be replaced. In addition, since the atoms sputtered from the target will diffuse in the chamber and some of them will inevitably fall on the chamber wall, it is necessary to clean the chamber on a regular basis.

The above simple diode sputtering system has limited application to the manufacture industry because the positive charges will gather on the cathode target, and once the cathode electrode is covered by the media, the glow discharge cannot be maintained. In addition, the simple diode sputtering system cannot be applied to in situ sputtering and etching because its process is opposite to the sputtering deposition process and the substrate, instead of the target, will be sputtered. In this case, the argon atoms will be used for cleaning the substrate, the polluted chamber as well as the etching remains.

4.8.3 RF and MC Sputtering Systems

In the RF sputtering system, the plasma is generated by the RF field. The RF is generally 13.56 GMz, which shall be applied to the target electrode back and coupled via the capacitor (see Figure 4.36). Both the electrons and the ions of the plasma are in the RF field, a lot of electron flow will be generated due to high frequency, resulting in negative charges depositing on the target electrode. When the argon ion sputters the target,

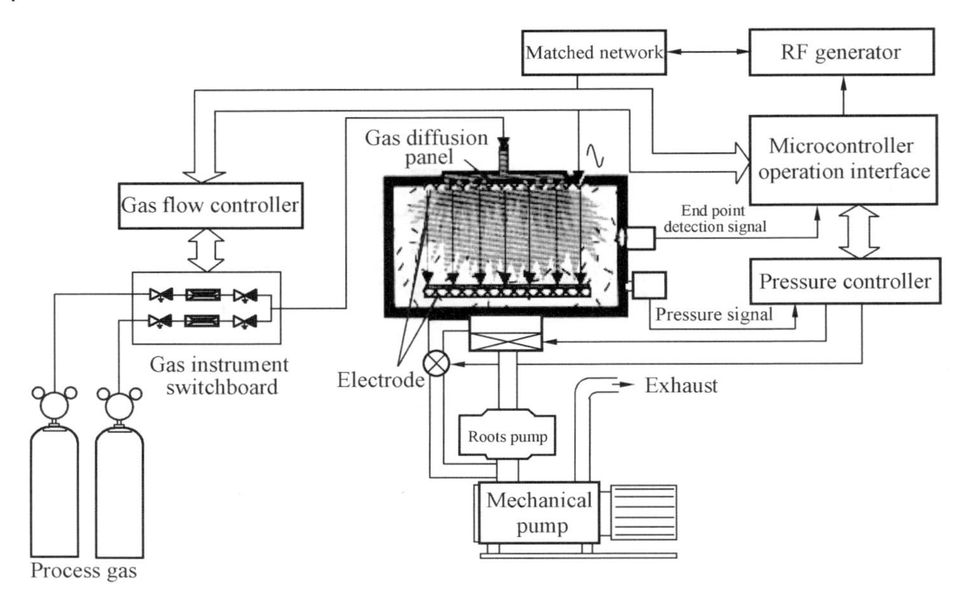

Figure 4.36 RF sputtering system.

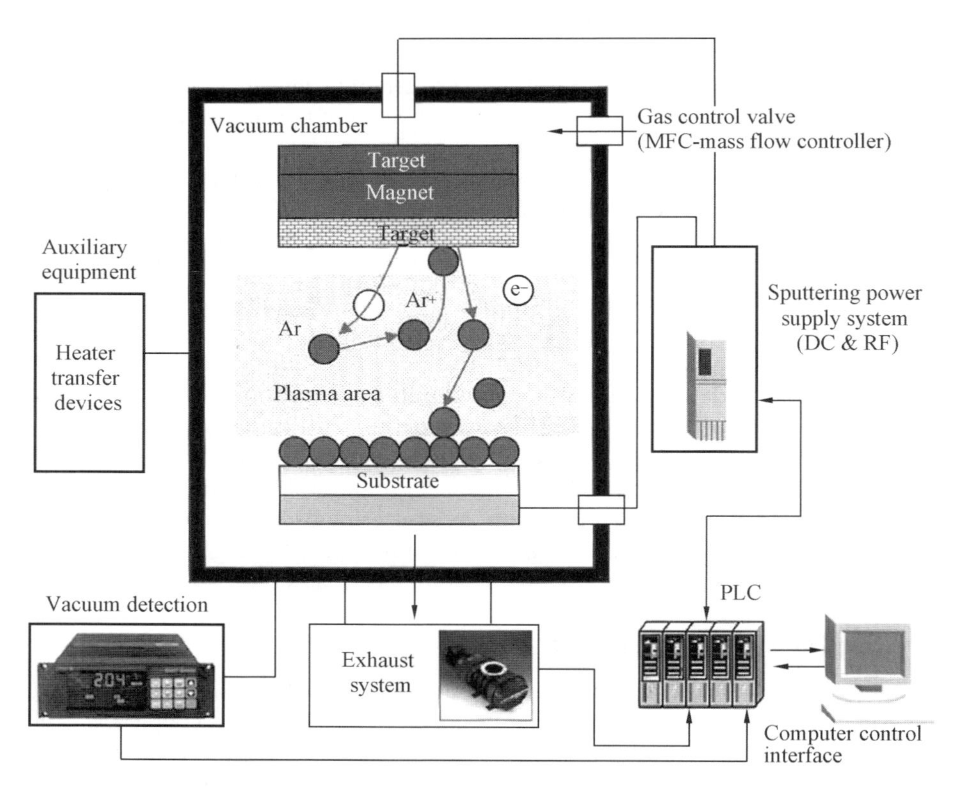

Figure 4.37 MC sputtering system.

these negative charges (generated by self bias) will attract some positive argon ions, resulting in some energy consumption in the collision with electrons. These electrons have no contributions to plasma generation, and thus they affect the sputtering rate of the RF sputtering system.

In MC sputtering, the permanent magnets are installed around and on the back of the target (as shown in Figure 4.37) to capture and restrain the electrons near the target, which increases the bombardment rate of ions on the target and generates more secondary ions, raising the ion rate in the plasma. And the result is that more ions produce more sputtering on the target, improving the system deposition rate.

Since the simple RF sputtering system has low sputtering efficiency, its application to the manufacture industry has been limited. The MC-based sputtering system is the currently mostly used one.

5

High-Efficiency Silicon Solar Cells and Non-Silicon-Based New Solar Cells

The past 20 years have witnessed great development in solar cells. Thanks to the new processes, new equipment and new design concepts, several kinds of high efficient silicon-based solar cells have been developed, and some non-silicon-based compound solar cells have also been widely used, and the next-generation promising solar cells have also emerged. In this chapter, we will introduce some representative efficient silicon solar cells, including the selective emitter ones, the passivation emitter ones, the interdigitated back contact (IBC) ones, the passivated emitter and rear locally diffused (PERL) contact ones as well as the crystalline silicon/amorphous silicon heterojunction (heterojunction with intrinsic thin-layer, HIT) ones and the like. As for the non-silicon-based compound solar cells, the focus will be III–V compound crystalline ones and II–VI compound thin film ones. And for the next-generation promising solar cells, we will introduce the organic ones, the dye-sensitised ones, the perovskite ones, the concentrator ones and the multiple quantum well (MQW) ones and so on.

5.1 High-Efficiency Crystalline Silicon Solar Cells (CSSCs)

In Chapter 3, some methods are introduced to improve the conversion efficiency of CSSCs, including:

1) Adopt the light trapping structure: The light irradiating surface is usually processed by the chemical texturing technique, and the reflectivity of the textured surface can be less than 10%. If the advanced reactive ion etching (RIE) or the laser texturing technology is used, the reflectivity of the textured surface can be less than 4% for the polycrystalline or monocrystalline silicon wafers, and it can be used to the thin wafers;
2) Produce the anti-reflection (AR) film: It is based on the principle that the film of certain thickness and refractivity will grow on the irradiating surface to make the reflections at all levels generated by the incident light interfere and offset with each other. To coat the AR film on the textured wafer surface can further reduce the reflectivity to below 2%;
3) Produce the passivation layer: The passivation process, including thermal oxidation passivation, atomic hydrogen passivation, phosphorus, boron, aluminum surface diffusion and so on, can reduce effectively recombination of photo-generated carriers

Technology, Manufacturing and Grid Connection of Photovoltaic Solar Cells, First Edition. Guangyu Wang.

in some areas. Take thermal oxidation passivation as an example, the silicon oxide film can be formed on the front and back of the cell, which can prevent effectively recombination of carriers on the cell surface. The silicon dangling bonds can be neutralised by the atomic hydrogen, reducing the recombination;

4) Increase the back surface field (BSF): For example, add a P^+ highly doped layer to the back of the cell made of P-type materials to establish P/P^+ structure. In this case, a built-in field pointing from P zone to P^+ will be generated. And the accumulation of the photo-generated carriers separated by the built-in field will increase the photo-generated voltage that has the same polarity with the original P/N junction photo-generated voltage, and further raise the open-circuit voltage Voc. At the same time, the presence of BSF can accelerate the photo-generated carriers, which is equivalent to increase the effective diffusion length of carriers, and thus increase the collection possibility of the minority carriers, raising the short-circuit current Jsc;
5) Improve the substrate materials and adopt quality wafers, for example, N silicon, which features such virtues as long carrier lifetime, few boron-oxygen complex, and small efficiency degradation and so on;
6) Design new structures of cells: For example, shallow junctions, selective emitters, and HIT of Sanyo, and the IBC electrodes of Sunpower, and so on.

It is gradually recognised in PV generation applications that if the PV conversion efficiency of the solar cell increases by 1%, the ultimate generation cost of the PV system can be reduced by 7–8%. In consequence, the above methods for improvement of cell efficiency have been fully applied to the PV industry and the scientific sector. It must be pointed out, however, that what the users are really concerned about is that how much it costs to generate 1 kWh, and how long the lifetime of the power generation system is. Accordingly, some cells may be not suitable for general commercial applications due to too complicated constructions and high process cost.

5.1.1 Selective Emitters and Buried Contact Silicon Solar Cells

As known, if the phosphorus diffusion concentration is increased in the N-type window of P-type silicon solar cells, this can promote good ohmic contact between the metal gate electrodes of the irradiating surface and the cells, however it will also cause it lower light absorption rate and higher defect density, resulting in decrease of the cell open-circuit voltage and short-circuit current. Although the low phosphorus diffusion concentration can improve the cell photo response, it may raise the contact resistance and reduce the filling factor. The two processes—the phosphorus diffusion and the metal gate electrode making—are mutually restrained. In the solar cell process with selective emitter junctions, the 'metal gate electrode contact area' and the 'light absorption area' can be independently diffused, which can effectively settle the conflict. Or in details, the high concentration diffusion can be made in the 'metal gate electrode contact area' to obtain good ohmic contact between the metal electrodes and the cell pieces, and the shallow diffusion can be made in the 'light absorption area', which is good for light absorption (Figure 5.1). Since the two areas have different sheet resistances, which can simultaneously achieve optimisation amongst the open-circuit voltage, the short-circuit current and the filling factor.

The position of the metal gate on the irradiating surface can be etched into gate slots by mechanical methods or laser, and the metal gate can be embedded into the narrow slots

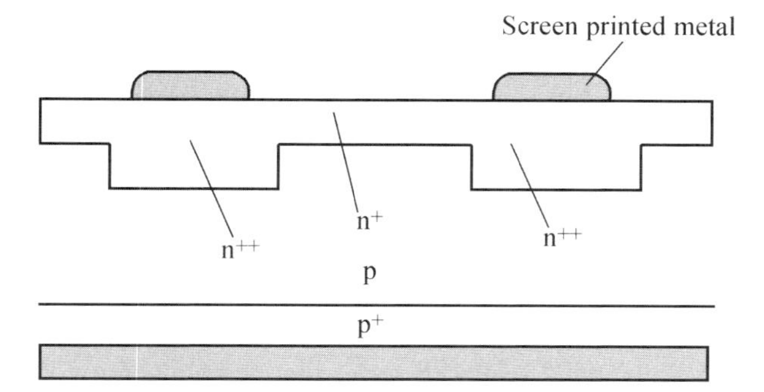

Figure 5.1 Different diffusion concentrations of the 'metal gate electrode contact area' and the 'light absorption area'.

to increase the height-width ratio of the gate electrodes and consequently reduce the shading area of the cell surface. The selective high-concentration phosphorus diffusion can be carried out in the gate slots, which can reduce the contact resistance between the metal gate and the cell and further increase the open-circuit voltage, effectively improving the cell filling factor. This is the basic principle on how to make the solar cells of buried contact selective emitter junction.

The minimum slot width can be approximately 40 μm by mechanical etching, and it can be 15–20 μm by laser etching. The principle of laser etching is to volatilise and remove the substances in the light area by means of the laser pulse energy, achieve selective etching and significantly raise the gate electrode thickness, reducing the contact resistance. Besides, selective doping can be realised by means of laser, that is, the gate slot can be melted by the thermal effect produced by the laser beam. The silicon beneath the gate slot in the liquid state is good for doping of phosphorus impurities of the diffusion layer and realise re-doping beneath the metal gate line (larger than the impurity concentration in the light absorption area), reducing the contact resistance of the emitter. This type of solar cells is also called as laser grooved buried contact (LGBC) cells. Figure 5.2 shows its structure.

Figure 5.2 Structure of LGBC cells.

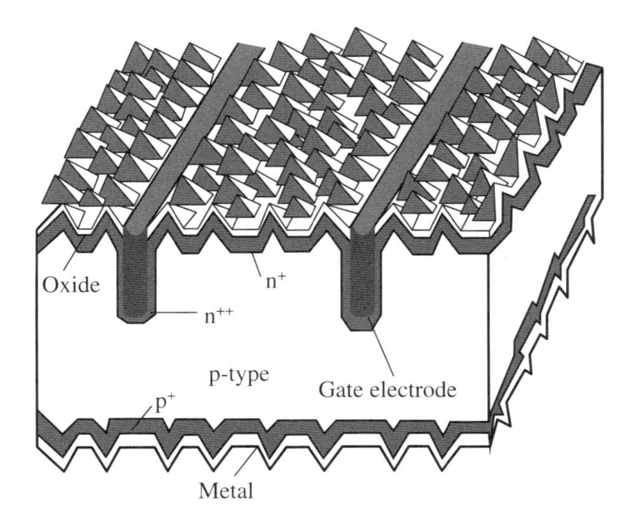

The metal emitter of the buried contact cells can be embedded into the narrow slots by means of screen printing and chemical plating methods, and the chemical plating method can achieve better metal contact effect. For the buried contact cell, the height-width ratio of the emitter can be up to 5:1, which significantly reduces the shading area of the metal gate and increases the irradiating area of the cells; and it can realise heavy-doping of the emitter area and reduce the contact resistance. As a result, the buried contact cell can significantly improve the performance of the solar cell.

5.1.2 Passivation Emitter Silicon Solar Cells

PESC, PERC, PERL solar cells are the advanced solar cell series that the University of New South Wales has studied for nearly 20 years. The first two letters 'PE' (passivated emitter) stand for passivation of the irradiating surface emitter and the latter two letters for diffusion and contact of the rear surface.

The PERL (passivated emitter and rear locally diffused) cells introduced in the section are short for 'passivated emitter and rear locally diffused solar cells'. In 1990, J. ZHAO from the University of New South Wales, based on the structure and technique of PERC back contact cells, adopted BBr_3 localised diffusion at the contact holes on the cell rear and produced the PERL cells, as shown in Figure 5.3. In 2001, the efficiency of PERL cells reached 24.7%, approximating to the theoretical value, which is the highest record in the P-type crystalline silicon solar cells.

The PERL cell has high efficiency because:

(1) Surface anti-reflection: The front of the cell is designed in a shape of an 'inverted pyramid', which has better absorption effect than the ordinary textured structure and has very low reflectivity, improving the cell J_{SC};
(2) Selective phosphorous diffusion and embedded gate: The phosphorous with high and low concentrations will be diffused in different zones; the phosphorous diffusion with high concentration beneath the finger electrode can reduce the contact resistance of the finger electrode; and the phosphorous diffusion with low concentration can meet the requirements on small transversal resistance power consumption and good short-wave response;

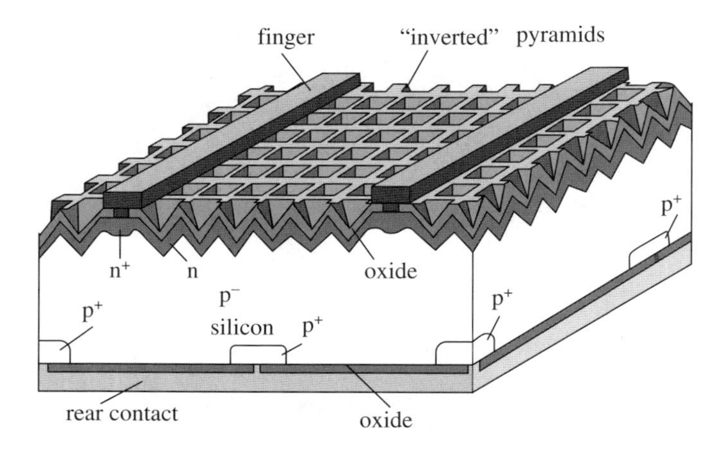

Figure 5.3 P-type PERL high efficient cells with passivated emitter, boron diffusion on the rear.

(3) Localisation on the rear, and boron diffusion P^+ zone of small area: It can reduce the contact resistance of the back electrode and increase the boron BSF. The steam aluminum back electrodes themselves are also good back reflectors, which can further improve the cell conversion efficiency;
(4) Double-surface passivation: Surface passivation of the emitter can reduce the surface state and recombination of the minority carriers on the cell front. The rear passivation can reduce the opposite saturation current density and improve the spectral response.

The production process of the cell is very complicated, which involves the preparation of the 'inverted pyramid' structure, dense phosphorous diffusion of the finger electrode, the light phosphorous diffusion on the front, SiO_2 passivation layer, back localised boron diffusion, the laser groove back electrode contact hole, the laser groove front finger electrode lead hole, front evaporation finger electrode, rear evaporated aluminum electrode, front silver plating and other procedures. It needs several laser groove processes, and thus its cost is not low.

The passivated emitter can be also applied to the N-type silicon solar cells. It is known that the lifetime of the N-type carriers is much longer than the P-type silicon, and the diffusion length of the carriers in the N-type silicon is usually larger than the thickness of the silicon wafer, and the carriers will have significantly less recombination in the wafer, and most of the carriers can pass through the wafer and be collected by the positive/negative electrodes. To further improve the PV conversion efficiency of the solar cells, it must reduce the recombination of the carriers at the electrodes of both the front and the rear. In this case, the electrodes on both sides shall be passivated. In the following N-PERT high efficiency solar cells, the front irradiating surface is provided with $SiNx/SiO_2$ double-layer passivated film protection, and the rear shady surface is designed with SiO_2 passivated film protection. Figure 5.4 shows the N-PERL solar cells (N-passivated emitter and rear locally diffused).

In the recent years, the production techniques of N-type silicon crystalline have been improved and the front/rear passivation technologies of the solar cell have been

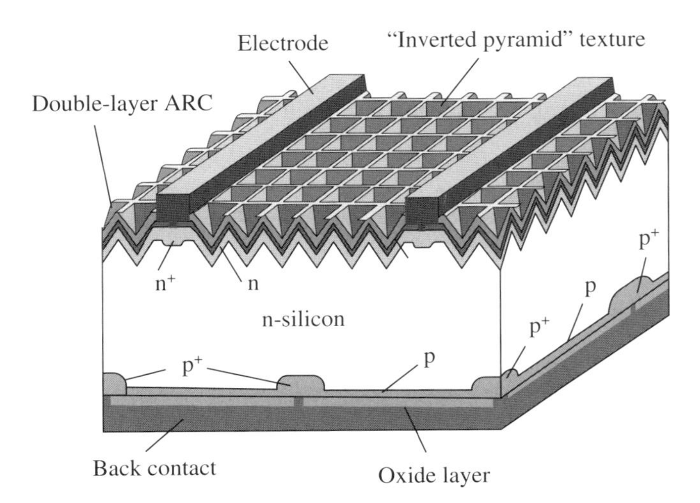

Figure 5.4 N-type PERL high efficient cells with passivated emitter and rear boron diffusion.

developed so that the N-type crystalline silicon solar cells have achieved profound development. The N-type crystalline silicon solar cells adopt boron diffusion, instead of phosphorous diffusion, on the cell rear to produce the P/N junction. Since the cell rear is protected by silicon oxide passivation and it shall penetrate the silicon oxide passivation layer and realise local contact with the substrate to reduce ohmic contact resistance. The silicon wafer used by the cells is very thin and the wafer with higher resistivity can be used.

5.1.3 Back Finger Electrodes and Boron Diffusion N-Type Crystalline Silicon (IBC) Solar Cells

The back finger electrodes and boron diffusion N-type crystalline silicon solar cells are also known as the IBC (interdigitated back-contact) cells, as shown in Figure 5.5. It is developed by Sunpower and it has the following characteristics:

1) The positive/negative electrodes are all arranged on the cell back, and the front is not provided with any metal gate electrodes. The irradiating surface of the cell shall receive the solar irradiation with the maximum area;
2) On the front, the SiO_2/SiNx composite film is combined with N^+ layer as the front surface field (FSF), and textured to reduce reflection;
3) Since the photo-generated carriers shall penetrate the whole cell and be collected by the P/N junction on the rear surface, the IBC cell is often provided with the N-type monocrystalline with longer lifetime of carriers as the substrate;
4) On the rear, the P^+ and N^+ are crossed by diffusion method, and metal contact holes are made on the SiO_2 to realise contact between electrodes and the emission zone or base zone;
5) The emitters (P-type gate zone) and the base electrodes (N-type gate zone) almost cover most of the rear surface and form crossed fingers. The metallisation of the rear can work as anti-reflection and offer low series resistance.

For IBC solar cells, the carrier diffusion length is larger than the cell thickness (e.g., 200 μm). To improve the conversion efficiency, the carrier recombination on the front/rear surface must be minimised. As a result, it is the key of the crystalline silicon solar cells to deposit silicon oxide film beneath the anti-reflection coating of the irradiating surface and on the back metal. The physical condition of the semiconductor

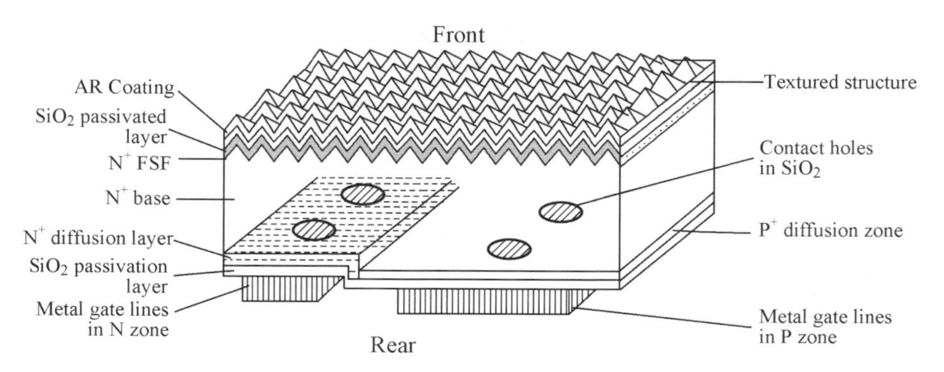

Figure 5.5 IBC solar cells, SunPower.

devices shows, the thermally oxidised Si/SiO_2 interface has very low defect density, which can effectively reduce recombination of carriers on the surface. Accordingly, the high-temperature oxidisation process is often used to prepare the SiO_2 passivation layer.

Because the thermally oxidised SiO_2 has high temperature and long oxidisation time, it will significantly reduce the carrier lifetime in low resistivity Si. As a result, the SiO_2 passivation film is mainly used to the silicon wafer with high resistivity and slightly doped with N type.

To make an ohmic contact on a lightly doped substrate, many small contact holes are made on the cell rear (Figure 5.3). At the contact holes, BBr_3 is used for localised boron doping and the back electrode will pass through the SiO_2 passivation film and realise local point contact with the substrate. It shall be noted that the spacing between these contact holes shall not be excessively large; otherwise, the carriers shall pass a long transverse distance to reach the contact holes and the transverse resistance will result in reduction of the filling factor. The total area of the contact holes shall be not too large because the holes cannot be passivated.

The cell has high efficiency because:

1. Double-surface passivation: The irradiating surface can be passivated, which can reduce minority carriers recombination through the surface state on the front surface; and the back electrodes can be passivated, which can reduce the reverse saturated current density and improve the spectral response;
2. The irradiating surface is designed in the shape of an 'inverted pyramid', which has better light trapping effect than the ordinary textured structure;
3. The localised boron diffusion can reduce the contact resistance of the back electrodes and improve the boron BSF.

The cells have a complicated production process, and the difficulties include P^+ diffusion, re-diffusion beneath the metal electrodes and laser sintering and the like. Moreover, for the IBC solar cells, the positive/negative electrodes shall be distributed on the cell rear in a crossed manner, it needs laser etching and thus the cost is high. The cell is often used to the application with high density incidence photon flows, for example, the concentrator solar cells.

All in all, for the high efficiency cells, special attention shall be paid in the production process to passivation of the surface dangling bond, including passivation of the contact zone of the back metal, and it shall, with all efforts, avoid impurities and other defects in the body and on the surface.

5.1.4 Metallisation Wrap-Through (MWT) Silicon Solar Cells

The MWT cell is short for 'metallisation wrap-through (MWT) silicon solar cell'. The MWT technique is developed by Solland Solar, the largest solar cell manufacturer of Holland. The technique adopts the P-type polycrystalline silicon, and transfers the energy collected on the irradiating surface to the rear of the cell by means of laser drilling.

In the traditional silicon-based cells, the emitters and the base electrodes are made on the irradiating surface and the rear, respectively. Accordingly, the irradiating surface shall be partially shaded by the metal gate lines of the emitters, causing some light losses. The emitters of the MWT cell are connected to the rear of the cell via the induction function in the base. In this way, it forms the back contact structure where both the emitters and the base electrodes are located on the rear. As a result, the irradiating surface of the MWT cell is increased, which effectively increases the cell short-circuit current. Its main advantages lie in the assembly on the same surface and reduction of light losses on the irradiating surface as well as massive production. JA Solar Holdings Co., Ltd., China, has realised massive production. Figure 5.6 shows the principle and Figure 5.7 shows the front/rear of MWT cells.

In the MWT cells, the contact electrodes of the front emitters are connected through the silicon wafer to the back by laser drilling and grouting printing techniques, which directly reduces the shady surface of the main gate and connects the positive/negative electrodes with the base by the conducting glue while the MWT cell assemblies are packaged. It is convenient and safe, and is especially suited to the solar cells with super-thin silicon wafer. The technique can reduce the consumption of the front silver slurry, and it also avoids the risk of cell damages caused by welding the main gates of the irradiating surface in series by the wide and thick welding strip.

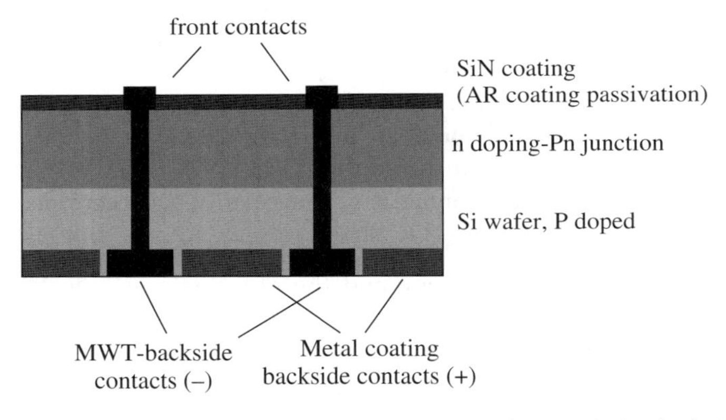

Figure 5.6 MWT cells: Emitter wrapped from the front to the backside (principle diagramme).

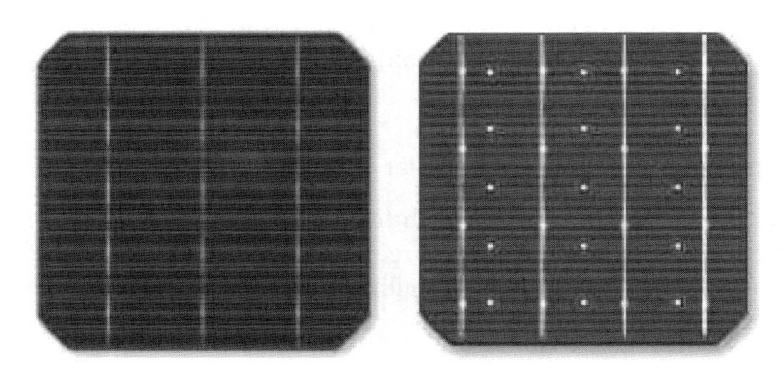

Figure 5.7 Front/backside of MWT cells.

Each silicon wafer, however, shall be drilled with about 200 through-holes during MWT production. The technical difficulties include how to drill the holes in an accurate and safe manner by laser and avoid leakage in and around the holes. In addition, attention shall also be paid to alignment and repeatability of laser drilling and grooving, size and shape of the holes, damages caused by laser and silicon substrate, filling of hole metal and the like.

It is predicted in ITRPV (SEMI, *International Technology Roadmap for Photovoltaic,* 5th edition), published by SEMI in 2014, that the development roadmap of the solar cell technologies and silicon materials in the upcoming 10 years, as shown in Figures 5.8 and 5.9. Obviously, the traditional BSF cells will gradually give way to the high efficiency cells such as the passivated film, IBC and HIT (see the next section) cells in the upcoming 10 years. At the same time, the proportion of the N-type silicon in the associated material will see dramatic increase.

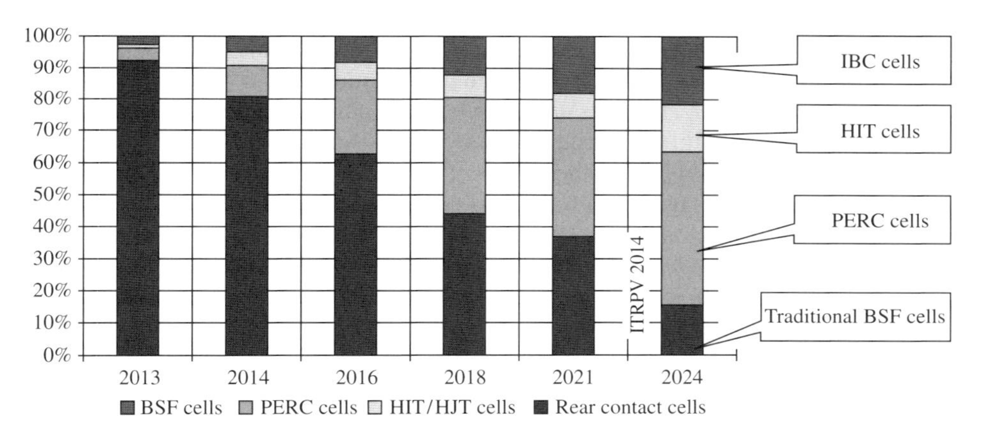

Figure 5.8 Development roadmap of the solar cell technologies in the upcoming 10 years.

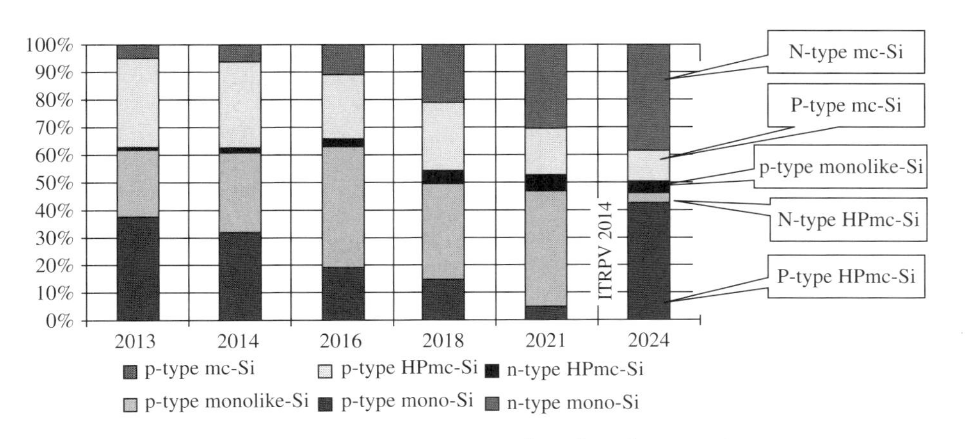

Figure 5.9 Estimated trend of silicon proportion in solar cells in the upcoming years.

5.2 Production Techniques of HIT High-Efficiency Solar Cells

5.2.1 a-Si/Monocrystalline Silicon Heterojunctions with Intrinsic Layer

In 1991, Sanyo proposed for the first time to fabricate the solar cell in the structure of a-Si:H (P-type layer)/a-Si:H (intrinsic layer)/C-Si (N-type layer) heterojunction with intrinsic thin-layer on the N-type crystalline silicon by the PECVD method, that is, an intrinsic layer a-Si:H (5–10 nm thick) shall be inserted between the P-type a-Si and the N-type monocrystalline silicon as a buffer. Since the buffer can passivate the silicon wafer surface and improve the interface characteristics, it can reduce recombination of carriers, enabling the solar cell with high open-circuit voltage and PV conversion efficiency.

5.2.2 Structure and Techniques of Double-Surface HIT Cells

The double-surface HIT cells are designed in the structure of Ag/TCO/a-Si(p)/a-Si(i)/c-Si(n)/a-Si(i)/a-Si(n)/TCO/Ag. It is combination of a-Si and monocrystalline silicon where i is the intrinsic a-Si. The HIT cell is shown in Figure 5.10.

The production steps of HIT cells are given below: First, the PECVD method can be used to deposit the very thin intrinsic a-Si:H layer and P-type a-Si:H layer on the front surface of the textured N-type CZ silicon wafer (200 μm or so); then, the intrinsic a-Si:H layer and N-type a-Si:H layer shall be deposited on the backside of the silicon wafer; finally, the sputtering technique can be used to deposit transparent conducting oxide coating (TCO) on both sides of the cell, and the silver electrodes can be produced on TCO. It is worth noting that all the steps can be carried out below 200 °C, which plays a significant role in excellent performance and energy saving.

During cell production, it is of great importance to control the thickness of the intrinsic a-Si in the HIT solar cell. On one hand, the open-circuit voltage increases with the intrinsic a-Si thickness; and on the other hand, the short-circuit current decreases with the intrinsic a-Si thickness. As a result, the intrinsic a-Si thickness shall be optimised to improve the inherent quality of the intrinsic a-Si layer.

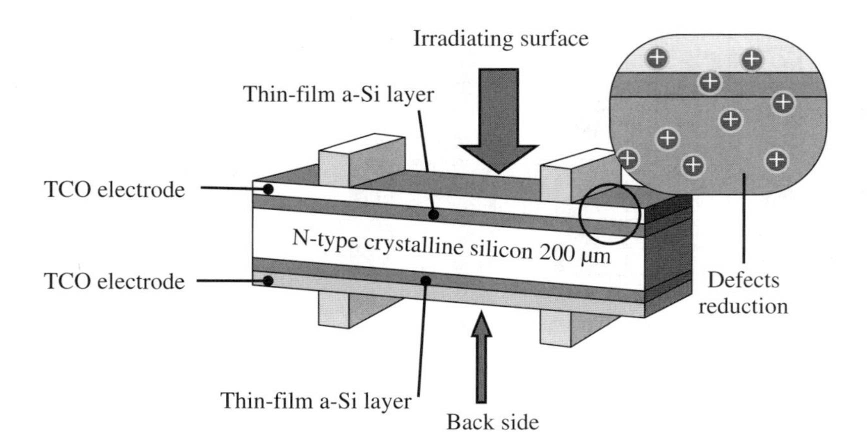

Figure 5.10 Structural diagramme of HIT solar cells.

5.2.3 Characteristics of HIT Cells

The HIT cells have the following characteristics:

1. The preparation techniques are simple;
2. Both sides can be provided with heterojunctions, which can absorb the light at any angle. In this case, the power generated can be raised by at least 10%, and even by 34% when vertically installed;
3. The electrodes on both sides are all TCOs; plated silver gate electrode;
4. Combination of a-Si/monocrystalline can absorb the solar spectrum in a wide range; the a-Si has a high solar absorption efficiency; the temperature coefficient of HIT cells is half that of the crystalline silicon solar cells, which is good for outdoor high-temperature environment;
5. The silicon wafer is thin, which is good for the carriers to pass through the substrate and be collected by the electrodes;
6. The heterojunction structure is good to raise the open-circuit voltage;
7. All the production procedures can be conducted at 200 °C, and the carrier lifetime will be not affected by impurity diffusion at high temperature;
8. The atomic hydrogen generated during PECVD deposition can passivate the crystal boundary;
9. The surface is textured, which can reduce reflection of solar rays.

All in all, the HIT cell adopts the low-temperature thin film deposition technique and it has the advantage of high mobility of crystalline silicon and the development outlook of the silicon solar cells with high efficiency and low cost. The HIT solar cells, produced by Sanyo, has achieved a high efficiency—approximating to 23%—at a large scale. The cells have realised rapid commercialisation. The HIT patent technique of Sanyo has expired in 2011. Since then, many companies have begun research on it.

5.3 Compound Semiconductor Solar Cells

5.3.1 Fabrication Methods of Compound Solar Cells

5.3.1.1 Vacuum Evaporation Technique

There are three types of vacuum coating: evaporation coating, sputtering coating and ion coating, all of which are physical vapour deposition techniques. The vacuum evaporation coating technique shall heat and evaporate a substance and then deposit it on the solid surface. In this technique, the evaporation substance will be put in a crucible or hung on the hot wire as the evaporation source, and the workpiece (base) will be put above the crucible. After the system is pumped down to a vacuum, the crucible or the hot wire will be heated to evaporate the substance. Then, the atoms or the molecules of the evaporation substance will deposit on the base surface in the condensation manner. The thickness of the thin film is ranged from several hundred angstrom to several μm. The distance from the evaporation source to the base shall be less than the average free travel length of the vapour molecule in the residual gas so as to avoid the chemical reaction resulted by collisions of the vapour molecules and the residual gas molecules. The average dynamic energy of the vapour molecule falls in a range of 0.1–0.2 eV. In the compound solar cells, the evaporation coating is used to produce CIGS cells, as shown in Figure 5.19.

The sputtering coating technique, which has been described in Section 8, Chapter 4, has been widely applied to deposition of back metal and ITO of the solar cells.

The ion coating technique, which can deposit ITO at the ambient temperature, has good coating performance but the deposition rate is slow with high cost.

5.3.1.2 Liquid Phase Epitaxy (LPE) Technique

For III–V compound solar cells, the early preparation method is LPE, which features simple equipment, safety and small toxicity. Ga and GaAs shall form saturated solution (the mother solution) at high temperature, and then cool down slowly. During cooling down, the mother solution will make contact with the GaAs monocrystalline substrate. Since the temperature is low, the mother solution will turn to oversaturated solution, and the surplus GaAs solute will be separated out on the GaAs monocrystalline substrate and grow new GaAs monocrystalline layer along the substrate lattice orientation, and get the epitaxy layer with good lattice integrity. Its shortcoming is that it is difficult to realise the cell structure with multiple layers, and that many heterojunctions cannot be grown by LPE technique. The LPE growth system is shown in Figure 5.11.

5.3.1.3 Metal Organic Chemical Vapour Deposition (MOCVD) Technique

So far, the metal organic chemical vapour deposition (MOCVD) technique is the main means to study and produce III–V compound solar cells. It is based on the principle that the organic metal compounds, for example, TMG, TMA and TMIn and so on, shall react with arsine, phosphine or other hydride in the vacuum chamber, and deposit GaAs, GaInP and AlInP III–V compounds on the GaAs or Ge substrate, realising epitaxy growth. Although the MOCVD technique needs costly equipment and gas sources, the doping concentration is controllable and the epitaxy layer has even thickness, and the solar cell has high efficiency and it is easy to obtain heterojunctions. The MOCVD system is shown in Figure 5.12.

5.3.1.4 Molecular Beam Epitaxy (MBE)

The molecular bean epitaxy (MBE) is another advanced epitaxy growth technique. With its principle similar to the vacuum coating technique, the MBE system has much higher

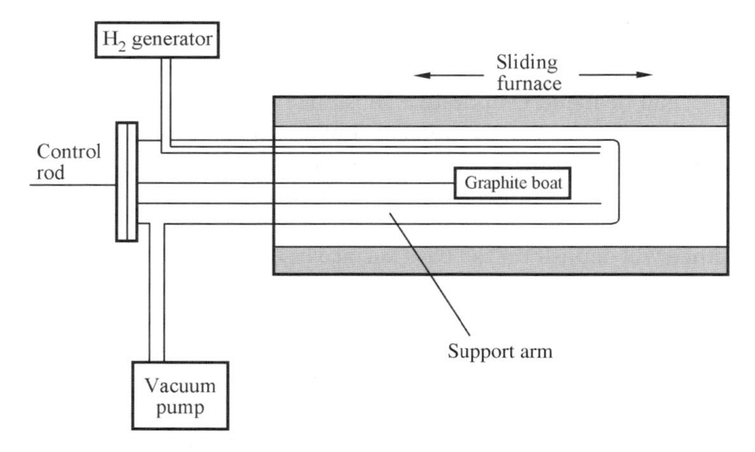

Figure 5.11 Diagramme of LPE growth system.

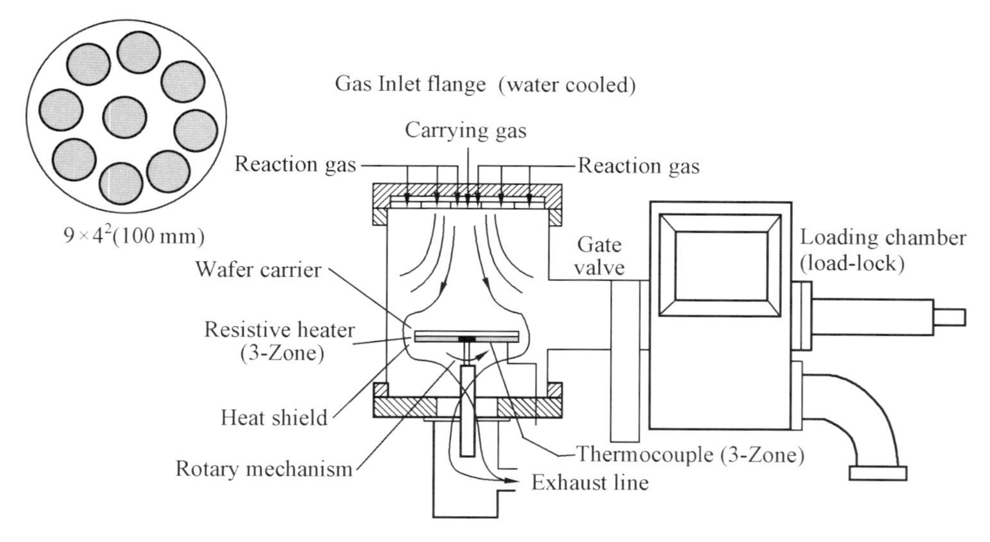

Figure 5.12 A schematic diagramme of MOCVD systems.

vacuum and much slower evaporation rate. The method can control the thickness of the epitaxy layer, the component and doping concentration and even the single molecular layer. Figure 5.13 shows the MBE system for the doped GaAlAs monocrystalline layer on the GaAs monocrystalline substrate. The effusion cell is provided with molecular beam source inside and the shutter control outside. When the effusion cell is heated to the given temperature at extra-high vacuum, the shutter shall be opened and the elements in the effusion cell will be emitted to the substrate in the form of molecular beam. When the substrate is heated to the given temperature, the molecules deposited on the substrate can migrate and grow crystalline layers by the substrate lattice order. The MBE technique can produce the compound monocrystalline coating of high purity and the desired stoichiometric ratio, and the slowest growth rate of the thin film can be controlled at 1 molecular layer/s. The control baffle can be used to produce the monocrystalline coating with the desired components and structure in an accurate manner. The MBE technique can be widely applied to produce various optoelectronic devices and various thin films with super-lattice structure. But the cost is high and it is difficult to realise industrialisation for the solar cells.

5.3.2 III–V Compound Multi-Junction Crystalline Solar Cells

III–V compound materials have many virtues, including the direct band gap structure, big light absorption coefficient, good anti-radiation performance and small temperature coefficient. And they are especially suited to prepare high efficiency and space-target solar cells. The band gap width of GaAs is 1.42 eV, which falls in the optimum range of the solar energy material. It has big light absorption coefficient and good anti-radiation performance. And the GaAs solar cell has low PV conversion temperature coefficient, which can be used to the sectors of high temperature and irradiation. In recent years, the high efficiency solar cells of tertiary and quaternary III–V compounds, especially GaAs, are generally designed with multi-junction stacked structure, for example, GaInP,

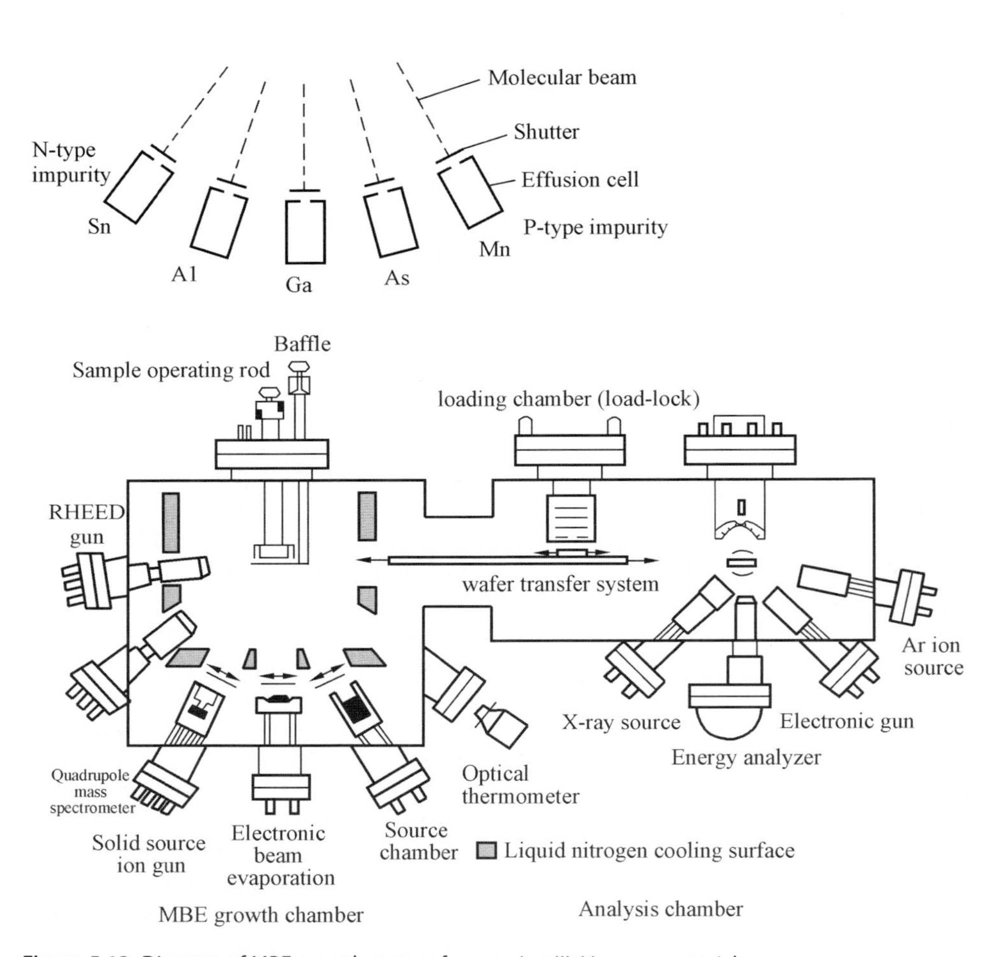

Figure 5.13 Diagram of MBE growth system for growing III–V group materials.

AlGaInP, InGaAs, GaInNAs and so on. Spectrolab, USA, in 2006, and Fraunhofer, Germany, in 2009, developed the three-junction concentrator GaInP/GaInAs/Ge stacked solar cell with efficiency up to 40.7% and 41.7%, respectively. The cost of the III–V compound solar cells, however, is far higher than the silicon solar cells, and thus most of them are applied to space sector and now they have begun to be used in the ground concentrator solar cell system.

The stacked solar cells are based on the following principle: The materials of different band gap Eg are made into several sub-cells, which, then shall be stacked together by Eg from wide to narrow and connected into a multi-junction solar cell. Each sub-cell will absorb and convert the light at different wavelengths in the solar spectrum, and thus the stacked cell can absorb more light than the single-junction cell, improving the PV conversion efficiency of the solar cell. The proportion of In and Ga in the tertiary III–V compound can be continuously adjusted to regulate the band gap width and crystal constant to make full use of the solar spectrum and realise lattice matching between sub-cells. The three-junction InGaP/InGaAs/Ge stacked solar cell adopts Ge instead of GaAs substrate, which has the following advantage: The Ge substrate is cheap and has better mechanical strength than the GaAs one. Accordingly, the Ge substrate can greatly reduce the thickness (about 140 μm). The structure of the three-junction GaAs cell is shown in Figure 5.14. Obviously, the absorption range of the three-junction GaAs cell on the solar spectrum is wider, as shown in Figure 5.15.

The three-junction stacked cell is generally grown by the MOCVD method. Two tunnel junctions are connected amongst the three sub-cells. The tunnel junction, a P^{++}/N^{++} junction grown by III–V compound, shall take into account the matching of top/bottom sub-cell lattice and work as the window for the bottom cell and as the BSF for the top cell.

5.3.3 Cadmium Telluride (CdTe) TFSCs

CdTe, a direct band gap material, has an ideal energy band gap (1.45 eV), high light absorption coefficient ($>10^4$ cm^{-1}), and a 2 μm-thick of CdTe polycrystalline layer can adequately absorb the solar rays. The theoretical PV conversion efficiency is 28%. It has such virtues as a good PV performance with a dim light condition and small reduction of cell efficiency at high ambient temperature and the like. In early 1990s, the II–VI

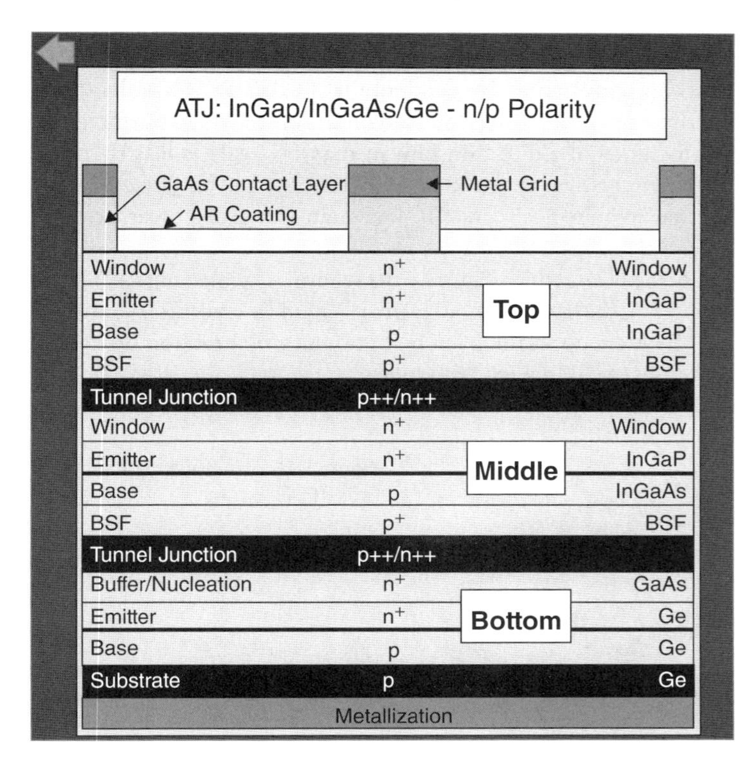

Figure 5.14 Structural diagramme of InGaP/InGaAS/Ge three-junction solar cells.

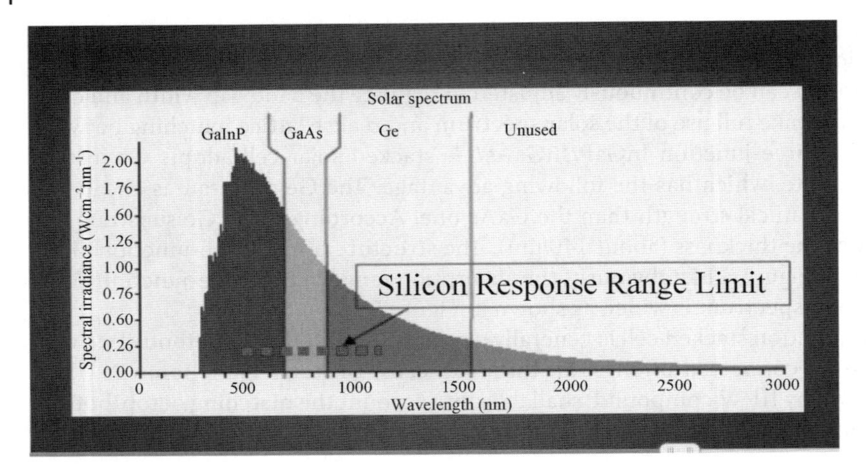

Figure 5.15 Utilisation of solar spectrum by GaInAs/GaAs/Ge three-junction solar cells (compared with silicon).

compound already adopted the thin film structure to realise commercialised production of PV power industry. Since the cost of II–VI compound semiconductors is much higher than III–V compound, such as GaAs, and it is more difficult to grow monocrystalline, the monocrystalline is seldom directly used to produce the solar cells.

CdTe TFSCs are the one with maximum yield in the TFSCs in the world. First Solar, USA, is the most successful company in massive production of CdTe TFSCs with output capacity ranked the first in the world. First Solar focused on the cost instead of efficiency. Although at present, the efficiency of CdTe thin film in massive scale is less than the crystalline silicon solar cells, First Solar declared in 2009 that the power price of solar power plant it contracted had reached 7.5 ¢/kWh, less than the grid power price of 9 ¢/kWh at that time.

The structure of CdTe TFSCs is shown in Figure 5.16 where, ① front glass; ② front contact TCO (0.5–1.5 μm); ③ CdS layer (0.03–0.2 μm); ④ CdTe layer (2.0–5.0 μm); ⑤ back metal electrode; ⑥ EVA encapsulant layer; ⑦ back glass; ⑧ junction box.

Below shows the process flow of CdTe TFSC modules:

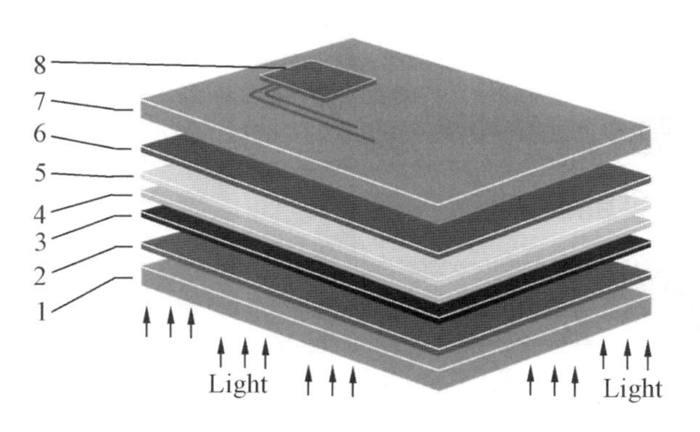

Figure 5.16 Structural diagramme of CdTe TFSCs.

Deposit and scribe TCO coating → deposit CdS/CdTe → atmosphere thermal treatment → laser scribe CdS/CdTe → deposit back contact layer → post-treatment → laser scribe back electrodes → package and test

There are many methods to deposit CdTe/CdS, including ① screen printing; ② spraying; ③ close-spaced sublimation; ④ electro-deposition; ⑤ vapour transport deposition; ⑥ sputtering; ⑦ evaporation. The key issues are discussed below:

1) The band gap of the window material CdS is 2.42 eV. Over the years, all the efforts to replace CdS by ZnS, ZnSe and so on have failed. CdS has an advantage: It will have mutual diffusion at the CdTe/CdS interface to generate $CdS_{1-x}Te_x$ ternary compound and improve the interface characteristics. To raise the output current of CdTe cells, CdS generally does not exceed 100 nm in thickness, and no pinholes are allowed (to prevent leakage current). The TCO layer consists of two layers with high/low resistance where the layer with high resistance can serve as the buffer layer to stop the likely leakage current at the CdS position. The buffer layer can be made of undoped SnO_2, Zn_2SnO_4 and Ga_2O_3;
2) After CdTe is deposited, it generally needs $CdCl_2$ thermal treatment, or the so-called activation. The procedure can significantly improve the CdTe cell efficiency;
3) Since the P-type CdTe has a large work function, it is often difficult to produce ohmic contact for CdTe. A high efficient method is to carry out Te-enrichment chemical treatment on the CdTe surface or form the telluride, for example, Cu_xTe, ZnTe:Cu and Sb_2Te_3, and then the sputtering or evaporation methods can be used to form the interface layer between CdTe and the metal electrodes by means of nickel, Mo or Ti.

Since a grown CdTe thin film has low concentration of carriers, it needs 380–450 °C thermal treatment in the chloride-containing atmosphere to promote CdS/CdTe interface diffusion and reduce lattice mismatching at the interface.

The 1.064 μgm of YAG:Nd laser is used to scribe the TCO coating; the 0.532 μm (by frequency doubling) of YAG:Nd laser is used to scribe CdS/CdTe absorption layer and the metal back electrode.

In summary, to further improve CdTe efficiency, the focus shall be put on defects (including interaction between intrinsic and extrinsic defects) and the interface quality.

As for the toxicity issue during cell preparation, some scientists have systematically studied the heavy metal emission during preparation of crystalline silicon solar cells and CdTe solar cells and compared it with that during power generation in the traditional power plants such as coal, oil, natural gas and so on. The results show, the oil power plant has maximum Cd emission, followed by coal, and the CdTe solar cell has minimum Cd emission. This is because 300–340 kg CdTe can produce 1 MW CdTe solar cell while it usually needs 7–8 t polycrystalline raw materials to produce 1 MW crystalline silicon solar cells. Although the experiments show, the modules of CdTe TFSCs are safe, it can increase the public confidence to build the recycling mechanism of rejected CdTe modules. The separated Cd, Te and other useful materials can be used to produce solar cell modules and carry out circulating production. The practise in the USA and Europe shows, recycling is technically feasible and its benefit is larger than the recycling cost. In fact, First Solar signs the recycling contract with the user during selling the CdTe solar cell modules where the recycling cost shall come to the solar power plant.

The A.N. Tiwari team from Empa (Swiss Federal Laboratories for Materials Science and Technology) started to study on producing CdTe solar cells with flexible and transparent Upilex-S polyimide plastic thin film substrate dozens of years ago, and the efficiency reached 11.4%. Compared with the glass substrate, the flexible plastic substrate has such virtues as super light and super thin, and roll to roll massive production can reduce greatly the cost. In June, 2011, Empa released A.N. Tiwari, which was used to produce single flexible CdTe TFSC modules on Kapton PV9100 transparent plastic thin film, produced by DuPont, USA, where the efficiency was 13.8%, representing international standard at that time.

5.3.4 CIGS TFSCs

CIGS and CIS have direct energy band gap, and their solar absorption coefficient $\alpha > 10^4 - 10^5\ \text{cm}^{-1}$, which is larger by two to three orders of magnitude than the crystalline silicon. With a different Ga ratio in CIGS, the band gap can be adjusted in a range of 1.02 eV–1.68 eV, which is good to design the multi-junction stacked cells. Figure 5.17 shows the solar absorption coefficient curve of various semiconductor materials. Obviously, CIGS has the maximum solar absorption coefficient in the wavelength range of 0.3–1.2 μm amongst the semiconductor materials. The absorption layer of CIGS TFSCs consists of 0.5–1.0 μm micro-crystalline, and it is about 2 μm thick. It has higher light absorption coefficient than the ordinary silicon. Thanks to such virtues as low-cost, high efficiency, attenuation-free, wide spectral response range and better output performance in rainy days than any other solar cells as well as variation of CIGS band gap with

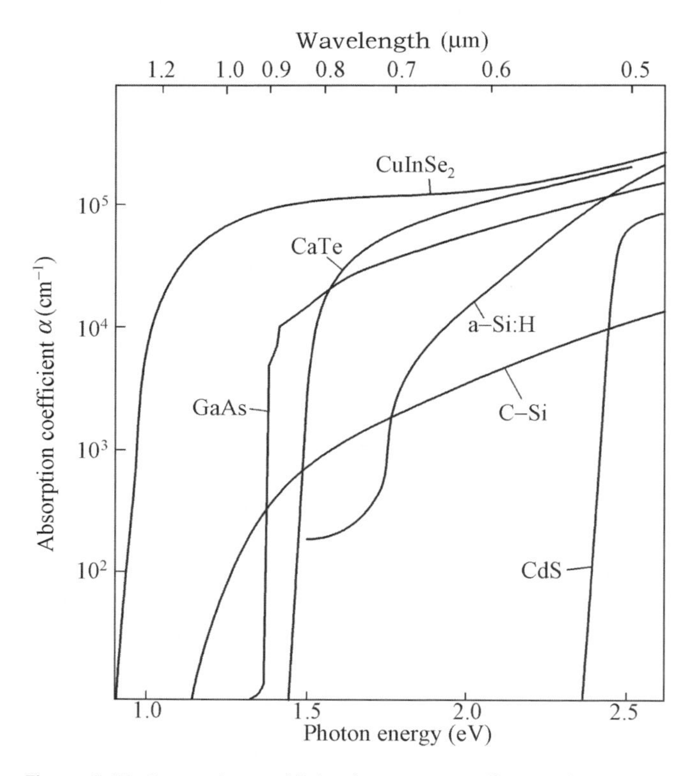

Figure 5.17 Comparisons of light absorption coefficient of various semiconductor materials.

composition changes, it is suited to produce multi-junction stacked solar cells and thus it has become one of the most promising PV devices. Here is an example. On October 3, 2012, MANZ, Germany, announced that the CIGS cell efficiency could reach 15.9% with module efficiency up to 14.6% at the international PV exposition, Taiwan (PV Taiwan), breaking the world record. ZSW, MANZ's partner, has achieved 20.3% efficiency in the lab conditions. The production cost of the CIGS production line with annual output of 200 MW is only \$0.55/Wp. It is reported, if the production line is expanded to GW level, the production cost can be further reduced to \$0.40/Wp, similar to that of the fossil-fuelled power generation. CIGS is considered as the thin film solar cells with the best prospects. The quaternary CIGS structure, however, is much complicated than the binary CdTe. At present, the countries in the world are focusing on the research and development and it is expected to realise industrialisation and massive production in the upcoming days.

The structure of CIGS TFSCs is shown in Figure 5.18 where ① substrate glass; ② back contact (Mo, 0.5–1.5 μm); ③ absorption layer P-type CIGS (1.5–2.0 μm)/N-type CdS (0.03–0.05 μm); ④ front contact (ZAO, 0.5–1.5 μm); ⑤ anti-reflection coating (MgF_2, 0.1 μm or so).

The core of CIGS cells is to deposit compound semiconductor thin film and metal coating on the substrate, which consists of the following process techniques:

5.3.4.1 Vacuum Co-Evaporation and Vacuum-Sputtering Methods

The core technology to fabricate CIGS cell lies in the growth of CIGS compound material, and the process is shown below:

Glass preparation→Mo deposition→laser scribe→CIGS/CdS deposition→laser scribe→ZAO coating→laser scribe→seal→test

At present, the CIGS solar cells are still based on the vacuum co-evaporation and vacuum sputtering techniques, that is, 'multiple stepped evaporation' and 'sputtering pre-set metal (Cu, In and Ga) and then selenylation' where the cell efficiency is high but it has to settle the issue of non-standard equipment for massive production.

The 'multiple stepped evaporation' method shall use Cu, In, Ga and Se as the source for evaporation step by step. It can obtain good crystal structure of the coating material

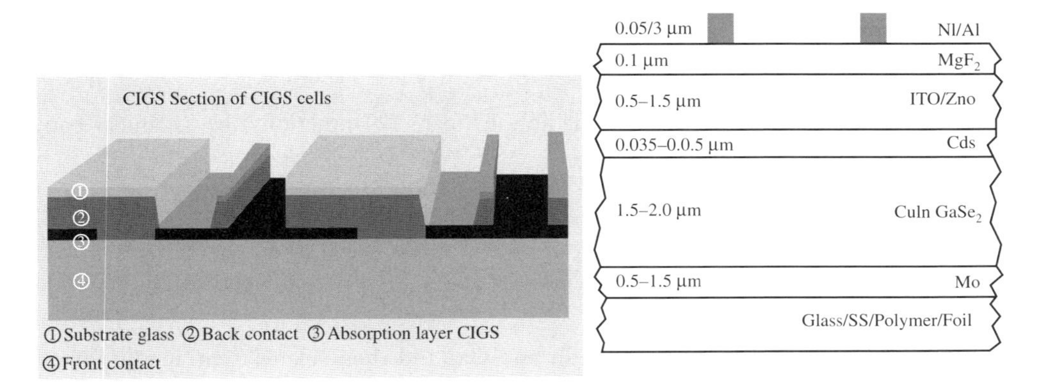

Figure 5.18 Structural diagramme of CIGS TFSCs.

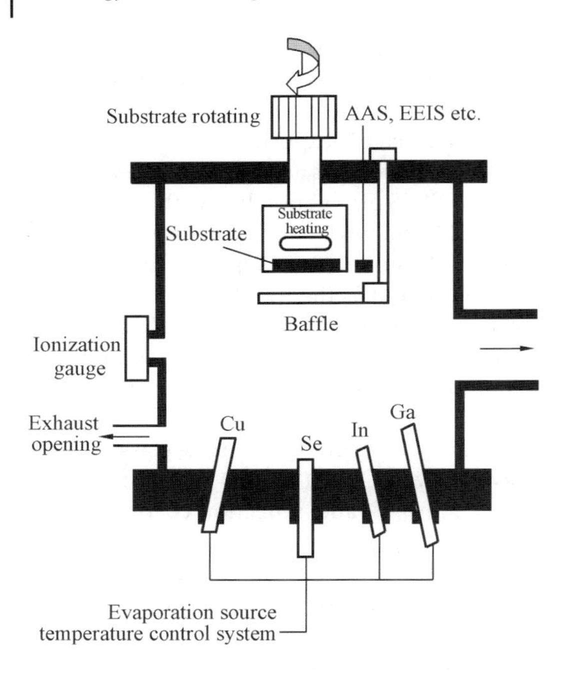

Figure 5.19 Diagramme of typical system for Cu, In, Ga and Se co-evaporation.

where the composition can be controlled during evaporation, especially the gradient band gap structure can be adjusted by longitudinal distribution of Ga in the absorption layer. The technique has short production period, less noble metal consumption, compact equipment, low cost and thus it has become a very promising technical route for preparation of CIGS thin film solar cells at large scale. The technical difficulties, however, lie in that the evaporation rate and quantity of each element shall be under accurate control and the evenness of massive deposition shall be improved to realise industrialisation. A typical co-evaporation system is shown in Figure 5.19.

Below shows the procedures for CIGS cell preparation by Cu, In, Ga and Se co-evaporation technique:

(1) Prepare the glass;
(2) Deposit Mo;
(3) Mo laser scribe;
(4) Sputter Cu, In, Ga, Se by vacuum co-evaporation method to generate $Cu(In,Ga)Se_2$;
(5) Deposit CdS (or other materials able to form PN junctions) by chemical bath deposition;
(6) Laser scribe;
(7) Coat ZAO;
(8) ZAO laser scribe;
(9) Coat the anti-reflection coating;
(10) Print the electrodes.

The 'sputtering pre-set metal (Cu, In and Ga) and then selenylation' method shall first sputter the preset layer of Cu, In and Ga on the substrate by proportion, then carry out selenylation and sulphuration at high temperature in the H_2Se atmosphere,

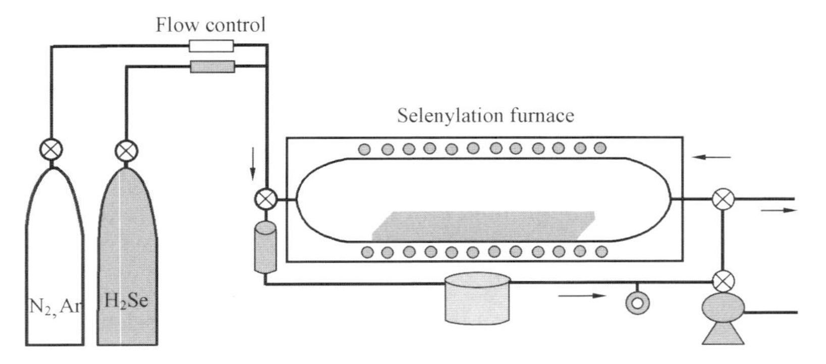

Figure 5.20 Diagram of H_2Se selenylation system.

and finally generate the CIGS polycrystalline thin film in the required proportion. The post-selenylation needs control of Se flow and substrate heater distribution etc. The methods have the following main problems: costly target and H_2Se gas, large consumption, long CIGS preparation time, huge equipment, larger cost of raw material consumption and production equipment than that of the co-evaporation method. The technical difficulty on composition and uniformity control, however, is less than that of the co-evaporation method. The H_2Se selenylation system is shown in Figure 5.20.

A typical process flow is as follows: The back electrodes (Mo) (0.5–1.5 μm thick) is sputtered on the glass substrate or other low-cost substrate by magnetic control; the CIGS absorption layer (about 2 μm thick) is produced by the 'sputtering pre-set metal (Cu, In and Ga) and then selenylation' method; the 30–50 nm CdS buffer layer is deposited by chemical bath deposition (CBD); the intrinsic ZnO (i-ZnO) and the Al-doped low-resistance ZnO (ZAO) is sputtered with thickness of 50 nm and 300–550 nm, respectively by magnetic control; the MgF_2 anti-reflection coating (about 100 nm thick) is deposited; and finally the Ni/Al grid top electrodes are generated by vacuum evaporation method.

Compared with the vacuum co-evaporation method, the sputtering and salinisation technique has low production process temperature, and it is easy to execute large-area deposition. The problems, however, include expensive target material and H_2Se gas, big consumption, long time to produce the CIGS film, huge equipment. Obviously, its cost on raw material consumption and production equipment is larger than the vacuum co-evaporation method. MANZ, Germany, adopted the linear co-evaporation source patent technique to realise element matching and improve significantly the evenness of large-area deposition. In 2014, it created the lab world record for conversion efficiency, which reached 21.7%, and it produced the 1200 × 600 mm turn-key module production line with conversion efficiency up to 14%.

5.3.4.2 Non-Vacuum Method

Since it is difficult to produce massive CIGS cells by the co-evaporation technique and the sputtering method has poor material utilisation, the non-vacuum methods, mainly including nano slurry printing and electrochemical deposition and so on, are under exploration to prepare CIGS cells. The non-vacuum methods have the following virtues: low equipment investment, controllable deposition rate, high material utilisation, and less noble metal raw materials. However, their cell efficiency is low, and the difficulty lies in selenylation and grain growth.

In the nano-level powder slurry printing and selenylation method, the Cu, In, Ga and Se shall be mixed to nano particle slurry by a certain proportion, which then shall be printed on the substrate by various techniques via annealing and selenylation. Or in details, the water-based nm mixed Cu, In, Ga and Se compound ink shall be first prepared, and then the slurry shall be printed on the Mo electrodes and reduced to Cu-In-Ga alloy at high temperature in hydrogen and finally form CIGS coating via selenylation. The process flow is given below:

Prepare the CIGS nano-level slurry→print on the substrate (2 μm thick)→selenylation annealing→cell fabrication

Since the nano powder slurry method prints the suspension liquid containing metal oxide particles on the substrate, the material utilisation can reach above 90%, and the equipment cost is much less than the vacuum process equipment. The printed selenylation method has the following virtues: simple and low-cost equipment, high printing speed and productivity. And the problem is that it is difficult to prepare the nano particles with sufficiently small size and thus the slurry must be added with several organic agents to make the nano particles distribute in the solvent in a stable and even manner. If the nm-level particle fluctuation is present, it will result in pores and loose defects in the coating. In addition, the oxide nano slurry shall be subject to twice heating: the first is used to reduce the oxide to the metal alloy and the second to heat selenylation or conduct selenylation in the toxic gas H_2Se, bringing difficulties for production.

The electrochemical method CIS is based on cathode deposition where the coating deposited is dependent on such factors as the concentration of the solute, the solution PH value, the electrolysed current density and the deposition time. The solution system used consists of two types: the chloride system and the sulphate system. The one-step or several-step deposition methods can be used. After deposition, it shall be annealed in argon at 400–600 °C for 30–60 min to carry out selenylation reaction. The electro-deposition is carried out in the acid water solution where the substrate serves as the cathode to deposit the preset layer of the CIGS coating. The electro-deposition can use two types of solution systems: the chloride system and the sulphate system. In addition, some auxiliary components shall be added to the solution (Formula 5.1). The electro-deposition reaction is given below:

$$Cu^{+2} + In^{+3} + 2H_2SeO_3 + 13e + 8H^{+1} = CuInSe_2 + 6H_2O \tag{5.1}$$

The CIGS deposited in the chloride solution system is generally copper-rich, which shall be subject to heat treatment in the Se gas atmosphere in vacuum, and In and Ga shall be simultaneously deposited to control the composition of the CIGS coating and then annealed to make the grain grow. Finally, the cell can be fabricated.

The electro-deposition can automatically purify and thus the raw material of low purity can be used, and the deposition potential can be used to change the composition of the preset layer to establish band gap gradient distribution. Compared with the print selenylation method, it can obtain the coating with dense preset layer but its deposition rate is low. The difficulty lies in that the coating contains a lot of oxygen, which can be only removed during H_2Se selenylation, resulting in difficulty for removal of toxic gases. In addition, the solution stability is also poor.

5.3.4.3 Roll to Roll Method

NuvoSun (Shanghai), USA, exhibited its production system for roll to roll flexible CIGS thin film solar cells at the exposition where the substrate is made of 1 m-wide stainless steel foil. The 'sputtering pre-set metal (Cu, In and Ga) and then selenylation' method is adopted to prepare CIGS absorption layer where the efficiency is about 12%.

It is the flexible CIGS thin film solar cells developed and produced on the plastic substrate by the roll to roll method that is more popular in the world. Since the polyimide thin film has smooth surface and good insulation performance, the metal impurities will not diffuse towards the CIGS absorption layer at the working temperature, and it is unnecessary to consider the roughness of the stainless steel substrate. It is an ideal material for substrate that is light and thin. In June 2011, Empa, Switzerland, announced that the efficiency of the CIGS produced on DuPont Kapton PV9100 plastic thin film reached

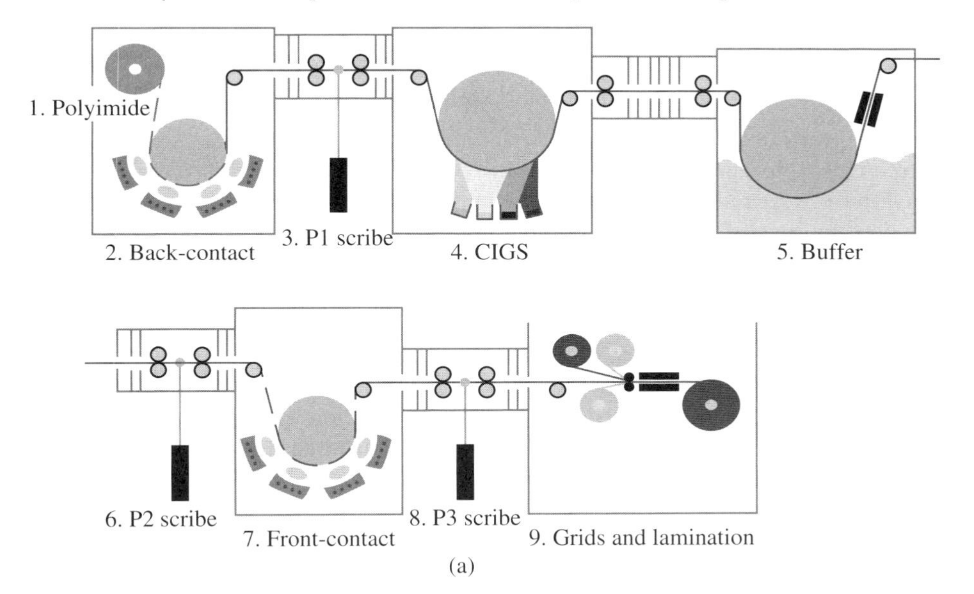

Figure 5.21(a) Flow diagramme of flexible CIGS solar cells.

Figure 5.21(b) Section of flexible CIGS solar cells.

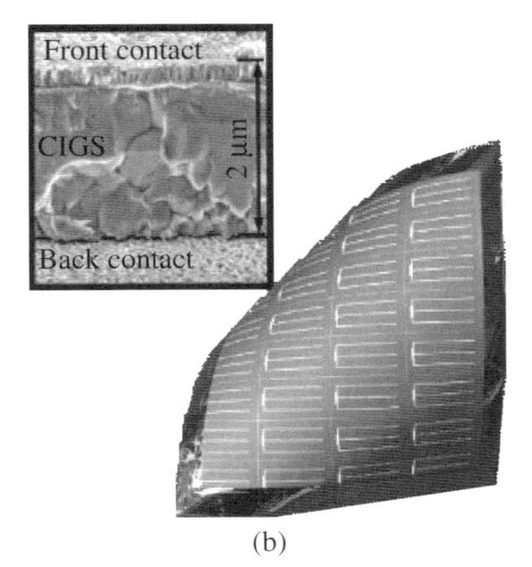

18.7%. In January 2013, Empa announced the efficiency reached up to 20.4%, close to the efficiency on the glass substrate. This indicates that the plastic thin film substrate has similar surface morphology and micro-structure to the glass. Empa, based on process simplification, has developed and optimised innovatively the low temperature process, laying foundation to massive production of efficiency and flexible CIGS solar cells.

Flisom, founded in 2005, has developed the industrial roll to roll production technique of flexible CIGS solar cells under technical cooperation and support of Empa. Flisom is building the production line with 15 MW annual output. Figure 5.21(a) shows the flow diagramme of flexible CIGS solar cells of Flisom and Figure 5.21(b) shows the section of CIGS cells.

5.4 Next-Generation Solar Cells

5.4.1 Organic Solar Cells

Based on the various PV materials and device structures, the solar cells can be divided into three types: inorganic, dye-sensitised and organic solar cells. The organic thin film solar cells have attracted more and more eyeballs thanks to such virtues as good flexibility, low production cost and easy to obtain massive and uniform thin film and the like.

The breakthrough on organic solar cells (OPV) came from the dye PV component with double-layer structure, which was reported by Dr. Deng Qingyun from Kodak, USA, in 1986. It uses the phthalocyanine derivative as the P-type semiconductor, and the tetra-carboxylic derivatives as the N-type semiconductor to form the double-layer heterojunction structure. Although the conversion efficiency was only 1% at that time, the study indicates for the first time that the introduction of OPV electron donor (P type)/electron acceptor (N type) organic double-layer heterojunction to the device can significantly raise the dissociation efficiency of the photo-generated excitons. Thanks to the efforts over 30 years, the component with efficiency of 7.4% was reported in 2010 (Liang Y, Xu Z, Xia J, *et al.* 2010. *Adv Mater,* **22**, E135).

For the organic solar cells, it generally needs five steps from light absorption to power generation, including exciton generation, exciton diffusion, exciton dissociation, charge transport and charge collection at the electrode.

The organic semiconductor (small molecules) material—phthalocyanine derivatives—is the typical P-type semiconductor, and the pyrene compound is the typical N-type semiconductor, which has high absorption coefficient ($>10^5$ cm^{-1}) in the spectrum zone of 600–800 nm and 400–600 nm, respectively. As a result, several hundred of nm organic materials can completely absorb the incidence light at the absorption peak. When the organic molecules absorb the solar rays, the electron will transit from the highest occupied molecular orbital (HOMO) to the lowest unoccupied molecular orbital (LUMO), and a hole will be formed on the HOMO, generating an exciton. Since the energy levels of HOMO and LUMO in the organic material are separate, the carriers are localised, which is different from the conduction band and the valence band of continuous levels in the inorganic semiconductor. Since the exciton (Frenkel exciton) electron-hole pair in the organic material has large Coulomb-binding force and the dielectric constant of the organic material is very small, the energy required for exciton dissociation is often larger than the thermal movement energy and it is difficult to generate free carriers. Due to the

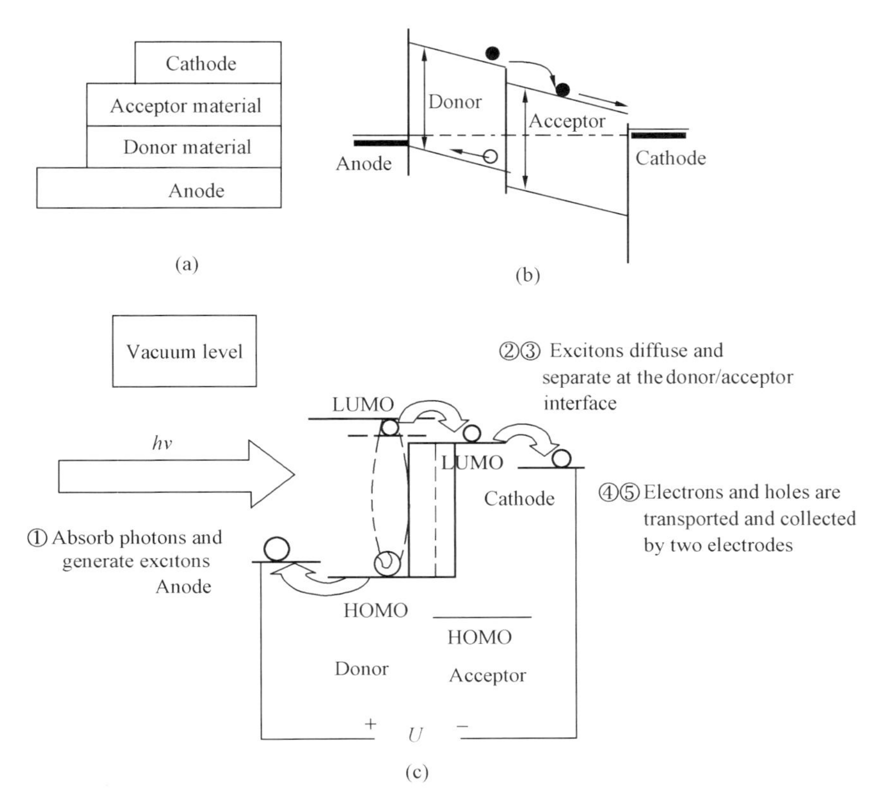

Figure 5.22 Double-layer heterojunction and basic physical process from photons to electrons in an organic solar cell: (a) Structure of double-layer heterojunction components; (b) diagramme of P/N junction built-in field formed at the electron donor (P)/organic acceptor (N) interface; (c) Basic physical process of photons to electrons.

above characteristics, the working principle of the organic solar cell is greatly different from the inorganic PN junction PV components.

In the double-layer heterojunction organic solar cells composed of two organic materials, the built-in field will be set up at the interface of the electron donor (P-type) and the electron acceptor (N-type) due to their LUMO energy differences, which drives the charges in the excitons to transfer and separate, and the transfer state of charges consist of the donor's HOMO and the acceptor's LUMO. After it receives the solar rays, the basic physical process from photon transferred to electron is described below and shown in Figure 5.22.

① When the energy absorbed by the material is larger than the incidence photon of the active substance Eg, the excitons will be generated;
② The excitons diffuse to the heterojunction;
③ Charge separation: The excitons are separated to free holes (on the donor) and free electrons (on the acceptor) near the heterojunction, which are the main carriers of the system. In the double-layer heterojunction solar cells, the drive force for charge separation is mainly the energy difference between the LUMOs of the donor and the acceptor;

④ Charge transfer: The separated free charges transfer to the associated electrode. Since the carriers in the organic material are localised and the excitons in the organic PV components have much shorter lifetime and diffusion length (about 5–20 nm) than the inorganic ones (≧1 μm), the charges can only be transported by leap or diffusion and reached and collected by the electrode and generate the photo-generated current. Since the oxygen and other wells are generally present in the organic material and the carriers in the organic material have excessively small mobility ($<10^{-4}$ cm^{-2}/Vs^{-1}), the thickness of the active layer often falls in a range of dozens of nm to several hundred nm, which limits the possibility to improve solar absorption by thickness growth, resulting in low short-circuit current of the organic solar cell.

⑤ Outlet of charges collected by the electrodes: Since the work function of the electrode material shall be matched with the HOMO/LUMO level of the organic material, and the ideal ohm contact in the organic solar cell shall be studied, the charge collection losses at the electrode is also a reason for low conversion efficiency.

In recent years, to expand the donor/acceptor interface effect, the bulk heterojunction organic solar cell, which is produced by the mixed solution containing donor/acceptor materials in the method of rotary coating (or evaporation), has been developed. Similar to the double-layer heterojunction solar cells, the bulk heterojunction organic solar cell is also based on the donor/acceptor interface effect to transfer the charges. The difference lies in that the charge separation is generated in the whole active layer in the bulk heterojunction solar cell while it only happens in the space charge zones (several nm) in the double-layer heterojunction solar cell. The transport rate of carriers to the electrode in the bulk heterojunction solar cell is lower than that of the double-layer heterojunction solar cell, which is because the bulk heterojunction particles are not continuous and the carriers are mainly transported to the electrode by the percolation function between particles.

The advantages of organic solar cells are described below:

① Big chemical variation, wide sources of raw materials;
② Light and thin, easy for carrying;
③ Easy to extend the spectrum absorption range;
④ Easy preparation, massive coating available by rotary and tape casting methods;
⑤ Low in price;
⑥ Degradable, less pollution to the environment.

The new high-performance organic solar cells have aroused general interests due to such virtues as light in weight, good flexibility, wet production available, easy for massive production and the like. The organic solar cells have some problems such as insufficient PV conversion efficiency, spectral response range and cell stability, which is mainly because the organic high polymer materials are mostly amorphous, and the photo-generated carriers after sunshine mainly move on the covalent bond and the mobility is generally low. It is an important approach to design and synthesise new conjugated polymer materials and optimise the photosensitive layer so as to improve the performance of the organic solar cell devices. The devices with small area can be generally prepared by rotary coating process, which, however, cannot prepare the modules of large area. As a result, it is of great importance to develop the solution preparation technique applicable to massive production.

On June 17, 2014, the project 'research on new high-efficiency organic solar cells', developed by Changchun Institute of Applied Chemistry of Chinese Academy of Sciences, achieved breakthroughs on design and synthesis of high-performance PV polymer, synthesis of new donor and interface materials and high-efficiency photosensitive thin film and preparation techniques of massive components, and synthesised many new polymer donor PV materials with strong and wide absorption, and high hole mobility (up to 1.1×10^{-2} cm^{-2}/Vs^{-1}), and the conversion efficiency of the PV modules with small area (0.04–0.1 cm^{-2}) developed in the lab by rotary coating technique could reach up to 8.79%. The solution preparation technique is adopted and further optimisation was conducted to post-treatment process of electronic ink and component photosensitive layers, and the conversion efficiency of the cell with single area of 10.2 cm^{-2} could reach 4.14%. The cell efficiency with effective area of 38.5 cm^{-2} could be 3.80%.

5.4.2 Dye-Sensitised Solar Cells

In 1991, M. Graumltzel published an article about dye nano crystalline solar cells (DSC) on *Nature*, where it says the PV conversion efficiency larger than 7% is achieved at low cost, opening up a new era in the history of solar cells and offering a new approach for solar utilisation. It has attracted many experts in the industry due to such virtues as cost advantage, relatively simple industrial production technology, nontoxic and pollution-free of all raw materials and production process.

The dye-sensitised solar cell refers to a photoelectric chemical liquid junction cell composed of nanoporous semiconductor film, dye-sensitised agent, redox electrolyte (I^{-3}/I^{-1}), Pt counter electrode and TCO glass and so on. The nanoporous semiconductor film is usually the metallic oxide (TiO_2, SnO_2, ZnO), and the sensitised dye will absorb on the surface of the nanoporous semiconductor film and condense on the glass plate with TCO to serve as the negative electrode of DSC. The other glass plate with TCO shall be plated with the catalyst Pt to serve as the counter electrode (+). Between the −/+ electrodes are filled with the electrolyte with redox counter electrodes, mostly I^{-3}/I^{-1}. The dye is generally N719 Ru dye, that is, di-tetrabutylammonium cis-bis(isothiocyanato)bis(2,2′-bipyridyl-4,4′-dicarboxylato)ruthenium (II).

The power generation principle of the dye cells are based on the electron transfer process of the dye molecules on the oxide semiconductor surface. Or in details, the dye molecules in the dye cell, after absorbing the solar rays, will jump from the ground state to the excited state and generate the photon-generated carriers. In this case, the dye-sensitised agent has the redox reaction. The photon-generated electrons will be rapidly injected to the conduction band of the adjacent TiO_2 nanoporous semiconductor and then flow to the external circuit; and the dye losing the electrons will be regenerated from I^- in the electrolyte, and I^{3-} in the electrolyte will simultaneously gain electrons from the Pt counter electrode. In this way, a cycle is finished.

It shows the structure and the working principle of the dye-sensitised nano solar cell in Figure 5.23.

Where

(1) After radiated by the solar rays, the dye molecules shall jump from the ground state to the excited state;
(2) The dye molecules in the excited state shall inject the electrons to the conduction band of the TiO_2 nano semiconductor;

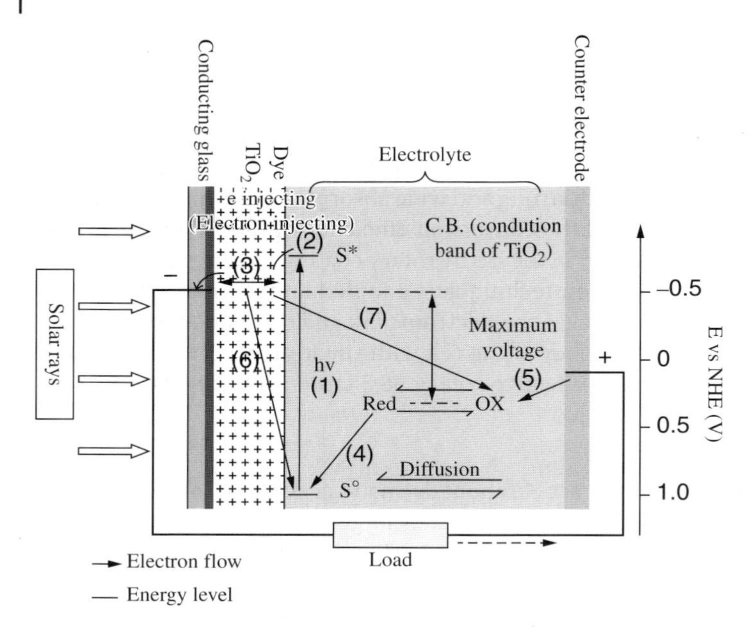

Figure 5.23 Structure and working principle diagramme of dye-sensitised nano solar cells.

(3) The electrons shall first diffuse to the bottom of the conduction band and then flow to the external circuit;

(4) The dye in the oxidation state will be reduced and regenerated by the electrolyte in the reduced state;

(5) The electrolyte in the oxidation state will be reduced after it accepts the electrons at the counter electrode, and in this way, a cycle is finished;

(6) and (7) show the recombination between the electrons injected to the TiO_2 conduction band and the dye in the oxidation state, and the recombination between the electrons in the conduction band and the electrolyte in the oxidation state.

The study results indicate only the sensitising agent molecules very close to the TiO_2 surface can smoothly inject the electrons to the TiO_2 conduction band, and the lifetime of the dye in the excited state is very short, and it must be closely bonded to the electrode, and it had better be chemically absorbed to the electrode. As a result, after absorbing the light, the dye molecules shall jump from the ground state to the excited state on the TiO_2 surface, generate the photon-generated electrons, and rapidly inject them to the TiO_2 conduction band. The dye molecules in the oxidation state that have lost the electrons shall be regenerated by the electrolyte in the reduced state during the redox reaction, that is, they can gain the electrons from the Pt counter electrodes. When the external circuit is connected, a solar cell is formed. The maximum open-circuit voltage is I^{-3}/I^{-1}, that is, the difference between the NHE potential in the redox reaction and the Fermi level of TiO_2. Obviously, the fundamental drive of the above power generation process is the photon-generated electron transfer where the electrons, after illuminated by the solar rays, shall jump from the ground state of the dye to the excited state and then injected to the conduction band of the semiconductor. The spectral response range of the dye molecules and the quantum generation rate are the key factors on DCS photon capture.

Although the semiconductor TiO_2 in DSC will not form PN junctions with the electrolyte, the redox potential in the electrolyte will form the liquid junction with the Fermi level of TiO_2 when it contacts with the electrolyte. When DSC is in the solar rays, it will form the open-circuit voltage, which can also realise the PV conversion process similar to the carrier generated by the electron.

The experiments show, for the large-area dye cells, the TCO glass surface resistance may have an obvious impact on the cell performance. This is because the electrons of the dye cells are mainly transported and collected by the TCO glass surface coating on both sides of the cell. In the preparation process of the dye cells, the TiO_2 nanoporous coating is usually prepared by the screen print technique and it is dye-sensitised by the simple soaking method. Since the liquid electrolyte in the dye-sensitised cell is apt to volatilise and leak, the cell must be sealed. In addition, the dye-sensitised cell must be coated with the anti-corrosion material for electrolyte on the print electrode surface (e.g., the low-temperature glass slurry, the porcelain material and the high polymer). In recent years, many scholars have added the organic gelling agent to the liquid electrolyte to solidify the electrolyte and form quasi-solid electrolyte so as to effectively prevent the liquid electrolyte from leakage. The study on the solid dye cells is also attracting more and more concerns.

After the study and optimisation for nearly two decades, Japan, Switzerland and Australia have created new DSC achievements, the efficiency of most DSCs is above 10% with maximum up to 12.3%, the life of the cell can reach about 15 years, and its production cost is only one-fifth the silicon solar cells. In 2009, Changchun Institute of Applied Chemistry of Chinese Academy of Sciences developed a dye-sensitised cell, whose efficiency reached up to 9.8%. The most important advantages of this cell are high efficiency, low-cost and simple preparation. Accordingly, it is expected to become a powerful competitor for the traditional silicon-based solar cells.

5.4.3 Perovskite Solar Cells

The perovskite solar cell is a new thin film solar cell developed on the basis of organic solar cells and the dye-sensitised solar cells, where the perovskite metal organic halide is used to replace the dye in the dye cell. In the perovskite structure (ABX_3), A is generally methylamino CH_3NH_3, B is Pb or Sn, X is Cl, Br or I. The most common perovskite is $CH_3NH_3PbI_3$, (also some I replaced by Cl), and the band gap is about 1.5 eV. The perovskite crystal structure is shown in Figure 5.24.

The perovskite solar cell consists of the glass, the electron transfer layer (ETM), the perovskite photo-sensitive layer, the hole transfer layer (HTM) and the metal electrodes from top to bottom where the ETM is generally dense TiO_2, as shown in Figure 5.25.

When illuminated by the solar rays, the perovskite layer will absorb photons and overcome the exciton-binding energy of the perovskite material to generate free carriers or excitons. From the perspective of crystallology, the perovskite is thermodynamically stable in the stereostructure where the PbI_3octahedron of long chain in order can facilitate electron transfer.

Compared with the traditional organic semiconductor materials, it has higher carrier mobility, larger carrier diffusion length (several hundred nm, and even above 1 μm)

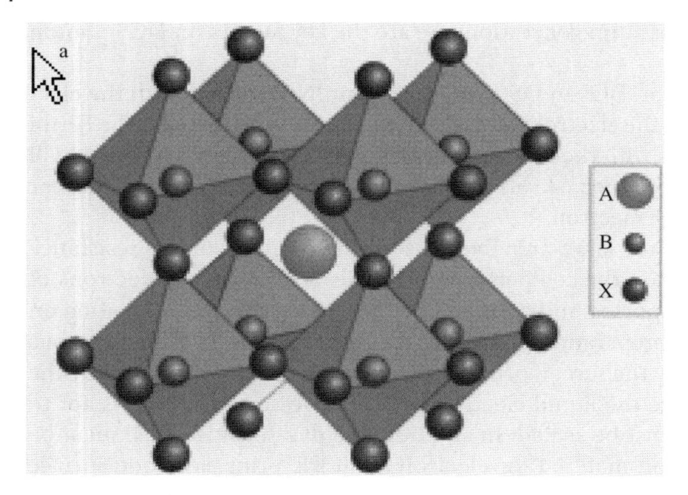

Figure 5.24 Typical structure of perovskite crystal.

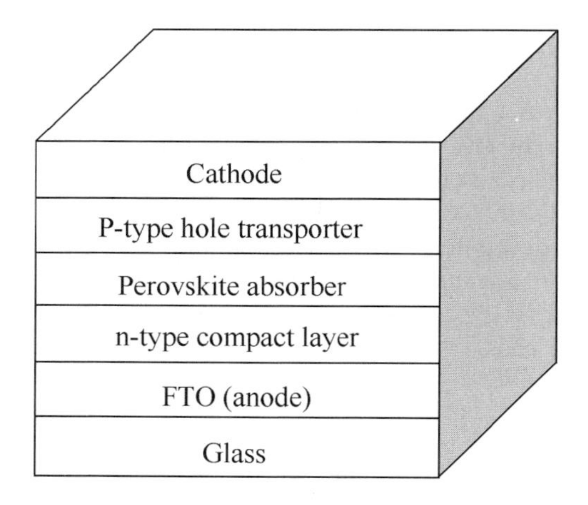

Figure 5.25 2D plane structure of perovskite solar cells.

and longer lifetime, larger by one to two orders of magnitude. Therefore, the probability of carrier recombination is low, and it can produce the solar cell with high efficiency. Compared with the dye-sensitised cell, the perovskite molecules don't need the porous TiO_2 film for absorption (the dense TiO_2, or even ZnO, Al_2O_3 can be used), and it is expected to replace the traditional liquid electrolyte by the solid organic material for convenient use. As the thin film preparation technologies develop, the 2D plane heterojunction structure, which is similar to the silicon components, is developed. Compared with the traditional silicon solar cell, the band gap of the perovskite material is about 1.5 eV (adjustable by doping substitute), and it has good matching with the solar spectrum; its absorption coefficient is big and the coating of several hundred nm thick can fully absorb the solar rays below 800 nm wavelength; the perovskite thin film solar cell has higher open-circuit voltage (>1 V); the synthetic method of the perovskite film material is simple where it can be produced either by the co-evaporation method or via the solution rotary coating method of low cost, and the production process has low energy consumption (<200 °C), and it has such advantages as flexible substrate available and so on, so that it can take both efficiency and cost into account.

The perovskite thin film solar cell has stood out from the numerous new solar cells developed in recent years, and it was elected as one of the 10 key scientific breakthroughs in 2013 by *Science*. Its conversion efficiency was raised from the original 3.8% to 19.3% in 2014. At present, it is regarded that the theoretical PV conversion efficiency of the perovskite solar cells still has a large room for improvement.

Since the perovskite solar cell has emerged for a short time, it needs further study, especially on such aspects as material selection, process preparation, cell structure and PV conversion mechanism etc. For example, the perovskite is apt to decompose with water and purify at high temperature, which may affect the service lifetime of the cell; the high efficient hole transfer material is costly, and it is difficult to prepare, which doesn't help reduce the cost; and the cell structure is under great influence of the deposition method of perovskite; in addition, the perovskite contains lead, which is toxic, and efforts are taken to replace it by tin. It is estimated that there is still a long way to go for the perovskite solar cell.

5.4.4 Concentrator Solar Cells

For the compound semiconductor solar cells, represented by CdTe (CdTe/Si) grown on GaAs and Si substrate, the conversion efficiency has reached 25–35%. They have large band gaps, so they can work in the outdoor applications with high temperature. These compound solar cells are costly. Since the power of the solar cell rises in proportion to the incidence irradiation, the concentrator photovoltaics (CPV) has been developed with target to the costly solar cells such as GaAs and so on. The ordinary solar cell can only make use of the natural solar rays. The CPV, however, can concentrate the outdoor solar rays of large area into a small range, forming the focal spot or focal strip, and the solar cell can be put on them to enhance the light intensity and achieve more electricity output. As a result, the CPV cell is more complicated than the ordinary ones.

The CPV usually consists of optical focus, dodging conduction, and solar cell tracking solar ray system. The optical focus structure can adopt the lens or the reflector. The lens mostly adopts the Fresnel lens, and the reflector often adopts the grooved plane concentrator and the parabolic concentrator. The sun tracking system is generally designed with PV automatic tracking function. The heat dissipating method can be air-cooled or combined with heaters so that it can obtain electricity and hot water. Compared with the compound materials (especially the III–V compound materials), the optical lens and the mechanical tracking system of the concentrator solar cell are less costly. The relatively low-cost optical system can be used to concentrate the solar rays on the costly semiconductor PV chip, which can reduce the total cost of the power generation system. The concentrator solar cell is an approach to reduce the total utilisation cost of solar cells.

Suppose N = solar concentration area/solar chip area is defined as the collection rate. For example, if the solar rays of 100 cm^{-2} are concentrated on a 1 cm^{-2} PV chip, the collection rate is 100 suns. Obviously, the plate solar cell without concentrators has only 1 sun. The CPV of 100 suns can be viewed as that the solar energy density is raised by 100 times. If the low-cost lens can focalise the solar rays by 500–1000 times on the high efficiency cell chip with small area (PV conversion efficiency up to 30–40%), it will have such virtues as less land coverage and costly material consumption and the like.

The collection ratio N, however, should not be too big: if N is larger than 1000 suns, the excessive focus may result in temperature rise on the PV chip and reduce dramatically the cell conversion efficiency, and the construction cost will rise again unless the complicated cooling system is provided.

The CPV optical components can concentrate and filter and dodge the light. Since the semiconductor materials of different types have different solar absorption spectrum, for example, the absorption of silicon mainly focuses on the near infrared zone while the III–V mainly absorbs the visible and near UV band. The long wavelength component in the solar spectrum larger than the absorption limits shall be removed to enhance the solar illuminated intensity of the semiconductor absorption waveband, reduce the chip temperature rise and keep the cell conversion efficiency. In addition, after concentration and filtration, the incidence light shall radiate evenly on the solar cell to avoid local overheating, hot spots and the subsequent solar cell damages. Different from the ordinary solar cells, it can still ensure good PV conversion performance at high junction temperature since the concentrator solar cell needs the solar illumination with high magnification. The ideal semiconductor material is GaAs, and generally N is 250–500. If the monocrystalline silicon is adopted, N is generally by the dozens. As for the cell structure, the P/N junction of the concentrator solar cells shall be deep, and the gate lines shall be dense, which accounts for about 10% of the cell irradiating surface.

The Fresnel lens is often used to produce the transmission solar generation system. Figure 5.26 shows the Fresnel lens and its solar cell system. The Fresnel lens has a characteristic that the optical energy losses transmitted will be the minimum when the angle between the incidence light and the prism top normal direction is equal to that between the emergent light and the prism bottom normal direction. The transmission concentrator is often made of polymethylmethacrylate (acrylic, or PMMA) of good transparency plastics by thermal pressing. To produce the Fresnel lens with such virtues as high concentration, good uniform performance, easy processing, light in weight and low cost, the special optical design shall be conducted. The system shall focus on how to reduce the losses after solar refraction, and how to avoid the heat island effect present in the cell chips due to uneven concentration, which will dramatically shorten the service life of the cell. At present, the Frenel lens is used for the solar cell system to concentrate the solar rays, as shown in Figure 5.26.

In the reflector concentrator, the solar rays will be reflected on the cell chips via dishes, as shown in Figure 5.27. The system covers more land space and it is heavy in weight. It shall focus on the concentrator evenness and the heat island effect caused by the shading of the concentrator on the cell board.

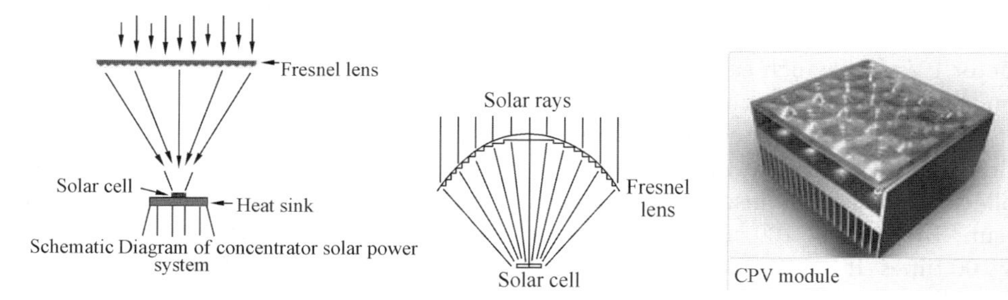

Figure 5.26 Transmission concentrator and PV cell chips.

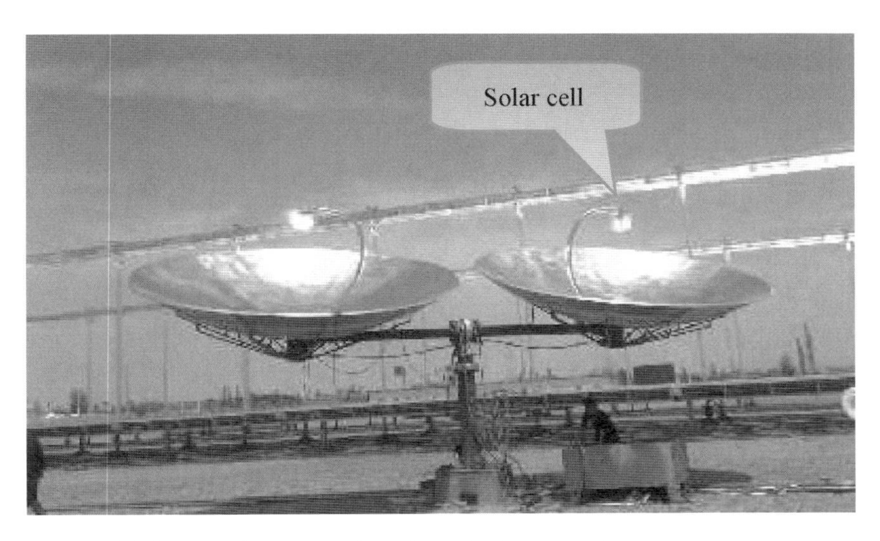

Figure 5.27 Principle diagramme of camber reflected concentrator.

The concentrator solar cell system has another characteristic: The concentrator and the tracking system are integrated together, that is, the PV chips must be aligned with the solar incidence direction so that the concentrator can work. This requires that the light intensity meter should be used as the sensor to control the concentrator and realise solar tracking, as shown in Figure 5.28.

5.4.5 Multiple Quantum Well (MQW) Solar Cells

The ultimate goals of the solar cells are to make full use of the solar spectrum as a whole, improve the PV conversion efficiency, dramatically reduce the cost and facilitate environmental protection and ecologic balance. The existing development trend of solar cell technologies show, the new-generation PV solar cells will be developed in the route of 'multiple films—stacked semiconductor materials—MQW materials—quantum dot materials'.

MQW solar cells are the PV cells composed of nano-level multiple-film micro-structure materials. It is expected to have high conversion efficiency due to the followings:

① The size of the nano-level grains is at the same order of magnitude with the scattering length of the carriers, and the scattering rate is reduced to raise the collection efficiency of carriers;
② Since the micro-structure has big electron density and strong absorption coefficient, the photons in a specific energy range can be absorbed by control of the size of the micro-structure;
③ The stacking layers can improve the light absorption capacity in a wide wavelength range;
④ In the nano-level multi-layer film, the electrons can be transported vertically to the super lattice interface and have resonant tunnelling, leap between local states and so on, significantly improving the PV conversion efficiency;

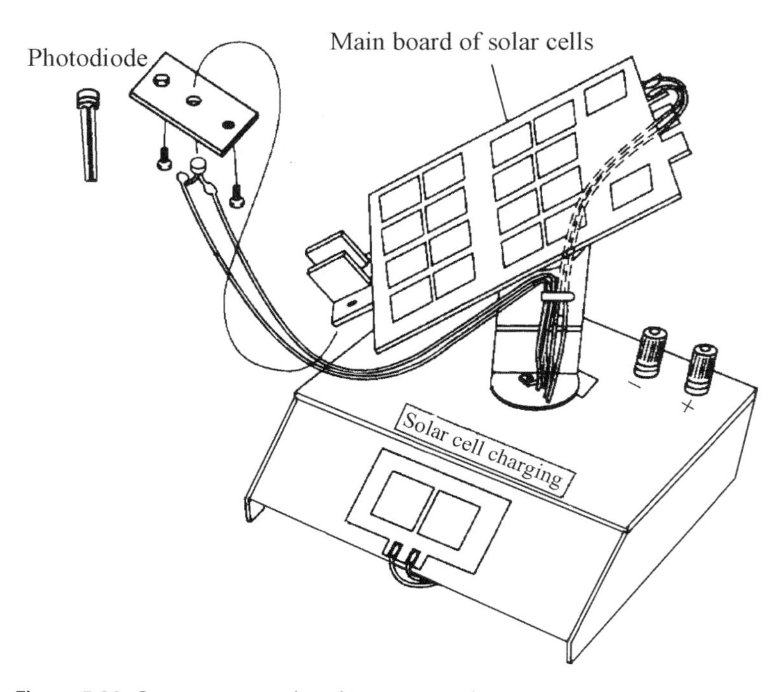

Figure 5.28 Concentrator and tracking system of concentrator solar cell system.

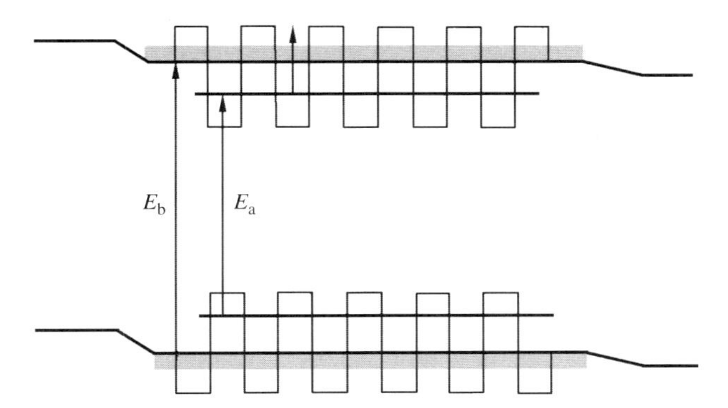

Figure 5.29 Diagram of quantum well superlattice energy band structure.

⑤ The base and barrier materials of the cell have wide band gap E_b; and the quantum well material has narrow effective band gap E_a. The value of E_a can be decided by the base state of the well quantum limiting level, and the gap of MQW absorption band can be adjusted by well material and quantum well width (barrier width L_b, well width L_z) to expand the absorption range of the solar spectrum wavelength and improve the photo-generated current. Figure 5.29 shows the diagramme of quantum well superlattice energy band structure.

The 'quantum dot' material has stronger quantum effect than the 'quantum well'. In the MQW solar cells, MQW is embedded to the intrinsic layer of the P-I-N solar cell, and in the quantum dot solar cells, the multiple quantum dot layer is embedded to the intrinsic layer of the P-I-N solar cell. In the quantum dot solar cell made of quantum dot materials, it can emit the energy of several wavelength when the energy of one photon is absorbed from the solar rays. The quantum dots are generally called as 'the artificial atoms', generally in dozens of nm, and the energy band structure is similar to the atom image. It has the quantum size effect in the 3D conditions, and as a result, it has the following functions:

1) The absorption coefficient is big: Since the quantum dot confinement effect can make the energy gap rise with particle size reduction, the quantum dot material can absorb the solar rays in a wider spectrum;
2) The quantum dot structure can form the sub-energy band so that several gaps can work. These sub-energy band transition can make the photons with incidence light energy less than the main band gap transferred to the carrier dynamic energy, generating electron-hole pairs;
3) The quantum dots have more obvious tunnelling effect than the quantum well, which can facilitate carrier transportation, and prevent the transversal photo-generated carriers in the quantum well from escape.

6

Modules and Arrays of Solar Cells

6.1 General

6.1.1 Modules and Arrays of Solar Cells

The voltage output by one single solar cell wafer is usually too low and the output current is also inappropriate. In addition, the wafer is fragile and apt to corrode. As a result, in practise, several solar cell wafers must be connected in series or shunt and then subjected to gluing, lamination, framework and external junction boxes and finally becoming the indispensable minimum unit capable to independently output DC current. The solar cell wafer shall be subjected to internal connections and external packaging to form a solar cell module.

As the actual PV power supply, several solar cell modules are generally mechanically fixed and electrically assembled according to the usage and design requirements to generate DC power after purchase. And this is called a solar cell array. The solar cell array is often called as the 'solar generator' in the foreign literature and the rest as the 'balance of system' (BOS) or the 'associated system'. In the solar cell array designed and built, the number of solar cell modules can fall in a range from two or three to several hundred, which is mainly dependent on the size of the PV system.

The solar cell modules can be classified by the following criteria:

(1) Cell materials: Crystalline cell modules, thin film cell modules and so on;
(2) Substrate rigidity: Rigid cell modules, flexible cell modules, semi-rigid cell modules and so on;
(3) Transmittance: Light-transmitting cell modules, non-light-transmitting cell modules and so on;
(4) Integration to the buildings: PV shingles (roof solar cell modules), PV screens (eave solar cell modules), glass curtain wall and other building integrated modules.

In this chapter, it will first introduce the structure, material, equipment, packaging process and tests after packaging of the crystalline silicon solar cell modules. Compared with the crystalline silicon solar cell modules, the thin film solar cell modules have simpler packaging process, which is similar to that of the crystalline silicon solar cell but usually integrated to the cell production.

Technology, Manufacturing and Grid Connection of Photovoltaic Solar Cells, First Edition. Guangyu Wang.

6.1.2 Packaging Techniques of Several Solar Cell Modules

At present, the PV modules mainly adopt EVA packaging. As the solar cell modules become diversified, other techniques such as UV packaging and vacuum packaging and the like have been developed.

Since EVA packaging is apt to become time-varying yellow, another, substitute—the UV solidification packaging technique is under development. The UV solidification is to pour the UV solidification glue to the two pieces of glass covering plate with solar cells and put into the UV solidification equipment for solidification. During UV radiation, the UV glue will have chemical reaction and change from liquid to solid in several seconds. The UV solidification packaging technique has the following unique virtues:

① Rapid packaging speed to improve the productivity;
② The UV solidification glue can be solidified at low temperature to avoid the damages due to high temperature on the solar cell performance in case of thermal solidification;
③ Unlike the EVA thermal solidification technique, which needs to heat the base materials, the UV solidification technique only needs the radiation energy of the photo-sensitive agent, saving energy. The UV solidification materials have high solid content, and the actual material consumption is few.

The main factors affecting the UV solidification technique are the viscosity of the packaging glue and the transmittance in the visible light range. If the UV packaging glue has excessively big viscosity, it is adverse to remove the bubbles generated during glue pouring; and it the UV packaging glue has excessively small viscosity, it is apt to result in big shrinkage during solidification and generate hollows in the glue and cause breaking of glue. The transmittance of the UV packaging glue can generally achieve high transmittance in the range of 400–1100 nm, up to about 90%.

Another packaging technique is vacuum glass packaging where the solar cell modules are packaged in the vacuum glass. The bonding interface of the glass and metal is inevitably to have air leakage during long-term service, which may reduce the conversion efficiency of the solar cell.

Because the EVA technique has been applied to solar cell packaging for over 20 years with satisfactory performance, it will mainly introduce it in this chapter.

6.1.3 Packaging Structure of Flat Plate Solar Cell Modules

Most of the solar cell modules are laminated to the flat plate structure, and the nonopaque, anti-ageing and hot melt EVA film of good bonding performance shall be packaged on the front and back after the solar cell wafer is connected in series or parallel. Then the low-iron tempered glass of high transmittance and anti-impact performance shall serve as the upper covering plate, and the wet/acid-resistant composite film or glass shall serve as the back plate. The vacuum lamination process shall be used to bond the cell wafer, the upper plate and the back plate into a whole by the EVA film. And then junction boxes shall be connected, and the aluminum alloy frames shall be used to fix. In this way, a practical solar cell module is built.

Figure 6.1 shows the structure of the flat plate solar cell modules, which consists of five layers of materials—the upper glass cover, the EVA film, the solar cell, the EVA film and the back plate. Figure 6.2 shows the diagramme of the flat plate solar cell modules after

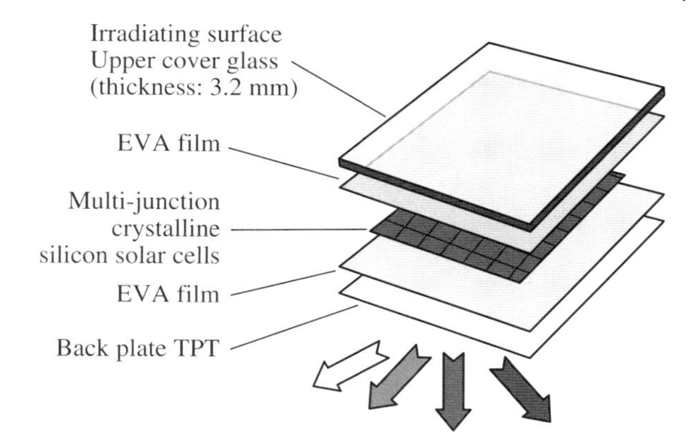

Figure 6.1 Materials of modules.

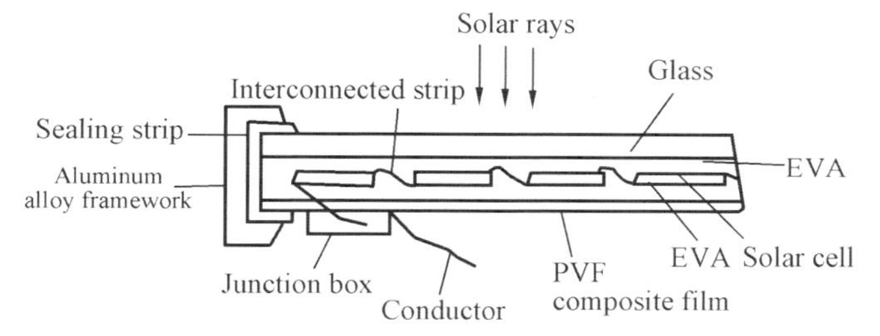

Figure 6.2 Structural diagramme of modules composed of solar cell wafers in series.

the crystalline cells are connected in series or parallel. Figure 6.3 shows the appearance of the flat plate solar cell module after frameworks and junction boxes are installed, as well as lamination and packaging.

To package the flat plate module, the glass of the upper cover shall have its transmittance larger than 90% (thickness: about 3 mm), and the two layers of anti-ageing EVA (ethylene vinyl acetate) film shall be added on the top/bottom of the cell wafer, and then the back plate is installed. The five layers of materials shall be laminated at high temperature and then installed with the aluminum alloy framework. After that, it shall be subject to epoxy and silicone sealing and other processes to finish the solar cell module.

6.1.4 Solar Cell Modules for Building Integrated PV (BIPV)

As the BIPV has been promoted in recent years, the module manufacturers have successively launched the solar cell modules with two-sided glass in addition to the flat plate solar cell modules for BIPV. The solar cell module with two-sided glass, with glass as the module back plate, consists of five layers—two layers of tempered glass (both the upper and the bottom cover plate), two layers of EVA film and the solar cell wafer. Figure 6.4 shows the structure. Figure 6.5 shows the picture of the nonopaque solar cell modules installed in the building.

The modules, being beautiful and nonopaque, have been widely applied to the PV buildings, for example, PV glass curtain wall, PV building roofs, PV louvers and PV screens and the like. It is expected the double-sided glass modules will witness further expanded commercial market. For the PV glass curtain wall modules, both of the two pieces of glass shall be tempered glass, and the glass facing the solar rays

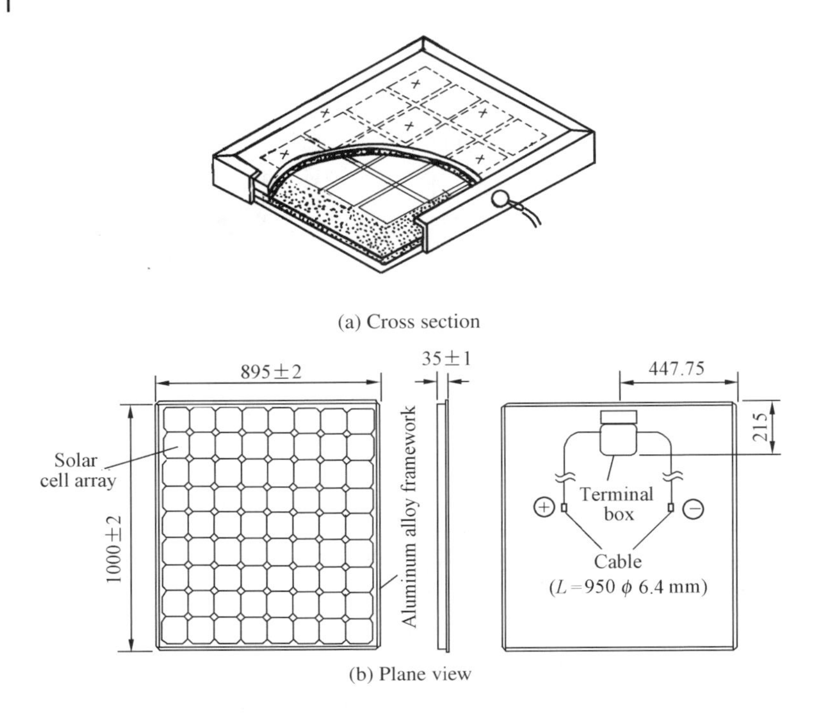

Figure 6.3 Appearance of flat plate solar cell modules.

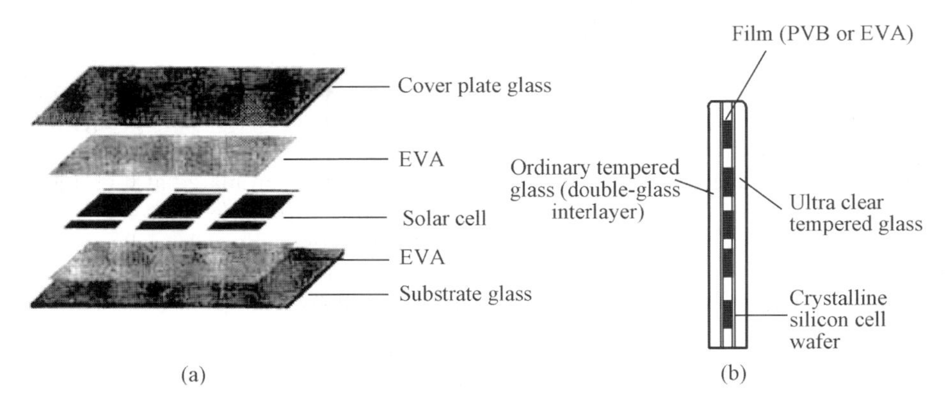

Figure 6.4 Structure of double-sided glass solar cell modules.

must be ultra clear glass, and the middle film shall be EVA, and the cell wafer can be mono/polycrystalline silicon or amorphous silicon. In addition, the PV glass curtain wall, as a part of the building materials, shall also take into account the available maximum wind pressure, anti-impact capacity and thermal conduction etc. The specifications are listed in Table 6.1, 6.2, and 6.3.

The building integrated PV (BIPV) modules have been developed towards big size, diversification, flexibility, modularisation and integration. Figure 6.6 shows some forms

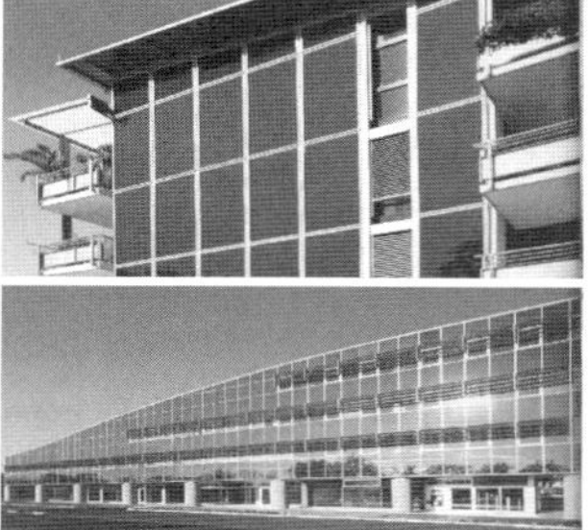

Figure 6.5 Nonopaque solar cell modules installed in the buildings.

Table 6.1 Wind pressure resistance level of building curtain walls.

Level	1	2	3	4	5	6	7	8	9
P(kPa)	1–1.5	1.5–2.0	2.0–2.5	2.5–3.0	3.0–3.5	3.5–4.0	4.0–4.5	4.5–5.0	≥5.0

Table 6.2 Impact resistance level of building curtain walls.

Level		1	2	3	4
Indoor side	E (N.m)	700	900	>900	
	H (mm)	1500	2000	>2000	
Outdoor side	E (N.m)	300	500	800	>800
	H (mm)	700	1100	1800	>1800

Table 6.3 Thermal conduction coefficient of building curtain walls.

Level	1	2	3	4	5	6	7	8
$K(W/\ m^{-2}\kappa)$	≥5.0	5.0–4.0	4.0–3.0	3.0–2.5	2.5–2.0	2.0–1.5	1.5–1.0	<1.0

Figure 6.6 Solar cell modules for BIPV (PV shingles, PV curtain walls and PV screens).

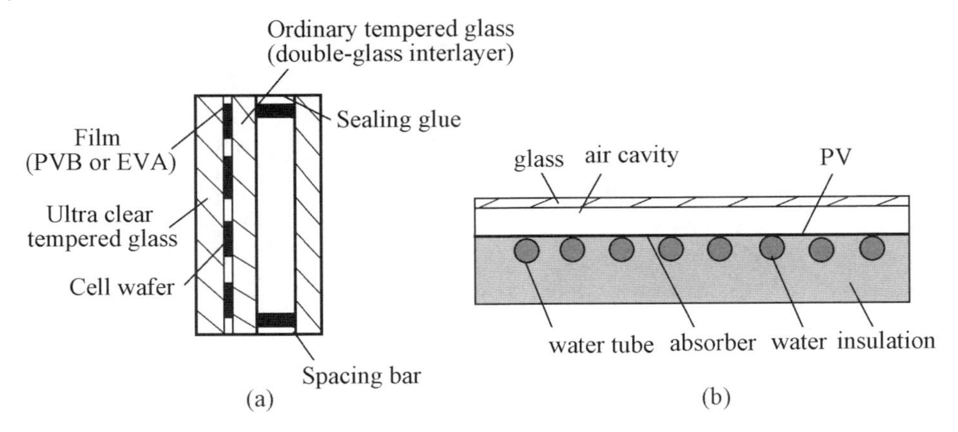

Figure 6.7 Hollow, double-sided glass PV modules.

of PV modules existed in BIPV, for example, PV roofing shingles, PV curtain walls and PV screens.

It shall be pointed out, if the hollow double-glass PV module (see Figure 6.7) is used, it shall pay special attention to expansion and shrinkage of gas in the hollow glass caused by temperature difference between day and night, which may shorten the lifetime of the sealing material and result in failure of hollow glass sealing. Since it can reach up to 50–80 °C outdoor in the daytime, and the air in the cavity will shrink as the cell wafer temperature falls down at night, the hollow glass will work in the alternating temperature. Moreover, in the crystalline silicon cell modules, it may have hot spot effect with local temperature likely up to 200 °C, which may result in excessive expansion of air in the hollow glass cavity or even breakage. It is proposed to make use of the heat and integrate the PV and photothermal energy and the heat absorption layer and water tube are installed in the hollow interlayer (see Figure 6.7(b)).

6.1.5 Double-Sided Cells and Modules

To make full use of the solar energy and save solar cell materials and site, the crystalline silicon solar cell wafers are built into double-sided cell modules with front/rear sides provided with PN junctions (e.g., HIT cells described in Section 5.4, Chapter 5). The module, also a PV shingle, made of light-transmitting double-sided glass, is installed with a certain angle with the ground so that the front can receive the direct solar rays and the rear can receive the solar rays scattered from each angle, improving the power generated of the whole solar cell. The energy of the light scattering towards all angles accounts for a certain proportion in the solar ray energy (Figure 6.8). The scattering light irradiating on the rear of the solar cell can be also used for power generation, and it can generate to the minimum extent another 10% power compared with the cells and modules with single-sided light receiving structure. Figure 6.9 shows the variation of power generation of HIT cells with double-sided structure as the installation angle changes, reported by Sanyo, Japan.

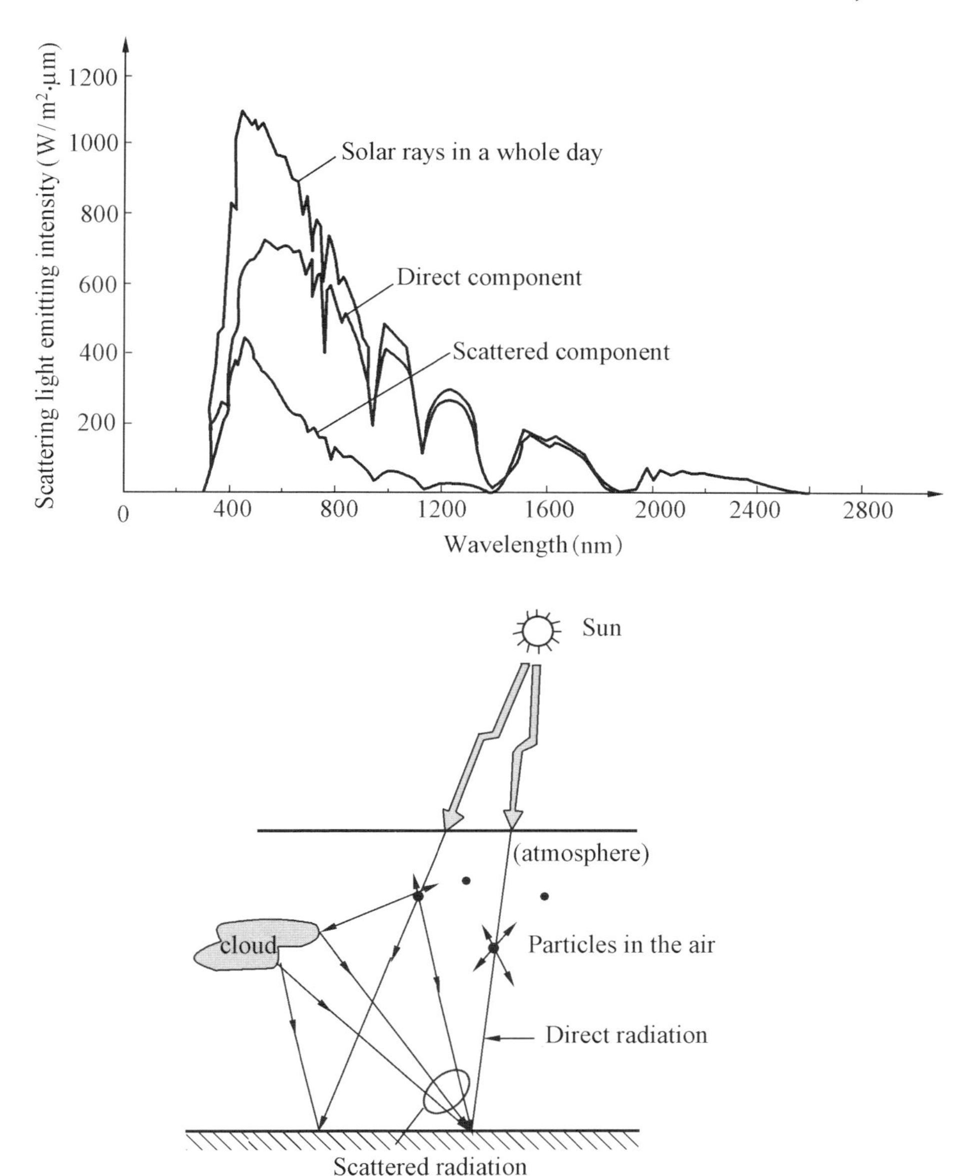

Figure 6.8 Solar ray intensity distribution in a whole day on the ground.

6.2 Module Packaging Materials

To ensure the service life of the module above 25 years, the quality of the module materials must be under strict control, including the cell wafer, the upper cover glass, the adhesive, the EVA film, the bottom plate TPT, the framework, the connecting lines, the junction boxes and the solder tin and so on.

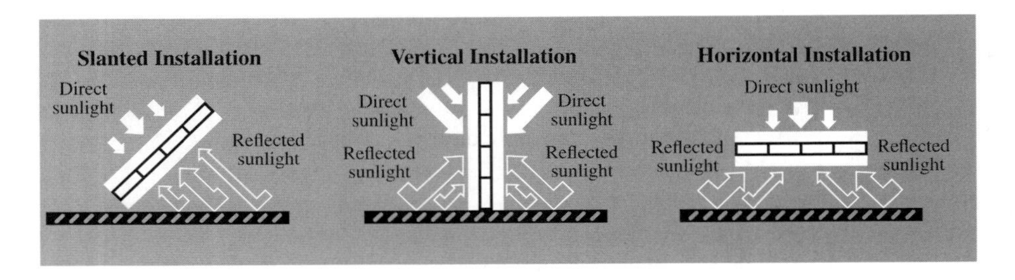

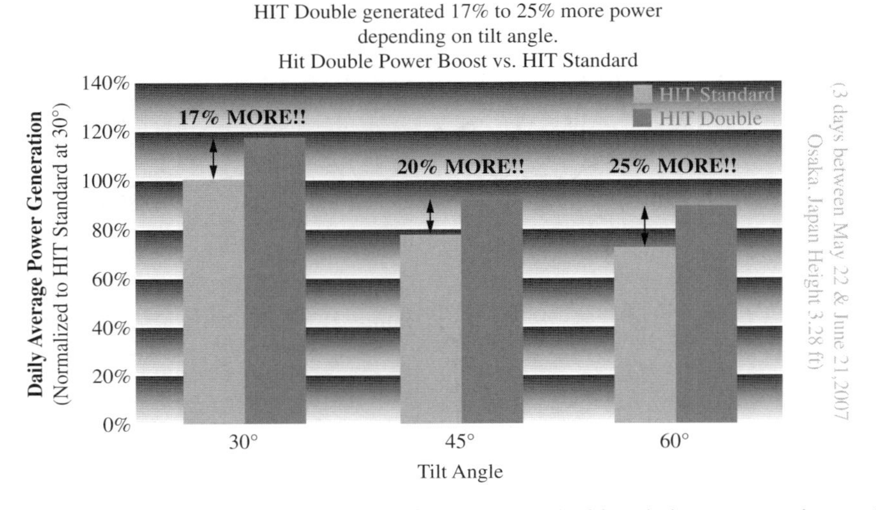

Figure 6.9 Variation of power generation of HIT cells with double-sided structure as the installation angle changes.

6.2.1 Inspection and Sorting of Cell Wafers

To ensure sufficient power output of the solar cell, the high efficient (>17.5%) monocrystalline silicon solar wafer packaging is adopted. Suppose one module consists of $m \times n$ solar cells (m rows and n columns) and if a cell of poor performance is mixed in a column, the current output of the column will be less than other columns in case of sunshine; and if a failed cell is mixed in a column, it almost has no current output in case of sunshine. The failed cell can only serve as the load and the cells of the other columns have to take bigger current (even larger than their permitted current capacity). The absolute voltage value at the working point of these cells may be larger than their permitted voltage, which may result in local heating of the originally normal cells, and rapid temperature rise and likely damages. The phenomenon where the local hot spot in the module comes into existence due to mixing of cells with poor performance is known as 'the hot spot effect'. (It shall be pointed out, it may also result in reduction of some cell output current and the hot spot effect when a local part of the module is shaded). As a result, strict inspection and sorting shall be conducted for cells to control the module quality.

The following items shall be included to solar cell inspection and sorting:

(1) The PV conversion efficiency test in the standard conditions;
(2) The open-circuit voltage and the short-circuit current sorting in the standard conditions;
(3) The reverse saturated current test;
(4) Hidden crack inspection;
(5) Classification of optical colour difference (colour);
(6) Consistency of mechanical performance (bending and thickness, etc.)

The sorting equipment for cell wafers has been described in Chapter 3.

6.2.2 Upper Cover Glass

Since the upper plate is the outer layer of the module, it shall be light-transmitting, firm, weather-proof and able to withstand sand and haze impact so as to protect the cell for a long term. The followings can be used as the top cover plate: tempered glass, polyacrylic acids, fluorinated ethylene propylene, nonopaque polyester and polycarbonate and so on. At present, the ultra clear tempered textured flat plate glass with low iron is generally used, which is in the micro-pyramid structure on the surface (pattern glass) and coated with anti-reflection coating with aim to improve absorption of scattered light and reduce optical pollution of glass surface on the environment. The detailed requirements are as follows: high transmittance, strong capacity of UV ray prevention, strong anti-impact capacity and good bonding strength with EVA film. It is generally 3.2 mm in thickness. The transmittance shall reach above 91% in the range of crystalline silicon solar cell response wavelength (320–1100 nm), and it shall have large reflectivity for the infrared rays with wavelength larger than 1200 nm. At the same time, it can resist radiation of solar UV rays with transmittance not falling down.

The glass manufacturers have been able to skillfully carry out tempering to the glass of 3 mm thick by physical or chemical methods in recent years where the transmittance can be kept a high level and the glass strength has been improved to three to four times of the ordinary glass.

To improve the glass transmittance and reflectivity, the quartz sand of low iron is adopted as the raw material (iron content: <40 PPM), and some materials are added during melting to turn the ferric iron to ferrous. During calendering roll process, the glass surface can be pressed with the patterns for anti-reflection. To reduce the reflectivity on the glass surface, the glass can be soaked in the sodium silicate solution and then dried naturally to produce 'the anti-reflection glass'. The coating and glass can be firmly bound together with good wear performance. And the final transmittance can be above 91%. The glass can be also applied to the glass curtain walls, building roofs, door/window glass of the UV-resistant BIPV.

All in all, the glass used as the upper cover plate of the solar cell module must have the following characteristics:

1. High solar ray transmittance and low absorption rate and reflectivity;
2. Resistant to wind pressure, accumulated snow, haze, thermal stress and etc., high mechanical strength;
3. Corrosion-resistant to the rainwater and the harmful gases in the environment;

4. No serious deterioration in case of long-term exposure to the atmosphere and sunshine;
5. Expansion coefficient matched with other structural materials.

6.2.3 Adhesives and Modified EVA Film

The adhesives are the key materials to fix the solar cell and ensure close bonding of the top/bottom cover plates. It is mainly used to fix the solar cells after series/parallel connections and seal the edges of the module framework. As a result, it consists of the film and the edge sealants. The solar cell module has the following requirements on the adhesives: nonopaque to the visible light, resistant to ageing due to UV rays; elastic to buffer the expansion between various materials and stop erosion of external humidity or harmful substances; good electrical insulation performance and chemical stability and not generating any harmful gas or liquid itself; and applicable to automatic module packaging.

The following adhesives can be used for vacuum lamination packaging: epoxy, organic silica gel and EVA film where EVA materials are of the most importance. Their performance is described below:

① Epoxy: Strong adhesive force and poor anti-ageing and apt to become yellow. In addition, epoxy is a high molecular material, and its molecular spacing is 50–200 nm, which is larger than the volume of the water molecule and thus it is easy to have water leakage. Besides, the expansion coefficient of epoxy is different from that of the silicon, which is apt to cause inner stress. Although the epoxy packaging process is simple with low material cost, it is only applied to the small-sized PV modules.

② Organic silica gel: It has the characteristics of both organic and inorganic materials such as high/low temperature resistance, anti-ageing, anti-oxidisation, electrical insulation, hydrophobicity and so on. The organic silica gel, an elastomer, can bond firmly the inorganic non-metal materials such as glass porcelain and so on and it has also strong adhesive force to the metal. In addition, the organic silicon is a nonopaque material with transmittance above 90%. The silica gel consists of the neutral and acid ones. The acid silica gel will corrode the silicon wafer and then the neutral silica gel is usually used. For some organic silica gel, however, the side group of polysiloxane is apt to oxidisation in heat, air, humidity and other ageing conditions, for instance, Si-O bond may react with the water in the air to break the chain and become ageing, and the physical performance will have obvious changes. As a result, the silica gel for solar cell module packaging needs appropriate agents to improve its anti-ageing performance.

③ Ordinary EVA film: EVA, the copolymer of ethylene and vinyl acetate, has higher transparency and better flexibility and thermal tightness, lower melting temperature (<80 °C) and good melting flow compared with polyethylene, and thus it is suited to solar cell packaging. the unmodified EVA has poor thermal performance, easy extension and poor elasticity and it is apt to crack, change colour and fall off from the glass or bottom plate (TPT) after long-time UV exposure and heating, and big thermal shrinkage may break the silicon wafer. As a result, the EVA shall be modified.

④ Modified EVA film: EVA shall be modified from two points: On one hand, such agents as the UV inhibitor, antioxidant and the solidification agent shall be added to improve the weather-proof and anti-ageing performance, and on the other hand, the chemical cross-link method shall be adopted to improve the thermal performance and reduce the thermal shrinkage of the EVA film. Generally, when the degree of cross linking is larger than 60%, the EVA film can withstand the variation of weather and be free from excessive thermal shrinkage.

At present, the solar cells mostly adopt the EVA as the raw material and add some appropriate modified agents to modify the EVA. At the ambient temperature, it is not cohesive (for easy cutting) and lamination packaging can be made to the solar cell modules during heating solidification, and after cooling down, it can produce permanent bonding sealing. Below shows the main performance data of the modified EVA solar cell coating:

① Two available solidification conditions: Rapid EVA, heating to 135 °C, and holding for 15–20 min at constant temperature; traditional EVA, heating to 145 °C, and holding for 30–40 min at constant temperature;
② Thickness: About 0.78 mm; width: 1100, 800, 600 mm available;
③ After solidification, transmittance: >91%; degree of crossing linking: >65%;
④ Peel strength of glass/ film: >30 N/cm^2; peel strength of TPT/coating: >20 N/cm^2;
⑤ Film thermal performance up to 80 °C and down to −40 °C, with stable dimension performance;
⑥ Good anti-UV ageing performance and anti-yellowing performance;
⑦ Not react with sealing silica gel.

EVA, the sealant of the solar cell, is used to bond and fix the glass, the back plate and the cell wafer, and it has high transmittance and anti-ageing performance. The EVA transparency has direct influence over the lifetime of the module. The EVA exposed in the air is apt to become yellow, which may affect seriously the module transmittance. If the viscosity is unqualified, and the bonding strength with the glass and the back plate is insufficient, it may result in early ageing of EVA and affect the module generation efficiency and lifetime.

The characteristics of typical solidification EVA sealing materials are shown in Table 6.4.

6.2.4 Back Plate and Localisation

The solar cell back plate, a protective material on the solar cell module, can protect the cell to work in the outdoor adverse environment for 25 years or even longer. In the main materials of PV modules, many, including glass, cell wafer and EVA film, have been localised except the back plate, which has to be imported. The detailed requirements on the back-plate material are: able to protect and support the cell; weather-proof, able to isolate the harmful gases entered from the rear; no any change at the lamination temperature; firm bonding with the binding materials, high peel strength amongst laminations, able to withstand high breakdown voltage. The overall requirements are insulation, waterproof, tightness and anti-ageing with 25 years of warranty period.

At present, there are two main types of solar cell back plate: One is to coat fluorine element on the PET-based surface, that is, to coat the back plate. The product has high

Table 6.4 Characteristics of typical solidification EVA sealing materials.

Characteristics	Unit	Testing method	Class	
			Traditional solidification	Rapid solidification
Tensile strength	Kg/cm^2		270	260
Rate elongation	%	JISK7113	430	410
Specific weight		ASTM D5105	0.95	0.95
Hardness	Shore A		79	82
Glass transition temperature	°C	DSC	−28	−28
Melting temperature	°C	DSC	76	76
Bulk resistivity	Ω cm		5.4×10^{15}	
Optical transmission coefficient	%		91.7	89.9
Haze			0.2	0.2
Reflectivity			1.191	
Specific heat	cal/g °C		0.55	
Thermal conductivity	Kcal/mh °C		0.1	
Thermal expansion coefficient	$°C^{-1}$		3.5×10^{-4}	
Hydroscopicity	20 °C, 24 hours	JISK7209	<0.01	
Water permeability	g/m^2, 24 hours		64.3	
Gel composition			95	95

productivity and low cost with market share rapidly growing. The other is to package the composite fluorine material on both sides of the PET base, that is, the composite back plate.

The coating product is to coat the paint, instead of fluorine coating, on both sides of the PET where the key lies in paint formulation technique and the coating technique. The typical quality data are shown in Table 6.5.

In recent years, many companies both at home and abroad have been developing the new back coating materials. For example, Suzhou Jolywood Solar Material Co., Ltd., adopted the plasma fluorosilicone coating technique, and developed the PTFE solar cell FFC back coating, which has good weather-proof, water/gas separation, insulation, thermal performance and low cost, and has won TUV and UL international certification. It is under accelerated localisation.

The composite back plate generally consists of three layers: PVF (fluorine-containing coating or film, about 40 μm thick)/PET (polyester, about 250 μm thick)/PVF (fluorine-containing coating or film, about 40 μm thick). Since it is fluorine-containing composite back coating, it is called as TPT composite coating. The fluorine film is formed by

Table 6.5 Main performance of back plate coated with fluorine element.

No.	Description	
1	Peel strength between laminations (N/cm)	≥ 4.0
2	Bonding degree with EVA (N/cm)	≥ 40
3	Steam permeability resistance (g/m^2.day)	< 2
4	Longitudinal or transversal size stability % (150 °C, 30 min)	≤ 1.0
5	Longitudinal or transversal tensile strength (350 mm/min)	≥ 110
6	Anti-ageing performance (hour)	≥ 2000
7	Anti-voltage strength (kV)	≥ 18
8	Local discharge test U_{DC}	≥ 1000

adhesives and PET, and the fluorine film can meet the basic requirements of the solar cell packaging materials such as anti-ageing, anti-corrosion, airtight and so on. TPT is a nonopaque material with high transmittance, and it can be produced in many colours such as white, blue, black, orange and and so on, as needed. It has such performance as anti-ageing, anti-corrosion, airtight, humidity-proof, fireproof, wet-resistant and self-cleaning and so on. And it is applicable to solar cell packaging. Generally, the back of the solar cell module is white, which is because white TPT can reflect the solar rays and improve the module efficiency. Besides, it has high reflectivity for the external infrared rays and thus it can reduce the working temperature of the module. The common white TPT specifications and performance are shown in Table 6.6.

Since TPT is costly in price, about 10 USD/m^{-2}, many packaging manufacturers are using TPE, instead of TPT, as the back material of the solar cell module. TPE consists of three layers: PVF/PET/EVA with colour close to deep blue. It is one-sided composite fluorine back film, and its weather-proof performance is less than the double-sided composite fluorine TPT, but its price is lower (about half of that of TPT). It has good bonding performance with EVA, and thus it has seeing more and more applications in module packaging. For the one-sided composite fluorine TPE, however, the side without composite fluorine is easy to become yellow, fragile and other defects.

6.2.5 Frameworks and Junction Boxes and Other Materials

The module is the part with highest value in the solar power system, and it works to convert the solar radiation energy to electricity. It shall have good corrosion-resistant and mechanical impact resistance performance. The frameworks of the solar cell module are mainly used to protect the modules and connect securely the modules with the array supports. The frameworks can be made of stainless steel, aluminum alloy, rubber and reinforced plastics and so on. For crystalline silicon solar cells, the main framework is usually made of aluminum alloy. The edge sealing of the solar cell module is realised by the frames and the adhesives.

The junction boxes, also viewed as the important materials of the module, shall have long service life and insulation to the earth. It shall be installed in secure bonding and good heat dissipation.

Table 6.6 Specifications and performance of white TPT.

Item	Unit	PET thickness in TPT		
		75 μm	250 μm	350 μm
Structural composition		T38/PET75/T38	T38/PET250/T38	T38/PET350/T38
Total thickness	μm	175 ± 5%	350 ± 5%	450 ± 5%
Specific weight	g/ m^2	240 ± 5%	485 ± 5%	625 ± 5%
Tensile strength MD	N/10 mm	150	380	500
Tensile strength TD	N/10 mm	180	480	600
Extensional rate MD	%	170	115	120
Extensional rate TD	%	120	95	100
Hot shrinkage rate MD	150 °C × 30 min, %	≦1.0	≦1.0	≦1.0
Hot shrinkage rate TD	150 °C × 30 min, %	≦0.7	≦0.7	≦0.7
Lamination peel strength	N/10 mm	5 (PVF falling off)	5(PVF falling off)	5(PVF falling off)
Degree of bonding with EVA	N/10 mm	> 40	> 40	> 40
Anti-permeability 38 °C, 90%RH	g/ m^2.day	5.1–5.5	2.6–2.7	2.2–2.3
Anti-permeability 23 °C, 50%RH	g/ m^2.day	1.0–1.2	0.6–0.7	0.4–0.6
Anti-voltage strength	kV	16	25	30
Local discharge test	V_{DC}	705	1040	1220

In addition to frames and junction boxes, other materials such as connecting bar (tin-dipped copper bar) and the soldering tin shall also meet the associated requirements.

All in all, special concerns shall come to the followings in the precondition of consistent cell wafer quality with aim to ensure PV module packaging quality and service life:

(1) Requirements on the top cover glass: High transmittance, low iron, tempered, able to filter UV rays;
(2) Requirements on the packaging materials: reflectivity matched with the solar rays, high bulk resistance, able to withstand breakdown voltage, low hygroscopicity, strong bonding force and good mechanical performance, difficult to break or damage, able to resist yellow deterioration caused by UV rays;
(3) Requirements on the back-plate materials: Good electrical insulation, mechanical strength, waterproof, weather-proof, bonding strength with packaging materials;
(4) Correct lamination processing technique, including temperature-pressure-time for lamination regulation and control;
(5) Module edges sealed to prevent foreign water, and tests of soaking/thermal circulation/leakage if necessary;
(6) Special concerns to anti-corrosion and mechanical strength of the frames;
(7) Special concerns to insulation and anti-ageing performance of the junction boxes.

6.3 Module Packaging Techniques

Since one crystalline silicon solar cell can only output voltage in a range of 0.45–0.60 V, and the output current of one 125 × 125 mm cell falls in a range of 4–5 A or so, several single cells must be connected in series and parallel and packaged into modules by the application so as to achieve the desired output voltage and current (power).

The front of the solar cell wafer serves as the irradiating surface where the PN junction is 0.3–0.5 μm in depth and the gate lines are 0.2–0.1 mm in width, and the bus bar is 2–3 mm in width. For series connections, the connecting wires on the front of the cell wafer (the negative electrode) shall be first welded, and then put the cell face down in order and weld them in series to form the cell set. Next, weld the bus bar together between the cell sets (pay attention to the polarity), and weld the leads of the modules. Both the leads and the bus bar are all on the back of the cell, which shall be insulated.

During production of crystalline silicon solar modules, the single crystalline silicon solar cells are usually welded into sets and then interconnected with each other to achieve the practical output voltage and current. During assembly, attention shall come to protect the electrode contact and prevent contacts from corrosion and the cells from damages. The packaging quality will have direct impact on the service life and reliability of the solar cells.

The crystalline silicon solar cell modules can be produced in the following steps: cell sorting—single cells welded to cell sets—interconnection of cell sets—composition and bus bar welding—lamination (glass-EVA-cell-EVA-TPT)—module tests—lamination—solidification—framework installation—electric junction box installation—insulation tests—module final tests.

The steps for crystalline silicon solar module production are described below:

1. Cell wafer sorting: Sort out the single wafer with similar open-circuit voltage and short-circuit current to form a module. Because one cell wafer with low power or a damaged wafer will reduce the output power of the whole module and even result in a 'hot spot'.
2. Single wafer welding: Weld the bus bar on the negative electrode of the cell where the welding joints shall be secure, straight and smooth and not fall off when pulling by hand in 45°. If the welding temperature is too low and the welding time is too short, it may result in dry joints while if the welding temperature is too high and the welding time is too long, it may result in accumulated tin or tin slag, causing wafer breaking in the vacuum lamination and other follow-up procedures. During single wafer connection, the contact resistivity of the welding points shall be controlled in a range of 20–80 μ Ω cm, and the welding firmness of gate lines shall be larger than 0.4 N/mm^2. The welding stress shall be low to prevent broken wafers and hidden cracks.
3. Interconnection of cell wafers: Based on the requirements of the module on voltage and current, the welded single cells shall be connected in series or parallel to finally form one positive electrode and one negative electrode. Connect the cells with identical short-circuit current in series. The solar cells in series shall have consistent performance and similar colour to achieve optimum PV conversion efficiency and output as well as good appearance of the modules. In addition, the soldering tin shall be smooth; otherwise, it may result in broken cells during lamination. During

interconnection, the contact resistivity of the welding points shall be controlled in a range of 20–80 μ Ω cm to prevent falling off. The firmness of the back electrodes shall be larger than 0.4 N/mm^2, the scaling powder shall be of good performance (the poor scaling powder will reduce the output characteristic FF or even fail the bypass diode in some cases, causing insulation breakdown). The welding stress shall be low to prevent broken cells and hidden cracks.

4. Composition and bus bar connection: Connect the welded cell sets with the bus bar. The layout shall be reasonable. Check the welding reliability and insulation amongst cell sets for safety concern.
5. Lamination sealing: It is aimed to conduct thermal compression and sealing for the connected cell module. After lamination, the cells in the module shall be free from breaking, hidden cracks, obvious displacement or impact; the bonding strength between layers shall be ensured and no bubbles shall be present or no layers shall fall off. Before lamination, the connected solar cell shall be subjected to short-circuit and breaking checks by a multimeter. It shall be noted that removal of the bubbles in EVA is the key to successful packaging during lamination because the oxygen reacted by the air entered in lamination with the EVA is the main reason for bubble formation. After lamination, check the sealing materials and the electrode welding.

During lamination, open the laminator, put the stacked cells into the laminator and close the top cover plate when the heating board reaches the given temperature (generally 100–120 °C when EVA is just in melting state). First, exhaust the air of the lower chamber to remove the air between the heated EVA and the cell, glass, TPT gaps. In case the exhaust time and lamination temperature is inappropriate, bubbles will be present beneath the module glass. Then, vacuumise the lower chamber and fill air to the rubber gas bag in the upper chamber (pressurise), and in this case, the melted EVA will flow under the action of compression and vacuumising to fill in the gaps between the glass, the cells and the TPT and exhaust the bubbles. In this way, the glass, cells and TPT will be bonded together by EVA. After lamination, open the cover, and fill air to the lower chamber and vacuumise the upper chamber to achieve balance between the lower chamber and the atmosphere, and finally open the top cover by the support to package the next solar module.

After taking out from the laminator, it must be put into the drying oven for solidification or to the laminator for direct solidification because the uncured EVA is apt to fall off from TPT or glass. Two solidification methods are available according to the types of EVA:

① For the rapid solidification EVA, the laminator shall be set as 100–120° and then put in the cell board, and exhaust for 3–5 min and then pressurise for 4–19 min and raise the temperature to 135° and solidify at constant temperature for 15 min, vacuumise the laminator for 30 sec, and then open the cover to take out the cell board for cooling down.

② For the traditional solidification EVA, the laminator shall be set as 100–120° and then put in the cell board, and exhaust for 3–5 min and then pressurise for 4–19 min and raise the temperature to 145–150° and solidify at constant temperature for 30 min, vacuumise the laminator for 30 sec, and then open the cover to take out the cell board for cooling down.

6. Module framework: The laminated cell module shall be provided with frameworks for outdoor installation. It shall pay attention to the bonding strength and elasticity of the silica gel and tightness of the framework. It shall avoid poor earth of the installation holes and framework corrosion.
7. HV insulation test: Connect the module lead in short-circuit manner and then to the positive electrode of the HV tester, and the exposed metal of the module to the negative electrode of the HV tester, and then apply the voltage at a rate no larger than 500 V/s until 1000 V plus double the system maximum voltage and then hold on for 1 min. It shall be free from insulation breakdown or surface damages.
8. Module test: Sort the cell module by process criteria. The IEC standard testing conditions are as follows: AM 1.5, 25 W, 1000 W/m^2. The following parameters shall be tested for the module: the open-circuit voltage, the short-circuit current, the working voltage, the working current, the maximum output power, the filling factor, the PV conversion efficiency, the series resistance, the parallel resistance and I-V curve and so on.
9. Installation of junction boxes: Install the junction boxes for the tested cell modules for electrical connections. Special attention shall come to electrical insulation, seal bonding and firmness during installation of junction boxes. In case of poor insulation of the connector/plugs and the framework or the earth connections, electric leak or even arc may occur.

6.4 Module Packaging System

6.4.1 Main Equipment in the Production Line of Solar Cell Modules

Figure 6.10 shows the layout of the automatic production line of 30–50 MW solar cell modules.

The production equipment of crystalline silicon solar cell modules mainly consists of the cell-sorting machine, the laser scriber, the cell welder, the cell series welding table, the laying work table, the turnover trolley for laminating modules, the laminator, the turnover trolley for assembling modules, the frame machine and the solar simulator (module tester) and so on. In addition, some auxiliary machines such as the glass washing machine, EVA cutting stand and the like. The cell-sorting machine has been described in Chapter 3.

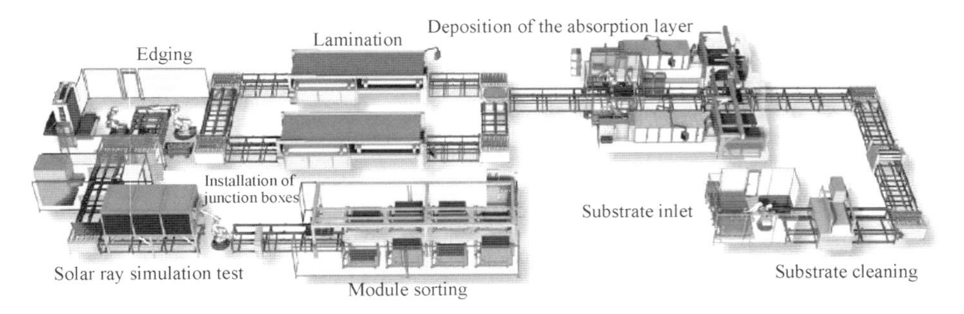

Figure 6.10 Layout of the automatic production line of 30–50 MW solar cell modules.

6.4.2 Laser Scribers

The solar cell module of small power can be scribed by laser to reduce the size. Laser scribing, non-contact processing, can scribe marks on the silicon wafer by the laser energy, which can reduce cell damages and prevent pollution, improving cell utilisation. The laser cutting depth is generally controlled as one-half to one-third of the cell thickness, which can be adjusted by the working current of the laser scriber. The laser scriber consists of the laser crystal (krypton flash lamp-pumped), the power system, the cooling system, the optical scanning system, the focus system, the vacuum pump, the control system, the work stand, and the computer and so on. Figure 6.11 shows the laser scriber for production of solar cell modules.

Compared with traditional mechanical scribing, the laser scriber has the following virtues:

① Controlled by computers with high speed and accuracy;
② Non-contact processing to reduce damages and prevent pollution;
③ Able to control the cutting depth (generally one-half to one-third of the cell thickness), applicable to cutting the thin, fragile and hard material such as silicon wafer;
④ Thinner laser beam, less processing material consumption, smaller heat-affected area;
⑤ Scribed grooves in good order, free from cracks and consistent depth;
⑥ Easy and simple operation, safe use, less labour and material cost.

6.4.3 Cell Welders

Manual welding is usually adopted in the factory. First of all, apply the soldering tin to the bus of the single cell on the welding table, and then weld the cells to cell sets on the series welding table, and finally weld in series and parallel the cell sets on the laying work table. See Figure 6.12 for single cell welding table, cell series welding table and the laying work table.

As for automatic welding of cells, two methods are available: hot air and infrared lamps, as shown in Figures 6.13 and 6.14.

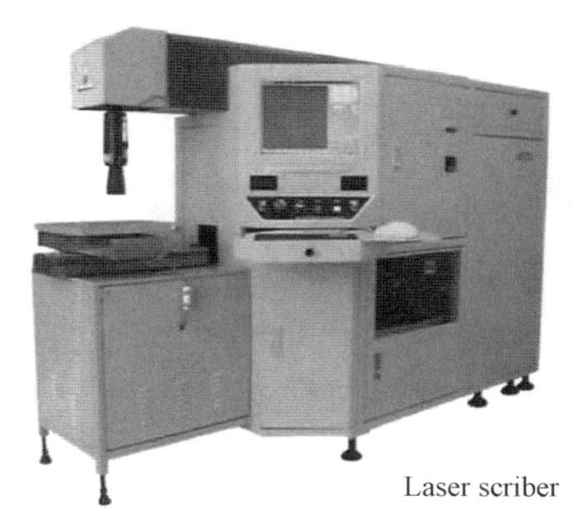

Figure 6.11 Laser scriber for production of solar cell modules.

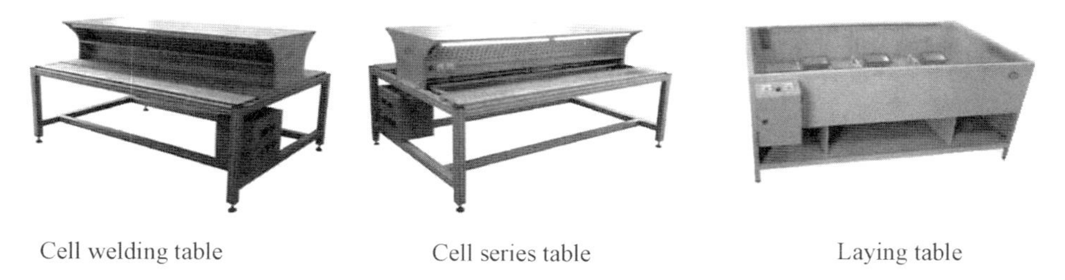

Figure 6.12 Cell welding table, series welding table and laying work table for production of solar cell modules.

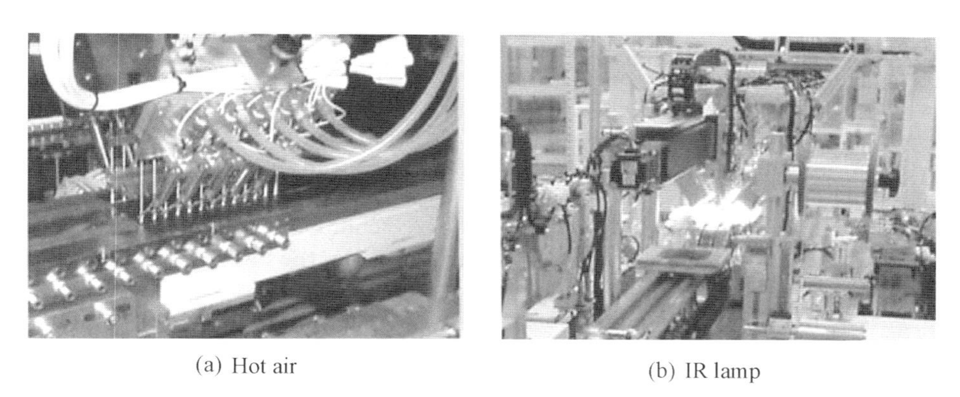

Figure 6.13 Automatic welding methods: (a) hot air; (b) infrared lamp.

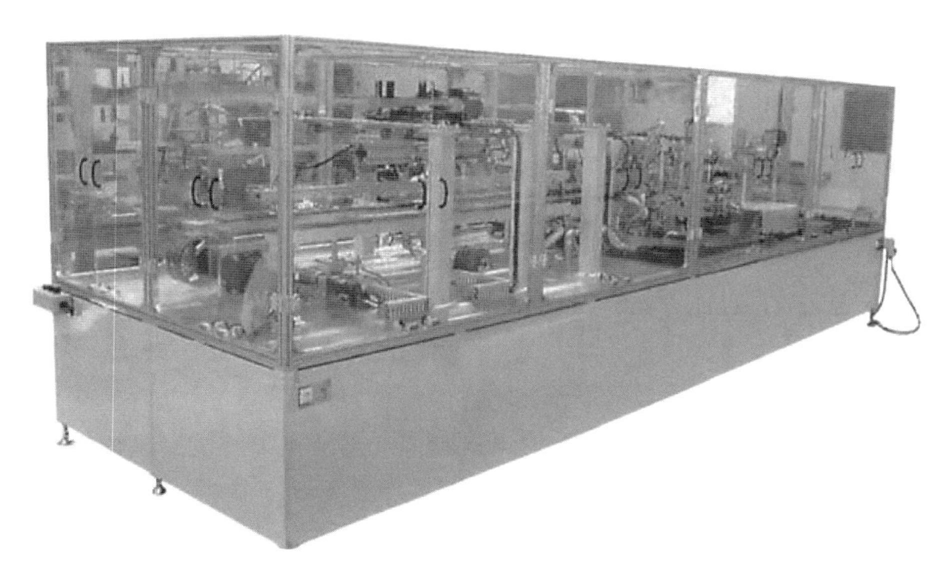

Figure 6.14 Automatic system for cell laying and series welding operations.

Figure 6.15 Manual laminator and auto-loading laminator.

6.4.4 Solar Cell Module Laminators

The solar cell laminator has integrated the vacuum, the pneumatic transmission and PID temperature control techniques. The following parameters can be set on the control table: the lamination temperature, the exhaust, lamination and air filling time and so on. Two control methods—automatic and manual—are available to control the vacuum in the upper chamber, air filling to the upper chamber, vacuum in the lower chamber and air filling to the lower chamber. When EVA is used to package the solar cell, it is generally set as 100–120°. At this temperature, EVA will be just in melting state and will not be solidified. Open the top cover of the laminator, you can see a rubber air ban on the inner side of the top cover, and a seal ring between the top cover and the lower cavity.

During lamination, the cells and the EVA, TPT, and the glass layer shall be stacked between the two layers of glass cloth. The glass cloth can reduce the temperature rise rate of EVA and generate less bubble and prevent the melted EVA from flowing out to pollute the heating board. The working temperature and exhaust lamination time shall be set based on the laminator and module size. Generally, the lamination time is set in a range of 5–10 min. Figure 6.15 shows the manual module laminator and the outline of the laminator with automatic loading/unloading functions.

6.4.5 Solar Simulators, Turnover Trolleys and Frame Machines

Similar to the principle of cell PV performance tester described in Chapter 3, the module tests also need a light source to simulate the solar ray AM1.5 in a larger testing area (see Figure 6.16).

Since the module has bigger size than the cell, some turnover trolleys are needed. And the frame machine shall be used to install the aluminum frameworks (see Figure 6.17).

6.5 Reliability of Solar Cell Modules and Inspection After Packaging

6.5.1 Module Packaging and PV System Reliability

It is internationally recognised that the solar cell module must ensure a service life of more than 25 years and the associated reliability. The design and production of PV

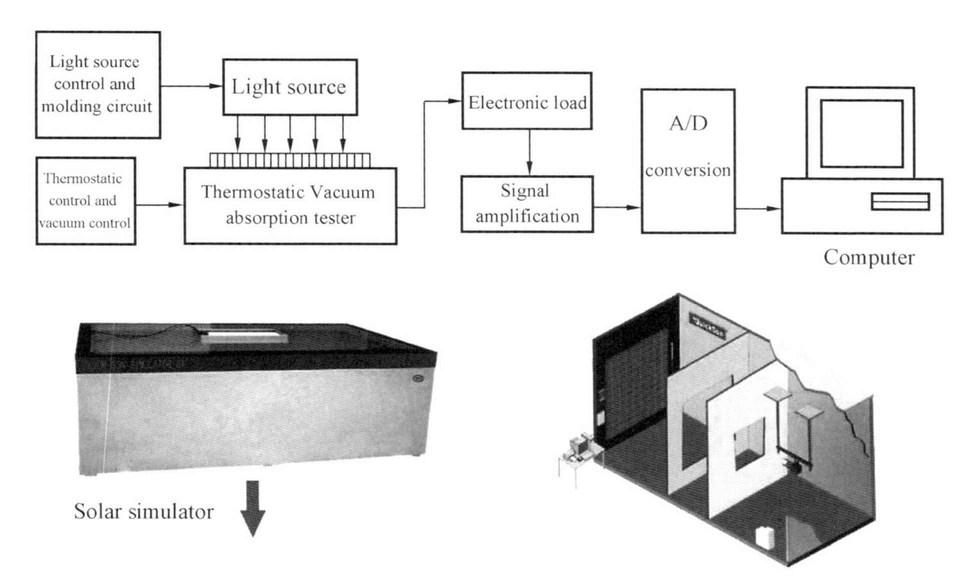

Figure 6.16 Solar simulator system for module tests.

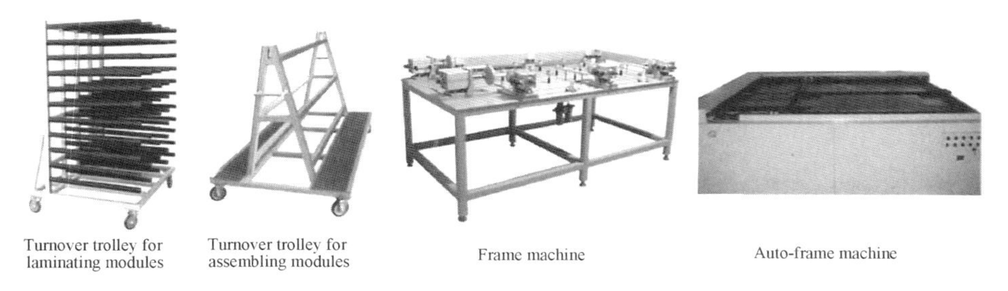

Figure 6.17 Turnover trolleys and frame machines for solar cell modules.

modules are closely related to the power generation and reliability of PV systems, and the module packaging reliability is usually viewed as the key reason for early failure of modules. The manufactures, the installation companies and the users often face conflicts on module quality, for example, it will reduce the generation efficiency of the solar cell when the EVA becomes yellow; the falling off of back boards can affect effective heat transfer, optical mismatch and water penetration, and further corrosion of tin welding points and circuit breaking and finally module failure; the glass will break with stress, resulting in more serious problems; one or more cells with poor quality will result in the hot spot effect; the different packaging materials will have direct impact on the module quality and performance; and the packaging process and processing conditions will have direct influence over the lamination quality and reliability of PV modules. As a result, it is of great importance to select the correct packaging material and carry out the tests after packaging. In conclusion, the overall reliability of the module is dependent on the material quality and processing technique of each component. These tests can help evaluate the expected performance of materials and improve the processed modules and reduce defect spreading.

After lamination, the EVA film, the back boards and glass shall be embedded with sealing rubber belt and coated with adhesives to bond with the aluminum framework, forming the sealing of the module edges. Since the bonded module shall bear its gravity, wind load and other mechanical loads, the adhesives shall have sufficient bonding strength. To ensure the framework sealing materials can withstand light, heat, water, oxygen (oxidisation) and salt and microoganism in the service life of 25 years, the adhesives shall be capable of rainwater prevention and foreign gas invasion to ensure the smooth, reliable and safe operation of the cell module. In addition, since the modules connected to the grid has relations to personal safety, it has certain requirements on the resistivity and breakdown voltage of EVA and adhesives. Besides, when the junction boxes are fixed on the back of the module by adhesives, the junction boxes shall be humidity-proof, dust-proof, air-tight and connected reliably.

During reliability tests, the followings shall be focused concerning defects and stress: (1) breakage at connecting points; (2) cell damage (considering heat and mechanical stress); (3) corrosion resulted by humid gas penetration at high temperature; (4) welding belt falling off by heat, mechanical or electrical actions; (5) colour change of packaging materials at high temperature and in UV radiation; (6) delamination due to poor bonding force; (7) hot spot (thermal effect due to electrical action); (8) poor contact of junction boxes.

The accelerated ageing test can be used to identify whether the external changes will have adverse impact on module reliability and service life, help predict the likely faults in case of outdoor application and remove these faults in advance, and facilitate develop the PV module of better performance.

The proportion of common module faults is listed in Table 6.7.

6.5.2 Objectives and Descriptions of Solar Cell and Module Tests

The solar cell module tester works on the following principle: when the solar rays simulated shine on the cell, the electronic load can be used to control the current variation in the solar cell to measure the voltage/current of V-I characteristics, temperature, light radiation intensity, and the tested data shall be processed, displayed and printed by computer.

Table 6.7 Proportion of common module faults.

Type of fault	proportion %
Corrosion	45.3
Damage at interconnecting point	40.7
Problem due to output	3.9
Junction boxes problems	3.6
Delamination	3.4
Overheating of electric wire, diode and connecting piece	1.5
Mechanical damage	1.4
Defect of bypass diode	0.2

The solar cell module tester is designed with the following functions:

6.5.2.1 Indoor Tests of PV Cells

Objectives:

- Determine the basic performance of the developed prototype and the finished product;
- Offer accurate performance parameters to module design;
- Sort the cell for current matching in case of series connection;
- Determine the cell level.

Descriptions:

(1) Light I-V characteristics (LIV) (basic performance test, classify);
(2) Dark I-V characteristics (DIV);
(3) Spectral response (infer the optical loss range);
(4) Reverse bias load test (determine the bypass diode compliant with the requirements);
(5) Reflection/transmittance (Evaluate the quality of anti-reflection coating);
(6) Others.

6.5.2.2 Indoor Tests of PV Modules

Objectives:

- Determine the basic performance of the developed prototype and the finished product;
- Offer accurate performance parameters for design of module specifications and arrays;
- Win customer's trust and marketing (conformity test, reliability test, and acceptance test).

Descriptions:

(1) Light I-V characteristics (LIV) (basic performance test, classify);
(2) Dark I-V characteristics (DIV) (evaluate the modules in shunt);
(3) Temperature coefficient of module conversion efficiency;
(4) Weather-proof test (humidity/heat, humidity/cold, thermal circulation) (for identification, product development, load test, reliability test).

6.5.2.3 Outdoor Tests of PV Modules

Objectives:

- Determine the basic performance of the developed prototype and the finished product;
- Offer accurate performance parameters for design of module specifications and arrays;
- Win customer's trust and marketing (quality test, and reliability test in extreme environment);
- Confirm the performance module;
- Measure in the expected operating conditions.

Descriptions:

(1) Electrical performance (remove the dust on the module surface before measurement, and the strong wind may result in unstable data);
(2) STC standard conditions (1000 W/cm^2, 25 W); NOCT normal operating conditions (800 W/cm^2, 20 W, wind velocity: 1 m/s);
(3) Temperature coefficient (in two hours before noon, wind velocity < 3 m/s, several thermal couples);
(4) Incidence angle (in two hours before noon, wind velocity <4.5 m/s, advance in steps from 0–90°);
(5) Long-term test (with appropriate inverter, measure the AC and DC performance on a regular basis, and such techniques as IR imaging, UV imaging, electroluminescence are used to take photos for the module on a regular basis to evaluate the change of performance, materials or appearance).

6.5.3 Testing Methods and Verification Standards of Solar Cell Modules

The testing methods of solar cells and modules are mainly based on GB/T9536–1998 'Crystalline silicon terrestrial photovoltaic (PV) modules—design qualification and type approval', GB/T14008–1992 'General specification for sea-use solar cell modules'.

1. Inspection items
   ① Performance test: Measure the open-circuit voltage, the short-circuit current, the maximum output power, V-I characteristic curve of the solar cell and modules at the specified spectrum light source, standard light intensity and cell temperature (25 W);
   ② Electrical insulation performance test: When the DC power of 1000 V + double system maximum voltage is applied to the module bottom plate and the leads, measure the insulation resistance, which shall be larger than 50 MΩ so as to prevent framework current leakage during usage;
   ③ Thermal cycle test: The module shall be subjected to the thermal cycle in the specified number in the range of 40–85 W and hold on for a certain time at the extreme temperature as specified; in this case, record and monitor the possible short-circuit, breaking circuit, appearance defect, electrical performance attenuation rate and insulation resistance rate to identify the thermal alternative performance of the module;
   ④ Humid hot/cold tests: The module shall be subjected to repeated cycles at the given temperature and humidity and keep some time for restoration to identify the performance of the module to withstand high temperature/humidity and low temperature/humidity;
   ⑤ Mechanical loading test: The module surface shall be gradually loaded to identify the performance of the module against wind, snow, ice and other static loads;
   ⑥ Hail impact test: The steel balls, instead of hails, shall be used to impact the module at different angles to identify the performance of the module against hail impact;
   ⑦ Ageing test: When the module is exposed in the site of high humidity and high UV radiation, observe the reduction of PV characteristics to identify its effective anti-attenuation capacity.

2. Codes and standards for inspection and verification of solar cell modules:

GB/T12637-90	General code of solar simulators
GB/T6495.1-1996	Measurement of PV I-V characteristics (IEC60904-1)
GB/T6495.2-1996	Specifications on standard solar cells (IEC60904-2)
GB/T6495.4-1996	Correction methods for temperature and irradiation of I-V measured characteristics of crystalline silicon PV modules (IEC60891)

3. Product marks of solar cell modules shall include the followings:
   ① Name and type;
   ② Main parameters, the short-circuit current, the open-circuit voltage, the optimum working current, the optimum working voltage, the maximum output power and the I-V curve;
   ③ Name of manufacturer, date and trademark.

6.5.4 Tests of PV Performance and Macro Defects of Solar Cell Modules

6.5.4.1 Tests of PV Performance of Solar Cell Modules

Beijing Delicacy Laser Optoelectronics Co., Ltd. has developed a solar cell module PV tester, which is designed with the following functions: I-V curve, the short-circuit current, the open-circuit voltage, peak power, voltage and current at the peak power, voltage and current at the working point, filling factor, conversion efficiency, series resistance, parallel resistance, solar cell temperature and light intensity. To ensure the accuracy of current measurement for solar cell modules, four-wire connections is used.

The solar cell module PV tester, the final testing equipment of solar cell power, consists of three optical structures: top lighting, bottom lighting or side lighting where top lighting and bottom lighting equipment is good for flow-line production, and the tester can be used for the tests before lamination, significantly improving the primary packaging product rates.

The infrared probe can be used to directly measure the temperature of the cells. To ensure accurate automatic corrections of solar cell temperature and light intensity, the temperature and light intensity shall be simultaneously measured.

Figure 6.18 shows the outlines of tester where the size of applicable solar cell modules are 1.2 m × 2 m. The high-energy pulse xenon lamp serves as the light source, the service life is larger than 300000 times, and the light evenness is 3%. The maximum testing voltage is 100 V, the maximum testing current is 20 A. The testing data can be saved, printed, transferred and the database can be organised.

6.5.4.2 Tests of Macro Defects of Solar Cells and the Modules

In addition to the above PV performance of the solar cells and modules, the macro defects, caused by thermal stress and mechanical stress, has attracted more attention in recent years. For instance, the mechanical pressure of the screen printer, the thermal stress in high-temperature sintering, the thermal stress during welding, the mechanical stress during lamination packaging, and the thermal stress during solidification, all of these may result in macro defects in the cell and the module, for example, the hidden cracks and damages of cells, breaking of gate lines on the cell surface and the insufficient solder of leads and so on. These will have a great influence over the module conversion

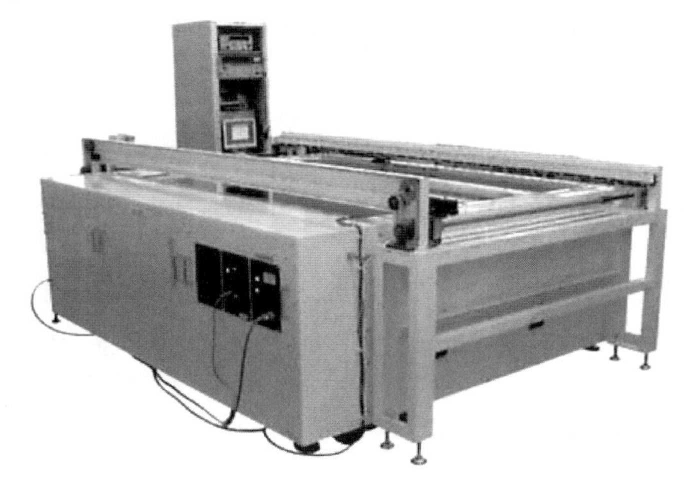

Figure 6.18 Automatic tester for solar cell modules.

efficiency, reliability and service life. The ultrasonic resonant test, photoluminescence and electroluminescence methods can be used to carry out intuitional tests on the macro defects of the solar cell and module.

(1) Ultrasonic resonant test: The frequency and amplitude of the ultrasonic wave can be changed to rapidly detect the cracks of the naked silicon wafer or cell on-line with each only in 2 s. The method, however, cannot detect the defects caused by uneven resistance or detect the module.
(2) Photoluminescence test: The external light source shall be used to trigger the solar cell, the filter and special photosensitive elements to collect the light signal of specific wavelength, and finally data processing to rapidly detect the cracks and resistance distribution data. The method is applicable to naked silicon wafer and cells, but not to modules.
(3) Electroluminescence test: Similar to photoluminescence, it excites the solar cell by means of the external field and collects the lighting signal with specific wavelength by filters and the special photosensitive element, and it can rapidly detect the cracks and resistance distribution by data processing. The method is applicable to detect cells and modules but not to naked silicon wafers.

The comparisons on the above three macro defect detection techniques are listed in Table 6.8.

Table 6.8 Comparisons on the above three macro defect detection techniques.

Detection method	UV resonance	Photoluminescence	Electroluminescence
Advantage	Non-destruction, rapid, on-line	Rapid, on-line, high sensitivity	Rapid, on-line, high sensitivity
Disadvantage	Low sensitivity	Filter required	Filter required
Applicable types	Naked silicon wafer, cell	Naked silicon wafer, cell	Naked silicon wafer, module

The comparisons show, the electroluminescence method has more advantages. It can discover the following defects of the module in an intuitional method: uneven resistance on the cell surface, cracks, fragments, broken gate lines, dry welding joints and so on. It has now become a research focus in the PV industry across the world to detect the defects of the cell and module by the electroluminescence method, and it has developed into a key subject in the research of PV product detection and production techniques.

6.5.4.3 Testing Principles of Electroluminescence

Since the evenness of resistivity has relations to impurity distribution and the surface of the minority carriers with more impurities will see serious recombination with reduced transition probability when the cell is excited by the external field, the light/dark patterns displayed on the electroluminescence picture can correspond to the resistance distribution on the cell.

The fragments and cracks may result in broken main gate lines, and poor screen print quality may result in discontinuous gate lines. The broken gate lines or discontinuity shall fail the external field to successfully reach the PN junction, or even make the whole series circuit open, failing to output the normal power. In this case, the photon density at this position is low, and the lighting picture is dark. As a result, the broken gate lines, cracks and fragments can be all displayed in the electroluminescence detection.

The dry welding joints are caused by welding failure, which can also fail the external field to reach the PN junction, and accordingly can be displayed in the electroluminescence graphs.

The uneven resistance on the cell surface may be caused by high-temperature sintering technique. During normal working of the module, the defect may result in current losses, reducing the conversion efficiency. The cell cracks in the module may be resulted from uneven pressure and bending of the silicon wafer, for example, inappropriate soldering tin operation of the cell, and unreasonable module lamination and solidification processes. These cracks may become fragments during module lamination and solidification techniques or outdoor service, resulting in electrical performance losses or even open circuit, and affecting seriously the service life and reliability of the module. As a result, uneven resistance, cracks, fragments, broken gate lines and dry welding joints must be replaced before solidification.

Figure 6.19 shows the diagramme of electroluminescence for solar cell and module detection. Figure 6.20 shows the picture of the hot spot effect during module detection by electroluminescence. Figure 6.21(a), (b), (c), (d) and (e) show uneven resistance, cracks, fragments, broken gate lines and dry welding joints during module detection by electroluminescence.

The ILMT3.2SI.V electroluminescence (infrared) tester, provided by Shanghai Solar Energy Research Centre, can be used for the possible macro defect detection

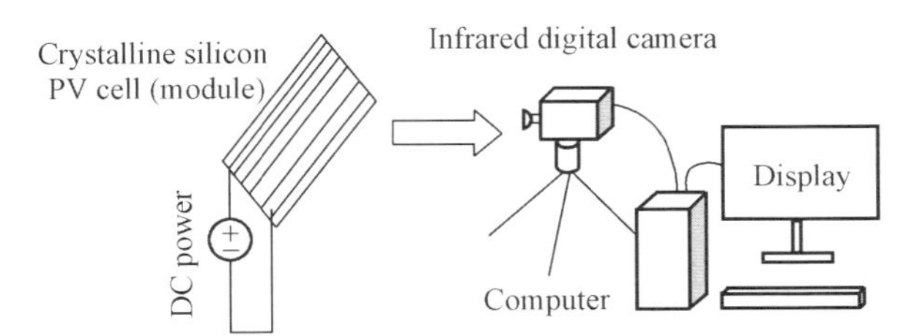

Figure 6.19 Principle diagramme of electroluminescence for solar cell and module detection.

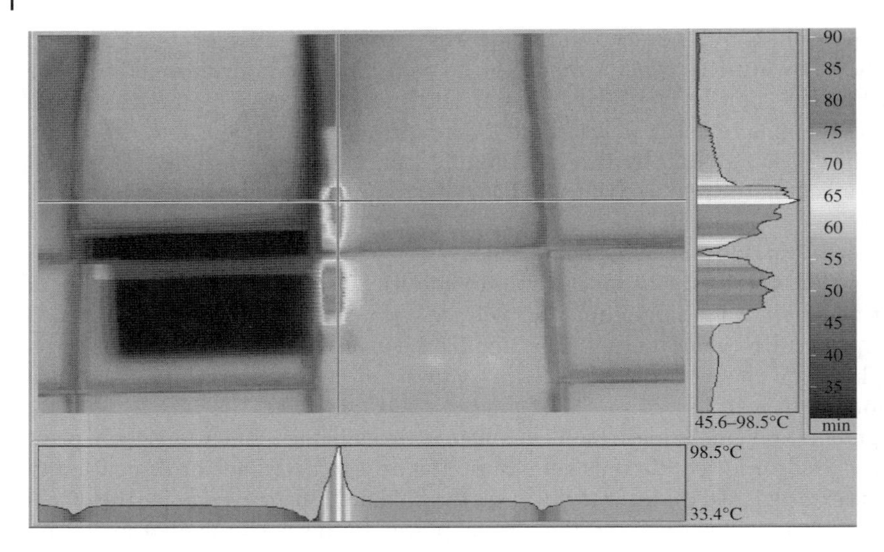

Figure 6.20 Hot spot effect during module detection by electroluminescence.

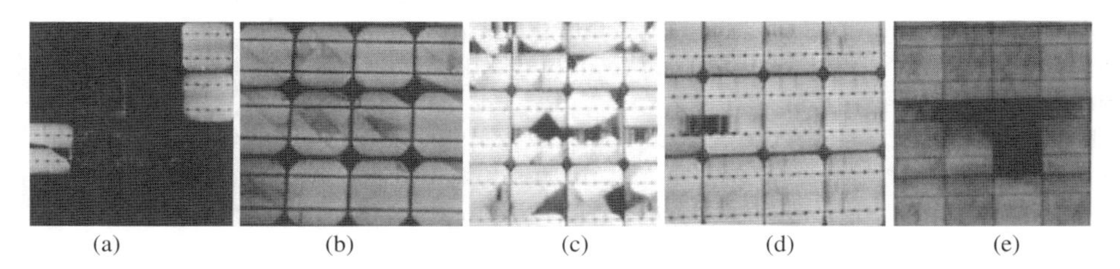

Figure 6.21 Silicon wafer defects during module detection by electroluminescence: (a) Uneven resistance; (b) cracks; (c) fragments; (d) broken gate lines; (e) dry welding joints

(e.g., cracks, fragments, broken gate lines, dry welding joints, uneven resistance) in each procedure of the flat plate crystalline silicon solar cell modules. After the module is powered on, the tester can collect the infrared of the module it sends out, which can be subjected to internal image processing and then automatic defect identification can be carried out. The effective testing area of the tester can reach 2000 × 1600 mm with detection time of 3–5 min, and the cracks with width less than 0.2 μm can be detected. The tested module can be manually loaded and tested or transported by the conveyor for automatic detection, automatic tapping and automatic serial number scanning.

6.6 Efficiency, Common Specifications and Market Development Trend of Solar Cell Modules

6.6.1 Estimates of Solar Cell Module Power and Efficiency

PV conversion efficiency of the solar cell module = peak power/incidence light power. Due to limits of glass and EVA transmittance, welding resistance losses between cells and junction box losses, the peak power total of the module is smaller than the peak power sum of all the cells in the module. In addition, there is ineffective area during cell composition in the module, for example, clearance at the rounded angle of the monocrystalline wafer, the composition clearance between the framework and the cell. As a result, the module conversion efficiency of the module is always smaller than the cell conversion efficiency.

Based on the two common composition methods available on the marked as shown in Figure 6.22—nine in series and four in shunt (36 cells) and 12 in series and 6 in shunt (72 cells), the estimation method of solar cell module power and conversion efficiency can be demonstrated.

For the 125×125 mm solar cell processed by the monocrystalline silicon bar of 150 mm in diameter, it shape takes on quasi-square (shown in Figure 6.23). If installed in the module with framework size of 158 mm(L) $\times$ 808 mm(W) $\times$ 35 mm(H) in the composition type of 12 in series and 6 in shunt (totally 72 cells), the actual irradiating surface of each cell shall be deducted with the clearance area, only 14858 mm^2, that is, <15625 mm^2 of 125×125 mm square area since the clearance is present between the rounded angles of the monocrystalline silicon wafers. Then, the actual effective area of the 72 cells in the module is only 14858 mm^2 (single cell area) $\times$ 72 pcs $=$ 1069776 mm^2 $=$ 1.069776 m^2. If the average conversion efficiency of these cells is 18% at the standard light intensity (STC, 1000 W/m^2), the peak power possibly generated by the module shall be $1.069776 \times 1000 \times 18\% = 192.55968$ W.

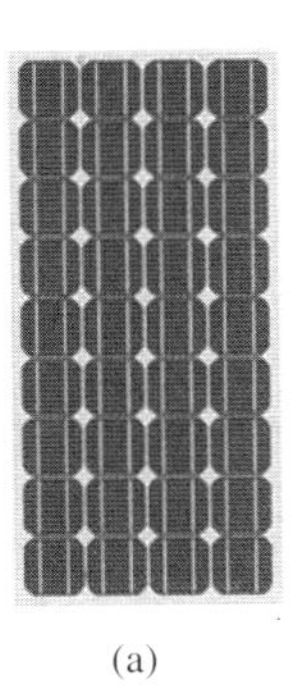
(a)

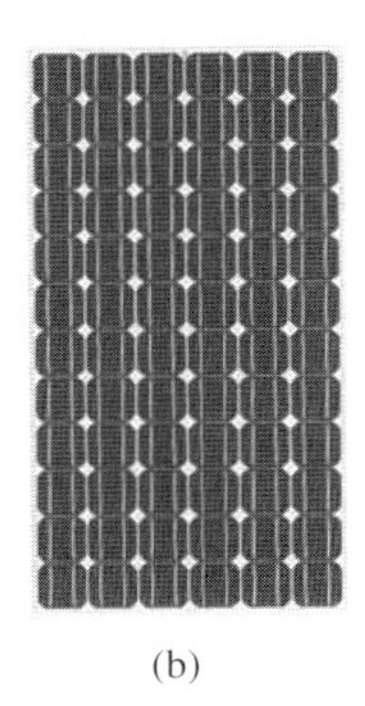
(b)

Figure 6.22 Two common crystalline silicon solar cell modules available in the market nine in series and four in shunt (125 × 125) and 12 in series and 6 in shunt (125 × 125).

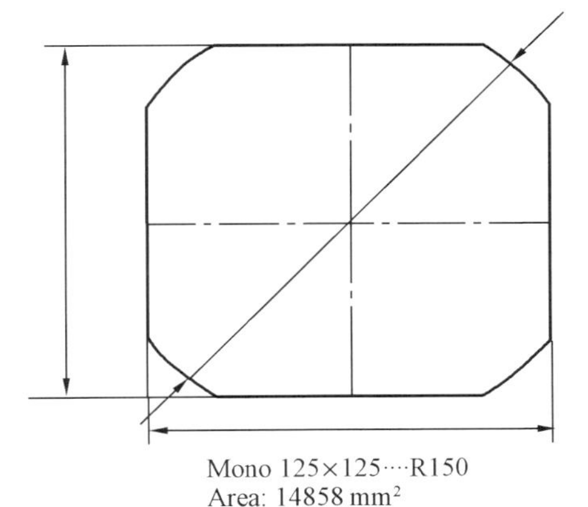

Figure 6.23 Effective area of 125 × 125 mm monocrystalline silicon wafer (deducting the rounded angle).

In addition, since the transmittance of both the top cover glass and the EVA is less than 100%, which, plus the welding resistance losses and the junction box power losses, will further reduce the module peak power. When estimated on 96%, the actual output electrical power of the module composed of 72×125 × 125 mm cells with efficiency of 18% shall be only 192.56 W × 96%=184.85 W. The nominal rated power of the module is 180 W in the market. If the framework area is 1580 × 808 mm=1.17664 m^2, the module efficiency can be worked out, that is, 15.3%.

Obviously, the module conversion efficiency is different from the cell efficiency in the module, and the conversion efficiency of the solar cell module is always lower than the conversion efficiency of the solar cell installed. In addition, since chamfering is present in the monocrystalline wafer, the area of the module cannot be fully used, and the rounded-angle monocrystalline silicon wafer has larger ineffective area than the square polycrystalline silicon module in the solar cell module.

6.6.2 Common Specifications in the Solar Cell Module Market

The solar cell module, the minimal component that can offer DC power supply in the market, has internal connections and packaging. The peak power of the crystalline silicon solar cell module can be 5 W, 10 W, … up to 240 W, or even larger. For easy selection, the common solar cell modules in the market are listed in Table 6.9.

Based on the peak power, rated voltage etc., the solar cell module can be classified into various types for easy selection. For the 125 × 125 mm crystalline silicon cell module (totally 36 cells) with composition type of nine in series and four in shunt, if the open-circuit voltage U_{OC} is 20 V or so, the voltage U_m at the optimum output power is only about 16 V, the rated voltage is 12 V, and when the 12 V lead-acid battery is charged, it can be directly charged in shunt; when the 24 V or 48 V lead-acid battery is charged, the cells in series and shunt shall be added. For DC appliance users such as the portable power charger, living power in the pasture and border post, the signal lamps for city transport and marine navigation, microwave relay stations and so on, the solar cell module can be directly selected. The ordinary user shall set up several modules to a solar cell array and convert them by DC/AC and then put them into use.

For the modules of the independent PV power system and the grid-connected PV power system, the modules are generally 36 pcs, 60 pcs, 72 pcs or 80 pcs of solar cells in series or shunt to form the 12 V, 24 V and 48 V DC output according to IEC: 1215: 1993. If it is used to supply power to the 220 V AC appliance, the DC-AC inverter shall be used to convert the DC power generated by the PV power system to the AC power.

Table 6.9 Specifications of common crystalline silicon solar cell modules.

Peak power (W)	Open-circuit voltage (V)	Short-circuit current (A)	Optimum working voltage (V)	Optimum working current (A)	Common size (mm^3)
5 W	21.48	0.36	16.8	0.3	250 × 300 × 25
50 W	21.77	3.58	17.5	2.86	840 × 535 × 30
100 W	21.77	5.62	18.3	5.48	1080 × 808 × 40
150 W	43.78	4.53	36.2	4.14	1580 × 808 × 43
180 W	44.58	5.21	37.37	4.82	1580 × 808 × 45

6.6.3 Attenuation of Solar Cell Module Power During Usage

We put our focus on how to improve the PV conversion efficiency in Chapters 3, 4 and 5. When the solar cells are installed and assembled into modules, the output power (efficiency) will attenuate with time, which is because:

1. Light degradation (Stabler-Wronski effect) of the solar cell: When the module is exposed to the solar rays in a long time, the solar cell performance will attenuate and finally reach the stable conversion efficiency. The different solar cells have different Stabler-Wronski effect. For instance, for the crystalline silicon solar cell, the reason for Stabler-Wronski effect is carrier lifetime reduction caused by boron-oxygen composite in the crystal. And the measures are to minimise the boron and oxygen impurity content in the cell substrate. For the thin film silicon, it is mainly cause by movement of metastable amorphous silicon atoms in the cell, and the module manufacturer shall deduct the attenuation. For the compound semiconductor, it almost has no Stabler-Wronski effect.
2. Hot spot: When the solar cell module is exposed in the solar rays and some cells in the module fail to work during shading, the shaded part will have a temperature rise far larger than that of the unshaded part, resulting in excessively high temperature rise and damaging the cell. In addition, the hot spot of the solar cell is also dependent on the internal resistance of the solar cell itself and the dark cell size of the solar cell. Because the solar cell is equivalent to a diode in case of no sunshine, the quality cell shall have good rectification ratio (i.e., the positive dark current is much larger than the reverse dark current). The solar cells with poor performance and big internal resistance and dark current are easier to have the hot spot effect and then damaged. To reduce the hot spot effect, the module manufacturer often shunts a bypass diode between the positive/negative poles of the solar cell module. When the cell with poor performance has the hot spot effect and fails to generate power, the diode can work as the bypass to flow the current generated by the cell through the diode and enable the module to continue power generation, and thus it will avoid blocking circuit due to a cell failure.
3. EVA ageing and yellowing: In the yellowing EVA zone, the transmittance of the incidence light will be reduced and the hot spot effect in the zone will be also deteriorated, resulting in reduction of module power.
4. Environmental vapour or corrosive gas invasion: If the module has poor tightness, it may result in broken welding joints or short-circuit faults once the outdoor rain humid gas enters.
5. Hidden cracks in the cell: The cell stress will be intensified due to big outdoor temperature change during long-term service, which may result in hidden cracks in the cell and finally cell breaking and failure.

6.6.4 Development Trend of Solar Cell Modules in China

In the recent years, the countries across the globe have carried out encouragement policies for PV and other new energy sources, and the PV companies have witnessed technical advance. Since 2008, both the silicon material and the solar cell module have seen significant price reduction. In 2013, the module price fell rapidly down to 0.7–0.8 USD/W from 4 USD/W in 2008. In 2013, China produced 25.1 GW modules, accounting

Table 6.10 Cost composition of a module factory, January 2014.

Item	Polycrystalline	Silicon wafer	Cell	Module
Proportion	15%	17%	23%	44%

Table 6.11 China PV parity development roadmap (generator side).

Year	2013	2014	2015	2016	2017	2018	2019	2020	2021
Desulfurization coal-fired power price (yuan/kWh)	0.42	0.44	0.45	0.47	0.49	0.51	0.53	0.55	0.57
PV power benchmark price (yuan/kWh)	1.00	0.95	0.90	0.86	0.81	0.77	0.74	0.70	0.66
Year	**2022**	**2023**	**2024**	**2025**	**2026**	**2027**	**2028**	**2029**	**2030**
Desulfurization coal-fired power price (yuan/kWh)	0.60	0.62	0.65	0.67	0.70	0.73	0.76	0.79	0.82
PV power benchmark price (yuan/kWh)	0.63	0.60	0.57	0.54	0.51	0.49	0.46	0.44	0.42

form 63% of the global PV market (global output: 43.3 GW). Six companies in China mainland, that is, Yingli Solar, Trina Solar, Arts, Jinko Solar, JA Solar and Hanwha Solar One, saw their output larger than 1 GW in 2013. BNEF predicts that the global module output in 2014 will reach 45–49 GW, and the six companies will still take the lead.

Table 6.10 shows the cost composition of a module factory based in China in 2014. Table 6.11 shows the PV parity development roadmap (generator side), China, given by China National Renewable Energy Centre.

6.7 Solar Cell Arrays

6.7.1 Design of Solar Cell Arrays

The solar cell array consists of the solar cell modules in series and/or parallel. The PV system users shall, based on the local solar irradiation and the actual load demands, design the output voltage, current and total power of the array, and determine the quantity of the solar cell modules and the serial/parallel connections so as to obtain appropriate output voltage and current and meet the demand of the user for normal power supply. To reduce the combined losses from serial/parallel connections, the output current and voltage between modules shall be identical. The arrays shall be designed to match the voltage of the module in series with the rated voltage of the inverter.

For the users with batteries, the solar cell array shall be designed to match the voltage of the module in series with the charging voltage of the battery. For the lead-acid battery, if the quantity of modules in series is too small, the series voltage will be less than the battery floating charge voltage and the array cannot charge the battery; and if the quantity of modules in series is too big, the voltage output by the array will much larger than the floating charge voltage and the charge current will not raise obviously. As a result, it can reach the optimum charging status only when the series voltage of the solar cell modules is the appropriate floating charge voltage. The optimum working voltage of the solar cell array shall be the sum of the battery floating charge voltage, the diode and the voltage drop caused by other factors. The battery floating charge voltage, however, has relations to the parameter of the battery, which shall equal to the maximum working voltage of the single battery cell multiplying by the quantity of batteries in series.

The inverters and batteries will be described in Chapter 7.

To help understand the relationship between the module (array) design and the inverter, the energy charging instrument and the local solar conditions, an example is given below. Suppose the average sunshine time in a place is 6 hr, the user's load power is 100 W with 5 hr in service each day, and the conversion efficiency of the inverter installed is 90%. In this case, the actual power demand is 100 W/90% = 111 W, and the power consumed each day is 111 W × 5 hr = 555 hw. For the system provided with batteries, the power consumption will be 70% off with consideration to the charging efficiency of the battery and the losses during charging, and thus the power consumed each day shall rise to 555 Wh/70% = 793 Wh, and the design output power of the module (array) shall be at least 793 Wh/6 h = 130 W. Taking the local weather into account, the working current, working voltage and output power of the solar cell array will become more complicated. In addition, based on the local meteorological data, the expected continuous sunny days between continuous gloomy and rainy days shall be worked out to figure out the power generated by the array, which shall be able to supply the loads and make up for the power of the cell lost in the continuous gloomy and rainy days. And it shall take into account the expected local continuous sunny days

The output power of the solar cell array shall be the product of the rated power of the solar cell module multiplying the quantity of modules in series and in parallel. For the large-power PV ground power station, dozens of or even hundreds of arrays shall be assembled to meet the demand, which shall be discussed in Chapter VII.

6.7.2 Array Electrical Connections and Hot Spot Effect

The solar cell array shall assemble many solar cell modules together. Since the array works outdoors, its output power and efficiency are influenced by temperature and solar irradiation. It shall be noted that the power losses of the array mainly come from inconsistent module characteristics, diode in series/parallel and connecting losses. To avoid hot spot, the solar modules of consistent conversion efficiency, open-circuit voltage and short-circuit current shall be selected to build the solar cell array. In addition, the cells had better have consistent colour and mechanical performance. In the severe cases, the temperature of the hot spot can reach 200°. The hot spot effect may fuse the welding joint, damage the packaging material and even fail the array.

When the modules are connected in series to build the array, the modules shall be of identical working current and each module shall be provided with bypass diode in parallel. And when the modules are connected in parallel to build the array, the modules shall be of identical working voltage and each parallel line shall be provided with blocking diode bypass diode in series. During building, attention shall be paid to prevent the solar module of poor performance from entering the solar cell array. In addition,

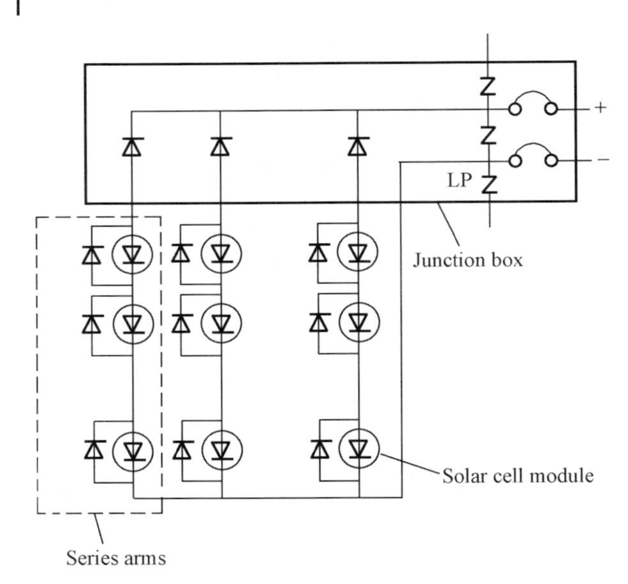

Figure 6.24 Single line diagramme of the solar cell array.

the joint box and the main terminal box shall be provided for the lightning arrester. In some cases, the bird repellent shall be mounted on the top of the array. The DC circuit of the array may be grounded according to the maximum output voltage of the array. See Figure 6.24 for the electrical connections of the solar cell array where the series part of the module is parallelled with bypass diode and the mains outlet is connected in series with blocking diode, which is aimed to prevent damages to the cell by the reverse current. LP refers to the lightning protector.

6.7.3 Installation and Measurement of Arrays

The support for installation of the solar cell array can be used to support the solar cell module, which can ensure secure connections between the module and the support, and facilitate module replacement. It can withstand the wind power of 120 km/h and be free from any damage. The tilt angle of the support plays an important role in local solar irradiation receiving. The tilt angle had better be adjustable so that the solar cell array can output maximum power in the design months (months with poorest average daily radiation).

When installed on the roof, the solar cell array must be connected with the main structure of the building. When installed on the ground, the minimum distance between the array and the ground shall be larger than 0.3 m to facilitate ventilation. The bottom of the column must be connected securely on the foundation to bear the weight of the array and the design wind velocity. For the portable small-power solar cell module, the support shall be also provided to offer reliable service.

The basic characteristics of the solar cell module and array are usually expressed by the following parameters in the standard conditions of IEC: the open-circuit voltage U_{OC}, the short-circuit current I_{SC}, the optimum working voltage U_m, the optimum working current I_m, the optimum output power Pm, the filling factor FF and the PV conversion efficiency η. Generally, a 10 kW array covers about an area of 70–80 m^2.

The capacity of the PV system is expressed by the standard solar cell array power, which refers to the maximum output power at the standard conditions of 1 kW/m^2, AM1.5, 25 W. In fact, it is difficult to measure the PV array in the standard conditions. Usually, the portable PV array measurement instrument is used in the natural solar rays and then the results shall be converted to the IEC standard conditions. The solar V-I characteristic curve and the basic parameters of the array shall be measured in the minimum time and then comparisons with the standard reference cell shall be carried out and the output power of the array can be finally worked out in the IEC standard conditions.

7

PV Systems and Grid-Connected Technologies

7.1 Overview on the PV System

7.1.1 Characteristics, Classifications and Compositions of the PV System

The PV system has the following characteristics:

① Inexhaustible, able to be installed in any place as long as the sun rises;
② Free from any rotary part and thus noise-free;
③ No combustion and thus fuel-free;
No medium consumption and thus reliable operation;
④ Simple maintenance, and low maintenance cost;
⑤ Free from air pollution, and no waste water discharged;
⑥ Long service life of the solar cell, which is the key part of the PV system;
⑦ Short time to acquire the energy, fast startup and thus especially suited for the distributed power stations;
⑧ Easy to install and transfer, and easy to expand or reduce the power size;
⑨ Flexible size (100 W–200 MW).

The PV system, however, has the following shortcomings:

① Since it needs the plane light sources, the plane density is small and the maximum solar irradiation energy on the ground is 1 kW/ m^2 (AM1.5);
② The power generated varies with the weather in a periodic or random manner, and thus it is complicated to regulate the power output;
③ The solar radiations are different in various regions.

The PV system can be classified by usage, as shown in Figure 7.1.

The PV system applications can be basically classified to two types: the stand-alone power system and the grid-connected power system. The former has a main characteristic: It uses the power generated by its own system, and it can purchase power from the utility grid in case of shortage. The system is generally installed with the inverter to supply power to the AC electric appliance, and some are provided with the energy storage device to store the surplus power generated by the solar cell array and supply power in case of insufficient solar rays. The stand-alone power system can be also called as the off-grid power system. See Section 7.6.1 for the detailed classification. The latter has a main characteristic: It can sell the surplus power generated by the PV system to

Technology, Manufacturing and Grid Connection of Photovoltaic Solar Cells, First Edition. Guangyu Wang.

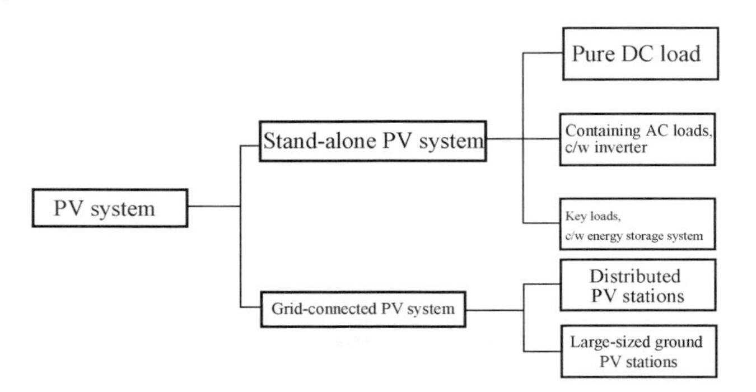

Figure 7.1 Classification of PV systems.

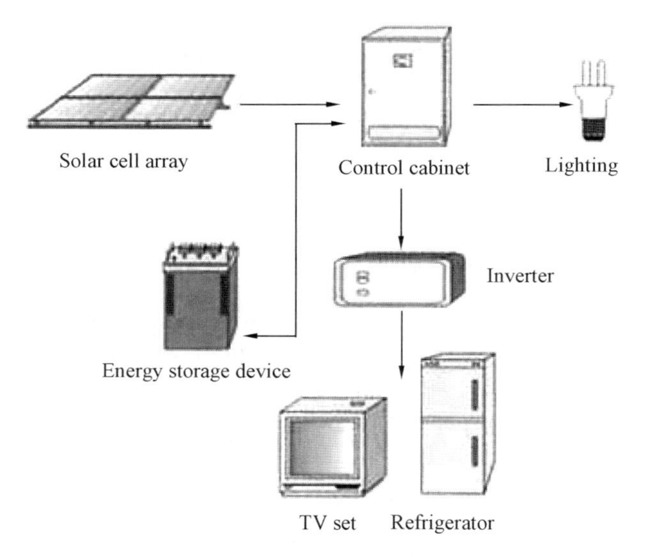

Figure 7.2 Diagram of a stand-alone power system.

the utility grid via the inverter, and it can also purchase power from the utility grid in case of power shortage. As a result, it shall be provided with the meters for power selling and purchase in practise.

Figure 7.2 and Figure 7.3 show the diagramme of the stand-alone power system and the grid-connected power system.

According to power feeding to the utility grid, the grid-connected PV system can be classified to two types: the reverse current type and the non-reverse current type. The former is allowed to feed to the utility grid, and thus it shall be provided with the grid-connected power metering instruments. The latter is not allowed to feed power to the utility grid, but it can be connected to the distribution side of the local grid and receive power from the utility grid. Based on the energy storage devices provided, the PV system can be divided into two types: energy storage type and non-energy storage type. The energy storage device of the former only stores the energy generated by its own PV system. Most of the non-reverse current PV systems are energy storage type, but

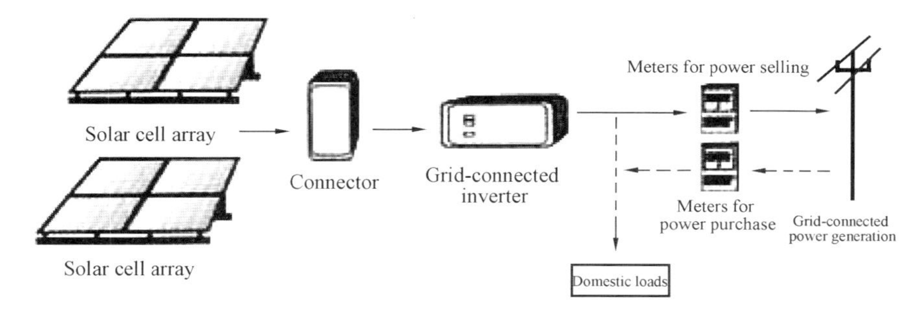

Figure 7.3 Diagram of a grid-connected power system.

the non-reverse current PV system can be not provided with energy storage devices to reduce investment and facilitate management. Usually, the reverse current PV systems are non-energy storage type, which can be also added with energy storage devices to carry out necessary switchover and ensure uninterrupted power supply to the key load in case of grid failure.

The PV system can be connected to the grid at the following voltage levels: to the 400 V LV grid (small-sized PV stations), to the 10~35 kV grid (medium-sized PV stations), and to the grid of 66 kV and above (large-sized PV stations).

For the reverse current PV system, its capacity is generally larger than the load; and for the non-reverse current type, its power is always less than or equal to the load and the power shall be supplied by the grid in case of power shortage, and thus it will never supply power to the grid and the power generated has to be abandoned even if some surplus power exists due to some reasons (see Figure 7.4 for details).

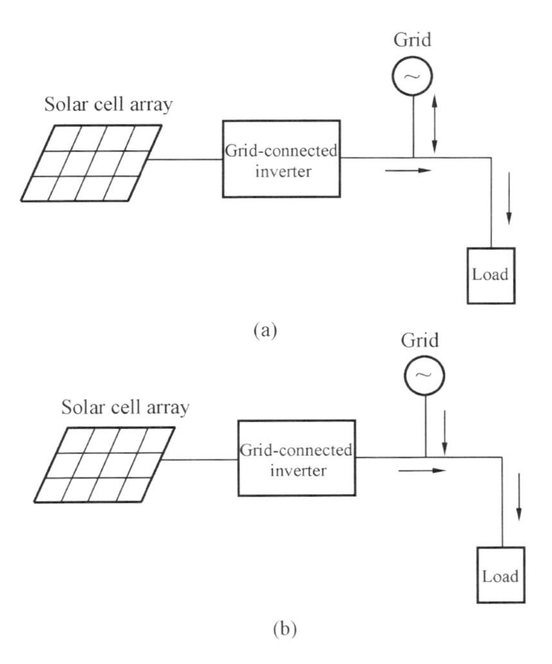

Figure 7.4 Power supply types of grid-connected PV systems (a) reverse current system; (b) non-reverse current system.

The basic working principle of a simple energy storage stand-alone PV system is that the power generated by the solar cell in the solar rays can be used to charge the energy storage device via the controller or supply power directly to the load. In case of insufficient solar rays, the energy storage device can supply power to the DC load under control of the controller. For the PV system containing AC loads, the inverter shall be added to convert the DC power to the AC power. Figure 7.5 shows the working principle of a simple energy storage stand-alone PV system.

Since 1990s, the PV power has seen rapid development. The research on the PV industry shows, the global PV industry developed at an annual speed larger than 30% over the past 10 years. The PV industry first developed in Europe, especially Germany, Spain and Italy and the like. In 2013, the total installed capacity of PV arrays in Germany was larger than 35.72 GW, accounting for 4.9% of the national total power. In 2013, the total installed capacity of PV arrays in was only 17.1 GW in China, accounting for 0.17% of the national total power. It is expected that the global renewable energy will account for more than 30% in the total energy structure as of 2030, where the PV power will account for more than 10% in the total power supply across the world; As of 2040, the renewable energy will account for more than 50% of the total energy consumption where the PV power will account for more than 20% in the total power supply across the world; and in the end of the twenty-first century, the renewable energy will account for more than 80% of the total energy structure, where the PV power will account for more than 60%. These figures show the development prospects of the PV industry and its important strategic role in the energy field.

At present, the grid-connected PV power has become the main stream in the global PV market and the grid-connected PV system has a trend of large scale. In the existing global PV market, the grid-connected PV capacity accounts for 85%, indicating that the PV power is changed from auxiliary energy gradually to alternative energy. In the PV application, China has witnessed rapid development in the recent years. In October 2012, a PV station with 200 MW installed capacity was built in Golmud, Qinghai. It was the maximum single PV system in the world with daily power generation larger than 1340 MWh. If the PV power accounts for one-tenth of the total power in China, additional 350 GW PV systems shall be installed. In the Chinese government plans, the total installed PV capacity shall reach 35 GW in 2015 and 100 GW in 2020. Soon, China will become the largest PV market in the world.

To develop the PV industry in a healthy and smooth manner, the State Council of China published 'Several Suggestions on promotion of healthy development of PV industry' in July 2013. In the document, it proposes to build PV stations in order and

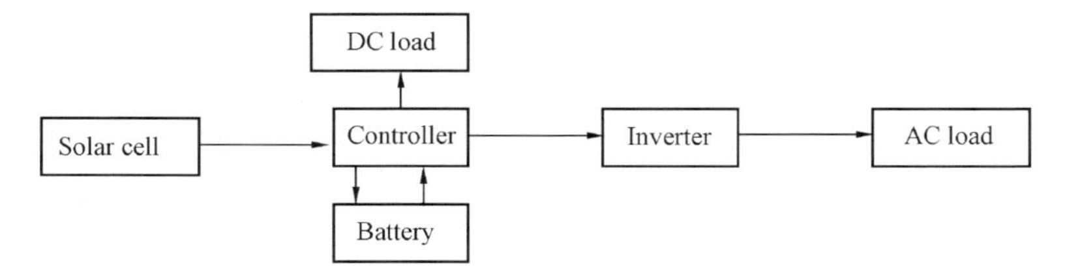

Figure 7.5 Working principle diagramme of a simple energy storage stand-alone PV system.

specified that the state grid must purchase all the power generated by the PV stations and settle according to the power generated, especially, the subsidies for distributed PV power projects are identified. In November of the same year, National Energy Board of China issued 'Temporary rule on PV power operation and management', further identifying the monitoring and management duties of the energy administrations and the associated institutions as well as the responsibilities of the operators and grid enterprises of the PV power project with aim to ensure smooth grid-connection and operation of PV power.

7.1.2 Composition and Simple Working Principles of the PV System

The PV system generally consists of the following parts: the solar cell module (array), the energy storage system, the electronic and electric equipment (the charging/discharging controller, the inverter, the testing instrument and computer monitoring system) and other grid-connected devices.

1. PV module array: The PV array consists of the solar cell modules in series and shunt according to the user's design and demand. It is the core of the PV system, and also the component with maximum cost proportion in the system. It is described in Chapter 6.
2. Energy storage equipment: The solar radiation is intermittent in a day and it is influenced by the weather. The energy storage equipment of the PV system can store the power generated by the solar cell module and release it to meet the demand of the load in case the solar rays are insufficient. For the grid-connected PV system, the energy storage equipment can be omitted to save investment since it can obtain power from the utility grid.
3. Inverter: It is a power converting device to convert the DC power to AC power. In the PV system, the inverter is used to convert the DC power released by the solar cell module or the battery to the AC power for the load. If the system does not contain AC load, no inverter is necessary. For the grid-connected PV system, the inverter shall be designed with regulation functions for voltage fluctuation, current waveform distortion, frequency fluctuation, and sudden blackout in the AC grid.
4. Controller and other power electronic equipment: The controller is the core control part of the whole PV system. It shall be designed with the following functions: 1) The output voltage and current will vary with changes of solar cell temperature and solar intensity. To ensure the PV system always works at the maximum efficiency, the controller shall be able to automatically regulate the working voltage of the solar cell to ensure that the PV system always operates at the maximum power point; 2) The PV system will run alone when the grid fails. In this case, incorrect judgment may be made for live lines. To protect the safety of maintenance staff and equipment, the controller shall be designed with protection functions (islanding protection); 3) The voltage, frequency and phase position shall be automatically regulated to make the output of the PV system consistent with the utility grid; 4) When the PV system or the inverter itself fails or no normal signal is detected, the PV system shall be disconnected from the utility grid in a timely and safe manner; 5) To protect the safety of the PV system and the utility grid, the controller shall be provided with overvoltage relay,

under-voltage relay and frequency up/down relay and so on; 6.) Other functions, for example, for the PV system with batteries, the charging/discharging conditions of the battery shall be specified and controlled, and the power output of the solar cell module and the energy storage system shall be controlled according to the load power demand. For the concentrator solar cells and some arrays with special requirements, the mechanical rotation system for solar ray tracking is often provided. For the large-/medium-sized PV station, the communication system shall be also provided with the utility grid. The multi-function objectives will definitely make the PV control system more complicated and it is difficult to meet such requirements as low price and small volume and so on. It has become the target subjects how to realise the above objectives.

According to the requirements of the users, the protection system between the PV system and the utility grid can be provided with or without insulating transformer. When the insulating transformer is installed between the inverter circuit and the utility grid, the efficiency may be reduced but it is safer (see Figure 7.6 for details).

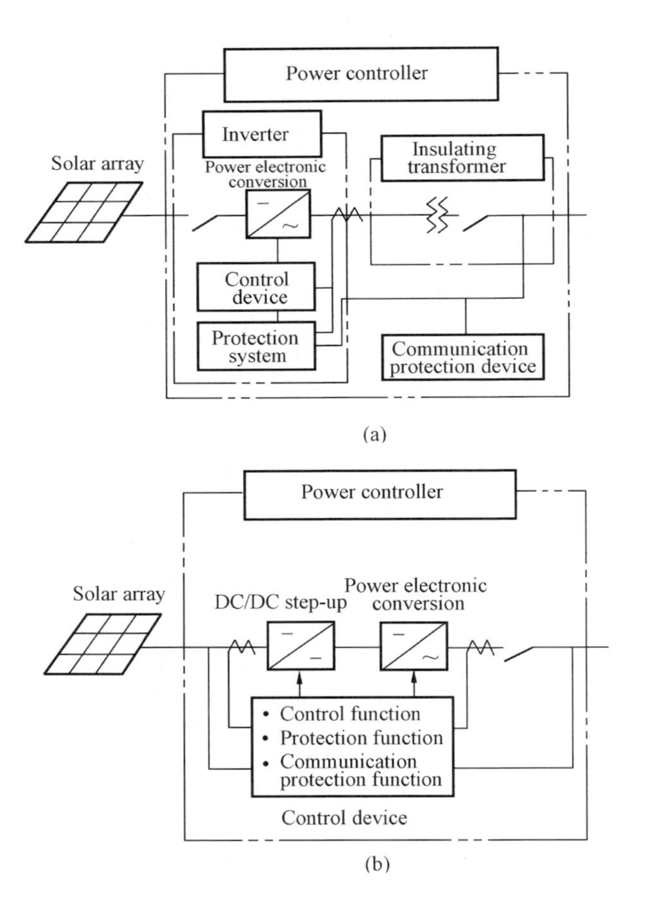

Figure 7.6 Protection circuit of PV system (a) c/w insulating transformer; (b) not c/w insulating transformer.

7.2 Energy Storage Batteries

7.2.1 Energy Storage Batteries and Their Application to PV System

Similar to wind power, the energy storage PV system shall be provided with the energy storage device, which will store the surplus power in the day and release it at night for the user to use (or converted to AC power and then transmitted to the grid). The stand-alone PV system shall be provided with the energy storage device as much as possible to supply power uninterruptedly to the key loads and make full use of the solar energy. For the grid-connected PV system, the energy storage device can work as a buffer and keep the voltage stable in case the utility grid fails.

The energy storage techniques can be divided into two types by their principles: physical and chemical energy storage. The former refers to convert the electric energy first to the potential or dynamic energy and then converted it back to the electric energy for use. The physical energy storage technique is mainly used to store pumped water energy, air compression energy and the flywheel energy storage (emerging in recent years). The physical energy storage technique has developed for a long time, but it has some limitations. For example, the pumped water energy technique can settle the problem and it has been applied to the PV system for many years, but it covers a large area of land and needs large-sized cost. The hydroelectric power station can be not built at any place. And the energy storage efficiency is low.

The chemical energy storage technique refers to convert first the electric energy to chemical energy, and then convert it back to the electric energy for use. The lead-acid battery, the nickel-cadmium battery and the lithium ion battery are all based on the chemical energy storage technique. The lead-acid battery has been widely used in the market thanks to its matured technique and low price, but it has low energy and power density, small number cycle times and it is easy to pollute the environment. At present, many stand-alone PV systems are provided with the sealed lead-acid battery, whose service life can reach 10–14 years. The nickel-cadmium battery has high energy and power density but its cost is high and the heavy metal cadmium is toxic. The lithium ion battery is a matured one. Its energy density is high, and its volume is 40–50% of the nickel-cadmium battery with the same capacity; the working voltage of a single cell can be 3.7 V, which is equivalent to three nickel-cadmium cells in series; it is pollution-free and does not contain any harmful metal such as cadmium, lead, mercury and so on. The number of cycle life is long, larger than 10000 times in normal conditions; it has no memory effect, and has been widely used in laptop, mobile phone and other wireless electronic devices. At present, it has been used as the energy storage battery of the electric vehicle and the PV industries.

The latest flow energy storage battery and the super capacitors are also being development nowadays. The flow energy storage battery, also a chemical one, has different working principle with the conventional chemical ones where the active substance of the flow battery is not accommodated in the solid electrode. The positive/negative active substances are mainly stored in the electrolyte, and they will be pumped to flow through the flow battery. The positive/negative electrolyte in the flow energy storage battery shall be separated by the ion exchange membrane. The flow battery has a big characteristic: it can carry out independent design for the storage energy and the discharge power. It can regulate the electrolyte in the storage tank to meet the demand on capacity of large-sized energy storage during solar or wind power generation, and it can regulate the pairs of positive/negative half cells and the electrode area to achieve the rated discharge power. Compared with physical energy storage techniques, the flow battery is not limited to space and regions since it covers a small area. As long as the new energy power equipment is installed, it can be installed, too. Compared with the physical energy storage technique and the traditional cells, the flow energy storage battery has the following virtues: longer theoretical cycle life, safer and more reliable, higher energy density, low primary investment and high efficiency and so on. It has wide application space in the small-/medium-sized PV systems and the communication systems and other emergency power supplies, and it is also suited to store large-sized power, playing a 'peak levelling and valley filling' role for the grid system during wind or solar power generation and improving the grid quality and reliability. The flow battery is a key development trend of the energy storage technologies, and it is expected to settle the key problems in the new energy power generation and grid connection. Table 7.1 shows the comparisons of the energy storage techniques:

The pumped water energy storage, flywheel energy storage, and compression air energy storage techniques have been applied to the power system, but they are limited to regions, construction period, or project investment, and thus they will not be described in this chapter. The focus shall come to the lead-acid battery, the lithium ion battery, the flow energy storage battery and the super capacitors and so on.

7.2.2 Lead-Acid Batteries

The off-grid solar cell power system is often provided with the lead-acid battery as the energy storage device. The lead-acid battery has low price and the widest application. For the PV system with high requirements, the deep-discharging closed valve-regulated lead-acid (VRLA) battery and the deep-discharging electrolyte-absorption lead-acid battery are often used.

The rated voltage and power of the battery selected for the PV system shall ensure the normal working of the load in gloomy and rainy days; otherwise, over-discharge for a long time will damage the battery. To achieve the working voltage of the load, the batteries can be connected in series according to the rated nominal voltage of the lead-acid batteries. To achieve the desired current for the load, the batteries can be connected in shunt according to the rated capacity. In fact, many factors will affect the capacity and service life of the lead-acid batteries. In addition, the capacity of the lead-acid batteries also changes, and two most important factors are the discharge rate and the ambient temperature. The capacity of the battery will vary with discharge rate, and it will reduce when the discharge rate rises. The capacity of the battery will vary with the battery temperature, and it will reduce when the battery temperature falls down. Usually, the capacity of the lead-acid battery is determined at 25 °, and the capacity at 0 ° will reduce to 90% of the rated capacity. In addition, the water generated during discharging will dilute the electrolyte, raising the condensation point of the electrolyte, and the electrolyte in the lead-acid battery may condensate at lower temperature, damaging the battery. The lead-acid batteries have the specified discharge depth, and even the deep-cycling lead-acid battery cannot make its discharge depth larger than 85%.

To achieve coordinated working of the PV modules (arrays) and the batteries, the batteries must be matched with the PV modules (arrays). It shall ensure the battery will

Table 7.1 Comparisons of energy storage techniques.

Technique		Advantage	Disadvantage
Chemical	The lead-acid battery	Matured technique (more than 100 years of application) cheap, accounting for 45–50% market	Low energy and power density Small cycle number, harmful to the environment
	Nickel-cadmium battery	Relatively mature technique, high energy density Longer cycle life than lead-acid batteries	More expensive than lead-acid batteries Toxic (Cd)
	Lithium (cobalt) battery	High energy and power density High charging/discharging efficiency	High cost (especially Co) Management required
	Sodium/sulphur storage battery	High charging/discharging efficiency Small volume, long cycle life	Expensive High working temperature (300–350°C)
	VRB	Big cycle charging/discharging number, long life Avoid self-discharging losses Applicable to large-sized application/ easy to upgrade	Expensive
Energy storage	Zinc-bromine battery	Low cost, working at the ambient temperature, applicable to large-sized application	Self-discharging
	Super capacitors	Fast charging Applicable to auxiliary peak levelling and brake energy storage of electric vehicles	Low energy storage density
	Fuel cells	React with the oxygen in the air to generate water for power generation without any combustion, and suited to energy storage of PV power generation	PV hydrogen production to be scaled Hydrogen storage technique to be developed
Physical energy storage	Flywheel	High power density Long cycle life, fast charging Power and energy capacities are independent from each other	Low energy density Big additional losses
	Pumped water energy storage	Huge power and energy capacity	Difficult site selection Big construction cost, long construction period

not be excessively discharged or changed at a rate larger than the battery's maximum charge rate, especially when the solar radiation is at peak, the PV modules cannot have excessively big charge rate at the battery; otherwise, it may damage the battery.

7.2.3 Lithium Ion Batteries

The lithium ion battery, a chemical one, has such advantages as high energy density, high working voltage, long cycle life, green and environmental protection. It uses two compounds that can reversely embed and get off the lithium ion as the positive/negative electrodes. The positive electrode adopts the transitional metal oxide of lithium, for example, $LiCoO_2$ and the negative electrode the carbon of special molecular structure, for example, graphite and carbon fibre. During charging, the potential applied to both electrodes of the battery will force the positive electrode compound to release lithium ion (getting off) and embed to the negative electrode carbon with molecules in laminated structure. During discharge, the lithium ion will separate out from the carbon in laminated structure and recombine with the compound of the positive electrode (embedding), and the current is generated by movement of lithium ions. The getting off and embedding reactions are reversible, the lithium ion can get off and embed repeatedly in the vacant lattice point with appropriate size, realising lithium ions moving between positive/negative electrodes. The structure and charging/discharging of lithium ion batteries are shown in Figure 7.7. The lithium ion batteries consist of the positive/negative electrodes, the electrolyte, the membrane, the positive/negative leads and etc. The membrane is polymer. The charging temperature of the lithium ion batteries falls in a range of 0–45 °, and the discharge temperature in a range of −20–60 °. To achieve repeated charging and discharging, the material of the positive electrode shall be added with appropriate additives, and the material of the negative electrode shall be designed with

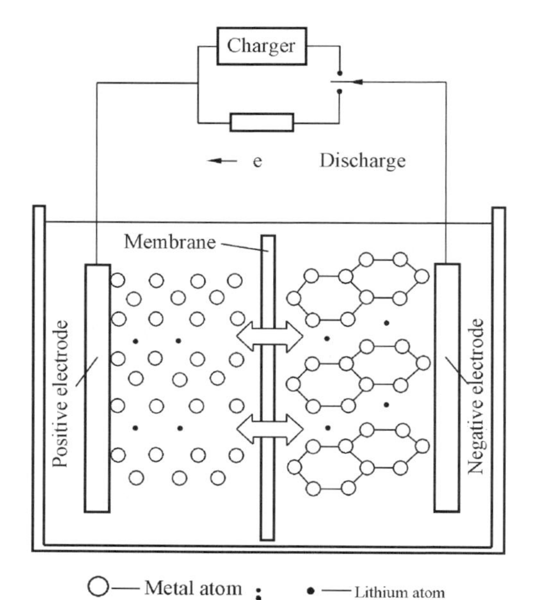

Figure 7.7 Structure and charge/discharge of lithium ion batteries.

proper molecular structure to accommodate more lithium ions. And the electrolyte between the positive/negative electrodes shall be of good stability and conductance.

At present, two factors limit the application of the lithium ion batteries to the energy storage field: the high price of single cell and the complicated power supply management system. Because the single lithium ion cell has limited capacity, and a lot of single lithium ion cells shall be connected in series/shunt to form the battery pack to meet the demand of energy storage field. As the quantity of cells in the battery pack rises, the difference of temperature, capacity and series impedance between cell units has become an important problem. Because the unevenness of the single cell nature will result in attenuation of battery pack capacity, the battery pack shall be under equalisation management.

7.2.4 Liquid Flow Energy Storage Batteries

See Figure 7.8 for the structure of the liquid flow energy storage batteries.

7.2.4.1 Sodium-Sulphur Batteries

The sodium-sulphur battery was invented and released by Ford in 1967. The sodium-sulphur battery consists of the positive/negative electrodes, electrolyte, membrane and the enclosure. Different from the ordinary traditional lead-acid batteries and nickel-cadmium batteries, the sodium-sulphur battery is composed of the electrodes in molten liquid state and the solid electrolyte instead of the solid electrodes and liquid electrolyte. The active substance of the negative electrode is the molten metal sodium, and that of the positive electrode is sulphur and sodium polysulfide molten salt. Since sulphur is insulated, it is often filled in the conductive porous carbon or graphite felt, and the solid electrolyte (and also as the membrane) is made of a ceramic material (Al_2O_3) to conduct sodions. The enclosure is usually made of stainless steel and other metal materials.

The sodium-sulphur battery has many characteristics: first, it has high specific energy (i.e., the effective electric energy possessed by the cell unit mass or unit volume) and the theoretical specific energy is 760 Wh/kg and it is larger than 300 Wh/kg in practise (the lithium ion battery is 100 Wh/kg). Second, it can discharge with big current and high power, and the discharge current density can often reach 200–300 mA/cm^{-2}.

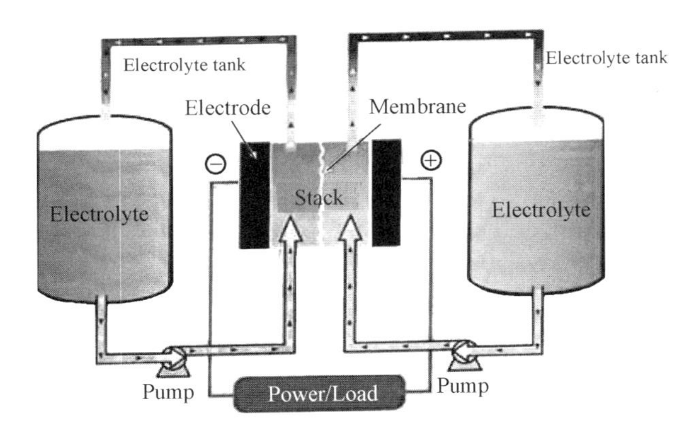

Figure 7.8 Structural diagramme of flow batteries.

Third, the charging/discharging efficiency is high. Since the solid electrolyte is used, it has no the self-discharging and other by-reaction like the energy storage battery with liquid electrolyte, and its charging/discharging current efficiency is almost 100%. The sodium-sulphur battery, of course, has its shortcomings. Its working temperature is 300–350 °, and it needs heating and insulation during service. The vacuum insulation technique of good performance can effectively settle the problem.

The sodium-sulphur energy storage battery was first applied to the electric vehicle. In late 1980s and early 1990s, the sodium-sulphur battery was mainly applied to some special fields such as the energy storage for power stations, UPS and transient compensation power supplies across the world, and it has entered commercialised implementation stage since 2002. At present, the 8 MW sodium-sulphur energy storage battery has been built in the globe. Shanghai Institute of Ceramics of Chinese Academy of Sciences, which cooperated with Shanghai Municipal Electric Power Company, achieved important breakthroughs on research and development of large-capacity sodium-sulphur batteries, and successfully developed the 650 Ah sodium-sulphur single cell with independent intellectual property, making China the second country with the core technique on large-capacity sodium-sulphur single cell in the world, following Japan. NGK, Japan, is the benchmark company on development and application of sodium-sulphur batteries in the world. Since the 1980s, Tokyo Electric Power Co., Japan and NGK have developed the sodium-sulphur battery as the energy storage battery. In 1992, the first sodium-sulphur battery energy storage system began pilot operations in Japan. In 2002, the sodium-sulphur battery began commercialised application. In 2005, NGK had its annual output of sodium-sulphur batteries larger than 100 MW and started exports overseas. At present, State Grid Corporation of China has attached great focus to the application of sodium-sulphur batteries to grid peak regulation and power supply reliability.

7.2.4.2 Vanadium Redox Batteries

The vanadium redox batteries (VRB) utilises electric energy storage and release via conversion of vanadium ions in different valence states, and it is made of the same elements, which is few in the numerous chemical power supplies. The positive/negative electrode reaction is finished in liquid phase, and only the chemical valence of the vanadium ion in the solution is changed during the charging/discharging process, and no external ions participated in the electrochemical reaction so that it can avoid the cross pollution caused by mutual penetration of the active substances of different groups. Since the active substances of the positive/negative half cell electrolyte are stored in different storage tanks, it can completely avoid self-discharging losses during electrolyte storage. Theoretically, it can conduct infinite charging/discharging cycles, and thus it can dramatically extend the cell service life. The VRB experimental stack has been built in the world, and it proves that its life is much longer than the existing lead-acid battery system. At present, the problem is that the proton isolation membrane is monopolised by other countries instead of China, and its cost is about a half the VRB. The Chinese Institutes are actively developing the technique.

7.2.4.3 Zinc-Bromine Flow Batteries

For the zinc-bromine power battery, the electrodes are made of zinc-bromine compound and carbon composite, the electrolyte is the $ZnBr_2$ water solution, and the membrane between electrodes is made of high polymer cellular materials. Since bromine has strong corrosion, the pH value of the solution shall be under strict control.

The theoretical open-circuit voltage of the zinc-bromine flow battery is 1.82 V, higher than that of the VRB. It can be charged rapidly and discharged at depth without any under-voltage protection. It is safe and reliable with long cycle life. It does not need complicated battery management system. It is pollution-free and its electric energy density is 3–5 times that of the lead-acid battery. A more important advantage is that the zinc-bromine battery can work near the ambient temperature, which makes possible to expand its application.

The total cost of the flow battery is to a great extent dependent on the price of the electrolyte. The electrolyte of the zinc-bromine battery is made of zinc and bromine, where zinc is a common metal, easy to acquire with low price and bromine is more common. As a result, the zinc-bromine battery has an inherent cost advantage. It is estimated by the insiders that the cost per kW of the sodium-sulphur battery and the vanadium battery is about USD500, while that of the zinc-bromine battery is about USD100, similar to that of the common lead-acid battery. For the zinc-bromine energy storage battery, the energy storage capacity and the discharging power can be independently designed, which makes it one of the options for large-sized energy storage batteries. The zinc-bromine battery is mainly applied to the energy storage battery for vehicles, the peak regulating and electric energy storage device for PV and wind power stations.

It is reported that the United States and Japan have assembled 5–45 kW zinc-bromine battery for the power supply of electric vehicles; ZBB Energy Corporation has designed a 400 kWh pilot stack; Japan has assembled a 1–4 MW zinc-bromine stack with overall energy efficiency about 65.9%.

All in all, compared with the lead-acid batteries, the flow energy storage battery has caught attractions in the wind power and PV power field and begun its pilot application thanks to such virtues as large capacity, large cycle times, long service life and so on. At present, the key technique of the flow battery is the membrane material, the preparation technique of electrodes and bipolar plates of the positive/negative half cells and the electrolyte preparation and storage as well as the testing and expression methods related to electrochemical materials and membrane production.

7.2.5 Super Capacitors

The super capacitors stores and releases the electric energy mainly via ion absorption and desorption in the electric double layers formed on the electrode/electrolyte interface. The maximum charging/discharging performance of the super capacitors is controlled by the ion and charge transfer speed on the active substance surface. It does not involve in the change of the electrode nature during the whole charging/discharging process, and theoretically, the super capacitors have high-power density and cycle life. Compared with the traditional batteries and the electrolyte capacitors, its power density is much larger than the lead-acid batteries, and its cycle life is longer with working temperature range in −40 °—+60 °. It is maintenance-free and it does not contain heavy metal or harmful substances and thus it is a green and environmentally friendly power supply. The super capacitor can work as the auxiliary power supply of the PV system. Because it can rapidly store the PV energy and release it as demanded, it is more applicable to the applications with big voltage fluctuation. At present, the standard super

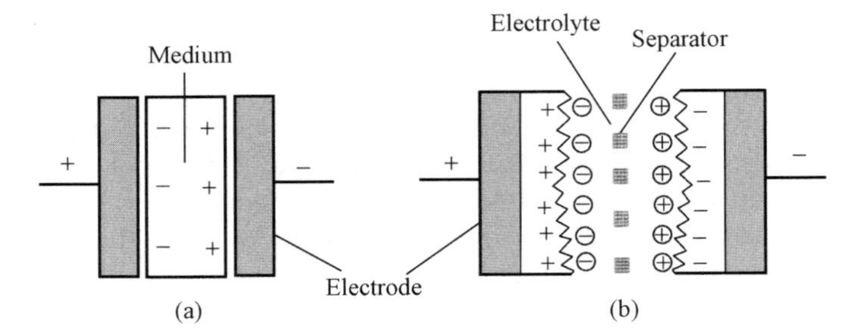

Figure 7.9 Structural comparisons between electric double-layer super capacitors and electrolyte capacitors. (a) Electrolyte capacitor; (b) Electric double layer super capacitor.

capacitors are available on the market, for example, the energy storage power supply for the PV beacon, the lawn and courtyard lamps and the street lamps. In addition, since it has fast charging/discharging speed, it is suited to the energy storage element for the LED flashlight of the camera phone.

Figure 7.9 shows the comparisons between the electric double layer super capacitors and the common electrolyte capacitors concerning structure. It stores energy via polarisation of the interface electric double layers between the electrodes and the electrolyte where no chemical reaction happens. The ions of the super capacitors can be very close, and if the active carbon with huge surface area is adopted, its electric capacitance can reach several thousand F, and the energy storage process is reversible. Other energy storage methods, including the lead-acid battery, the nickel-cadmium battery and the lithium ion battery, are all related to chemical reactions.

EEStor, USA, has produced the new multi-layer ceramic super capacitor by high-purity $BaTiO_3$ powder where the high-purity $BaTiO_3$ powder is coated on the surface of the resin film and it will peel off after drying, and then cut and processed to the multi-layer ceramic capacitor in the extra clean environment. No electrochemical reaction happens during charging/discharging process, and there is no organic solvent electrolyte in the capacitor, and it has no material decomposition problem during usage, and its cycle life is up to 10×10^6 times, and it has no the safety problems of the lithium ion batteries such as fire or explosion due to overheating.

7.2.6 Fuel Cells

7.2.6.1 General

The fuel cells used hydrogen, natural gas, coal gas, alcohol, methanol, methane or petroleum as the fuel where the hydrogen (or hydrogen in the natural gas, the coal gas or petroleum), instead of combustion (i.e., free from the complicated process from chemical energy to heat, and then to mechanical and electric energy), directly reacts with the electrolyte and the oxygen in the air to generate water and power. The fuel cells have the following advantages:

① Since no combustion is necessary, it has few pollution and the emission of harmful gases (NO_X, SO_X and CO_2) is nearly zero;
② The conversion efficiency is high, and the electric conversion efficiency can reach 40%, and the electro thermal overall conversion efficiency can reach 90%;

③ The fuel cells have no rotary parts, and thus the noise is small;
④ It can continuously supply the reactant (fuel) and discharge the product (water) and thus output electric power in a continuous manner;
⑤ The modular system design is available to build the cells of various specifications and power, and it is expected to connect to the grid;
⑥ It is free from regional limitations and thus has caught wide attention.

If the power generated by the PV system is used to electrolyse water to generate hydrogen and oxygen, the fuel cell can discharge power for this purpose. For the PV power, it is an ideal energy storage battery.

In essence, the fuel cells are based on chemical energy power generation (see Figure 7.10 for the structure).

The fuel electrode and the air electrode of the fuel cells have the following reactions:

Reaction of the fuel electrode:

$$H_2 = 2H^+ + 2e \text{ (under the action of the catalyst)}$$

(H^+ shall move towards the air electrode via the electrolyte while the electrons shall flow towards the air electrode via the external load)

Reaction of the air electrode:

$$2H^+ + 2e + O_2 = H_2O \text{ (under the action of the catalyst)}$$

The overall reaction of the oxidation-reduction reaction that happens in the fuel cell is as follows:

$$2H_2 + O_2 = 2H_2O$$

The balanced voltage of the fuel cells can be figured out by the Nernst equation, and the open-circuit voltage of the fuel cell at the normal pressure is 1.23 V.

For the fuel cell for energy storage purpose of PV power, the key technique lies in solar cell scale hydrogen producing technique and the hydrogen storage technique after production as well as the production technique of hydrogen fuel cell (e.g., catalyst and cell membrane).

It shall be pointed out that the fuel cells can be designed with many fuels in addition to hydrogen. By the types of electrolyte, the fuel cells can be divided to alkaline fuel cells,

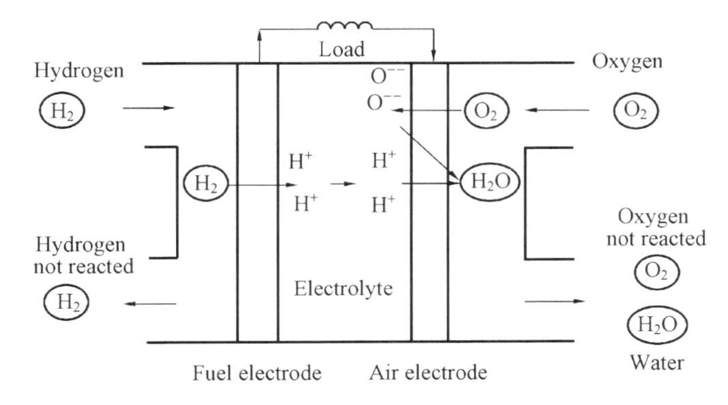

Figure 7.10 Basic structure of fuel cells.

phosphate fuel cells, molten carbonate fuel cells, solid electrolyte fuel cells, methanol fuel cells and proton exchange membrane (PEM) fuel cells. Of them, the PEM fuel cell is very promising because its separation material between the fuel electrode and the air electrode is the electrically insulated high molecular thin film, and it can avoid the troubles of electrolyte. Moreover, it is airtight and it can work at low temperature. The other fuels except hydrogen have all a restructuring system, that is, these fuels shall be converted to the fuel with hydrogen as the main content that can be used for the fuel cell.

The fuel cell for the PV system adopts the hydrogen directly electrolysed by the water that is generated by the PV system and thus no fuel restructuring system is needed. As a result, the fuel cell with hydrogen as the raw material is especially suited to work as the energy storage battery of the PV system.

Since the fuel cells are free from space limitations, it can serve as the distributed power supply (e.g., electric vehicles) and be connected to the grid. Japan has installed 10,000 sets of fuel cell systems. It is reported that all the gas service stations in the United States must be provided with the alternative fuels for users (hydrogen or the natural gas) as of 2018. At present, most of the fuel cell vehicles use the compression hydrogen as the fuel, and the research and development are on the way for the power vehicle using the natural gas as the direct fuel. And some large natural gas and power companies, together with the fuel cell manufacturers, are developing the fuel cell power station system. The fuel cells have been applied as the portable and emergency power supplies, the UPS, the domestic power stations, the backup power supplies of the vehicle and military facilities.

7.2.7 Capacity Design of Battery Packs

At present, the PV system is mostly designed with the following energy storage power supplies: the ordinary lead-acid batteries, the lead-acid maintenance-free batteries and the alkaline nickel batteries. Although the nickel-cadmium battery has such virtues as good low temperature, overcharging and over-discharging performance, it has high price and thus is only applied to some special occasions. In China, the PV systems are mostly designed with the lead-acid maintenance-free batteries. During design of battery pack capacity and composition, it shall take into account that the battery often works at float charging and its voltage will vary with array power generation and load power consumption. In addition, the power provided by the battery is also under the influence of the ambient temperature and the gloomy and rainy days. To achieve longer service life, the capacity of the battery pack shall be matched with the associated PV system.

The capacity of the PV system battery shall be based on the insufficiency of the PV power generated for the load, and special attention shall be paid to the load power consumption during continuous gloomy and rainy days; otherwise, the battery may be damaged due to long-term over-discharging. As a result, the daily average power consumption of the load and the power consumption during gloomy and rainy days as well as the discharge depth are the key factors to determine the battery capacity. Below is the equation for battery capacity (Formula 7.1):

$$B_C = KQ_LN_LT_O/C_C(\mathrm{Ah}) \tag{7.1}$$

Where, K is the safety coefficient, 1.1~1.4; Q_L is the daily average power consumption of the load, working current×daily working time; N_L is the number of maximum continuous gloomy and rainy days; T_O is the temperature correction coefficient, generally 1 for above 0 °, 1.1 for above −10 °, 1.2 for below −10 °; C_C is the discharging depth, generally 0.75 for the lead-acid batteries, and 0.85 for the alkaline nickel-cadmium battery.

In addition, the battery of larger capacity shall be selected in the actual applications to reduce the number of cells in shunt, which is aimed to avoid the effect of imbalance between cells in shunt. The number of cells in shunt is often not larger than 4. Nowadays, many PV systems are provided with two sets of cells in shunt. In this way, it can be disconnected from service for maintenance or repair in case one set fails.

7.3 Core of the Inverter—Power-Switching Devices

7.3.1 MOSFET and IGBT and Other Power Electronic Power-Switching Devices

The modern inverter technology is based on many subjects, including the power electronic technology, the semiconductor material and component technology, the modern control technology, pulse width modulation (PWM) technology and industrial electronic technology. Its core is the fully controlled power device with switching characteristics, that is, the switching control signal will be transmitted periodically by the main control circuit to the power device according to the given control logic, and stepped up or down by the transformer and filtered to achieve the desired AC current.

Generally, the small-power inverter adopts the power-metal oxide semiconductor field effect transistor (MOSFET), the medium-power inverter adopts the insulated gate bipolar transistor (IGBT), and the large-power inverter adopts the gate turn-off thyristor (GTO). The IGBT of HV-withstanding and big current performance has witnessed great rise in application to the PV ground station inverters. It has been listed to the national key projects by the Chinese Ministry of Science and Technology. This is because IGBT has the virtues of power- MOSFET such as high input impedance, fast opening, simple driving circuit, small power for driving and so on and the virtues of the bipolar junction transistor (BJT) such as conductivity modulation in the drift zone, low conduction losses and the like. Compared with the power-MOS device, it has bigger current density, larger power capacity; and compared with the BJT, it has higher switching frequency, wider safety working zone, which makes it the leading power switching device in the voltage applications above 600 V. At present, the inverters of the PV ground stations are mostly provided with the large-power IGBT of 3000 A/1700 V and above, and it is developing towards higher voltage and bigger current, and it is taking the market share of SCR and GTO. MOSFET is usually applied to the occasions with low power and high frequency (<1000 W, <100 kHz; see Figure 7.11). At present, the power electronic device for PV power and wind power grid-connecting inverters has developed from the MOSFET to the third, fourth and fifth generations (silicon IGBT), and the silicon carbide IGBT will be developed in the future.

7.3.2 Structure and Working Principles of IGBT

We will focus on the working principles of IGBT. Figure 7.12 shows the structural diagramme of IGBT.

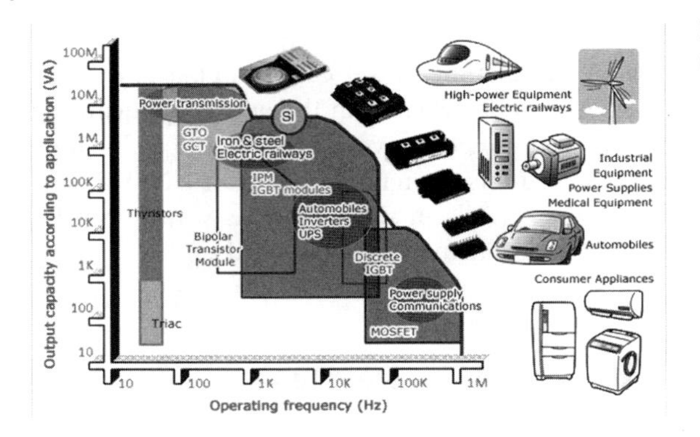

Figure 7.11 Applications of IGBT (frequency and power range).

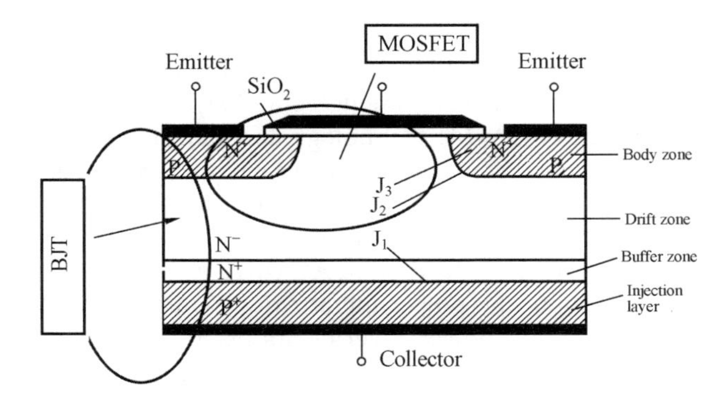

Figure 7.12 Structural diagramme of IGBT.

IGBT, a hybrid device of BJT and MOSFET, introduces the conductivity modulation effect of the BJT to the high-impedance drift zone of the MOS component, significantly improving the conduction characteristics. Although both of them are voltage-driven, the conduction mechanisms are different.

The structural diagramme of IGBT shows, the structure of IGBT is similar to the power-MOSFET except that a P^+ zone is added between the drain and the substrate of the original power-MOSFET. In the structure of IGBT, the P^+ zone is called as the collector zone, and the N^+ zone above is the buffer zone, and the N^- layer is the drift zone; the N^+ layer on the top is called as the source zone, and the electrode on it is the emitter; the control zone is the gate zone, and the electrode on it is called as the gate.

In the structural diagramme of IGBT, the P zone connected with the emitter, the N^- zone in the drift zone, the N^+ in the buffer zone and the P^+ zone connected with the collector form a PNP-type BJT. The P^+ zone connecting the collector is a functional zone only possessed by IGBT, and it can inject holes to the N^- zone in the drift zone to make it have high carrier concentration and reduce the making voltage when IGBT is switched on. However, the switching on/off of IGBT is controlled by the gate voltage. When the gate voltage is larger than the threshold voltage of MOS tube ($V_G > V_{th}$), and electronic conduction trench will be formed in the P zone beneath the gate, and the electron will be injected to the N^- drift zone via the conduction trench from the N^+ zone of the emitter, that is, it will offer base current for the PNP transistor in the IGBT, so as to realise PNP tube conduction, or IGBT conduction. In this case, to keep electrical balance in the N^- drift zone, the P^+ zone will inject holes-carriers to the N^- drift zone to keep the N^- drift zone high carrier concentration (i.e., carry out conductivity modulation to the N^- drift zone), reduce the conduction resistance in the drift zone, and make the high voltage-withstanding IGBT with long drift zone also have low conduction voltage drop. When the gate voltage is less than the threshold voltage of MOS tube ($V_G < V_{th}$), the conduction trench of MOSFET will disappear, the base current of the PNP transistor will be disconnected and the IGBT will be switched off.

The followings can be drawn:

1. IGBT is a hybrid component with BJT as the core and MOSFET as the drive. In IGBT, the P zone connected with the emitter, the drift zone, the N zone in the buffer zone and the P^+ zone connected with the collector form a PNP transistor. The P^+ zone can inject holes to the N^- zone in the drift zone to make it have high carrier concentration and reduce the making voltage when IGBT is switched on.
2. The switching on/off of IGBT is controlled by the gate voltage. When the gate voltage is larger than the threshold voltage of MOS tube ($V_G > V_{th}$), an electronic conduction trench will be formed beneath the gate, and it will offer base current for the PNP transistor in the IGBT, so as to realise PNP tube conduction, or IGBT conduction. When the gate voltage is less than the threshold voltage of MOS tube ($V_G < V_{th}$), the conduction trench of MOSFET will disappear, the base current of the PNP transistor will be disconnected and the IGBT will be switched off.
3. IGBT can be equivalent to the Darlington, which uses the MOSFET of N trenches as the input and the PNP transistor as the output. IGBT is equivalent to a PNP transistor with thick base zone driven by MOSFET. It has the fast switching characteristics of power-MOSFET and the low conduction voltage drop of BJT (see Figure 7.13 for the equivalent circuit of IGBT).

IGBT integrates the advantages of power-MOSFET and BJT with voltage withstanding capacity up to 600–6500 V, and it is suited to the applications from several kW to several MW. It has been widely applied to inverters, UPSs, general-purpose frequency converter, transmission of electric locomotives, power supplies of large-power switches, induction heating power supplies and etc. As the voltage withstanding capacity of IGBT is improved, it will cover wider power applications.

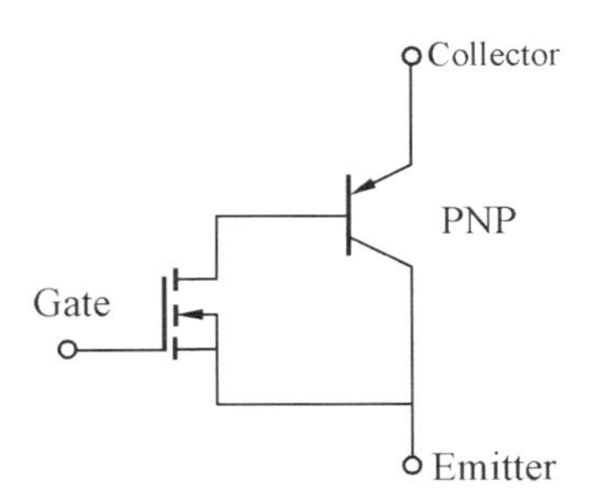

Figure 7.13 Equivalent circuit of IGBT.

7.3.3 Development History of IGBT

The development history of IGBT shows that the subject on performance and reliability improvement of IGBT has always focused on saturated voltage drop and switching characteristics. The IGBT of first through third generation is designed with the CZ epitaxial silicon wafers and the basic concept of 'high injection, low transmission efficiency'. To achieve low conduction voltage drop and fast switching off, the IGBT collector has to inject a lot of carriers to the buffer zone and the drift zone, and the lifetime of carriers has to be very short, which will definitely reduce the transmission efficiency of carriers and be adverse to improve the conduction characteristics of IGBT, and tail current will occur when IGBT is switched off. The IGBT after the fourth generation has important improvement on three aspects: design concept, chip material and doping technique. The design concept is changed to 'low injection, high transmission efficiency' from 'high injection, low transmission efficiency'; the chip material is changed to FZ monocrystalline silicon from CZ monocrystalline silicon; and the epitaxial doping is replaced by accurate ion injection doping technique. More importantly, the trench gate technique, the non-punch-through (NPT) technique and filed stop (FS) technique have been developed (see Figure 7.14).

7.3.3.1 Trench Gate Technology

Figure 7.15 shows the structural comparisons of planar gate IGBT and trench gate IGBT. Since it is inevitable to have a junction transistor J_{FET} beneath the planar gate IGBT, and the trench gate structure can be realised by trench digging, side wall oxidation and vertical gate production, which can greatly reduce the J_{FET} resistance beneath the planar gate IGBT, and reduce the series resistance, make carriers evenly distributed on the sectional area. As a result, the IGBT can have smaller conduction losses and switching losses as well as smaller chip size.

7.3.3.2 Non-Punch-Through (NPT) Technique

The non-punch-through (NPT) technique can accurately control the P^+ collector doping by the ion injection method where the N^+ buffer layer in the punch-through (PT) IGBT can be omitted. The collector is often called as 'the transparent collector',

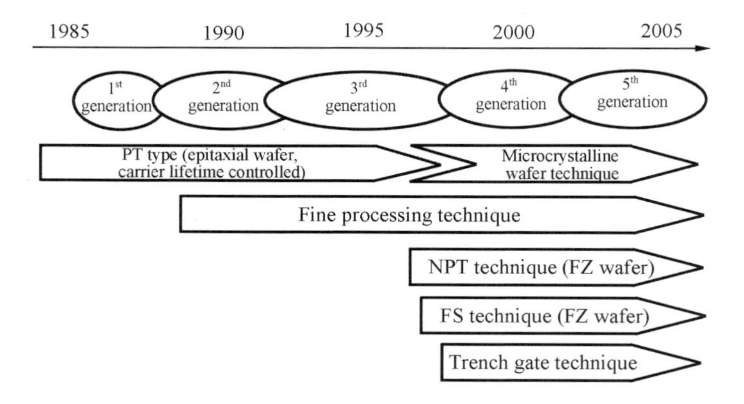

Figure 7.14 Development history of IGBT.

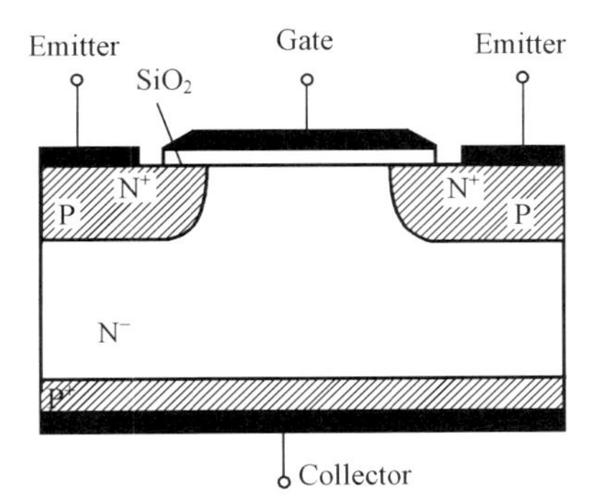

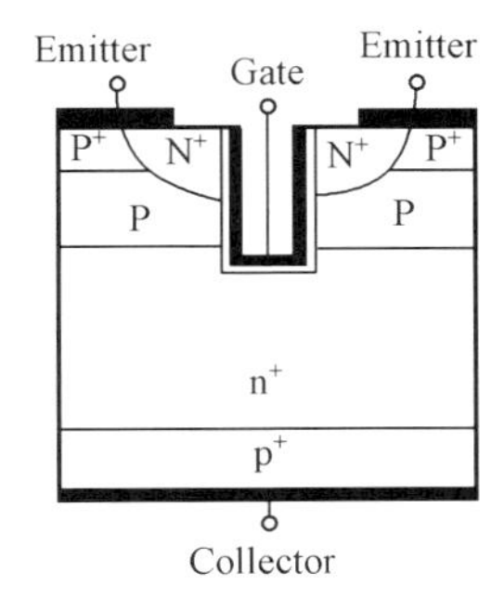

Figure 7.15 Structural comparisons of planar gate IGBT and trench gate IGBT.

which is defined as one with thin P^+ layer, thickness in submicrons and low hole injection concentration. In this way, slight doping will not make the drift zone produce excessively surplus carriers, which can improve the breaking speed and achieve low $V_{EC(ON)}$, reducing the tail current in case of IGBT breaking. Since the carriers in the NPT IGBT have long lifetime and the switching speed is insensitive to temperature change, it can keep stable performance in the working temperature range. Its conduction voltage drop rises with growth of working temperature, the IGBT device has positive temperature coefficient, and several NPT IGBTs can be connected in shunt to raise the capacity and reduce the switching losses. Since the transparent collector technology is used, the thickness of the P^+ layer is in submicron level, the chip thickness of the NPT IGBT is usually two-fifths of the PT IGBT chip, reducing the heat resistance in case of conduction. Its voltage impulse withstanding capacity is better than the PT IGBT, and it can produce the switching components with lower conduction voltage drop in the same voltage withstanding conditions. In this way, it can raise the switching speed and reduce the losses in case of breaking. More importantly, the NPT IGBT has positive conduction voltage temperature coefficient after the transparent collector is adopted, which is good for IGBT in shunt and can produce the module with high voltage and big current. The IGBT after the fourth generation is mostly provided with 'the transparent collector', which can be used to the applications of above 600 V (see Figure 7.16 for comparisons between PT IGBT and NPT IGBT).

7.3.3.3 Filed Stop (FS) Technology

The field in the fourth-generation NPT-IGBT is in delta. To make the chip thinner, the fifth-generation IGBT is added with a FS layer, which can make the field in trapezoidal distribution, and the chip becomes thinner, further reducing the losses during conduction and breaking. Figure 7.17 shows the comparison between the fourth-generation NPT IGBT and the fifth-generation FS IGBT.

Figure 7.18 shows the sixth-generation IGBT that has the trench gate and FS structure.

From the first-generation IGBT to the sixth-generation IGBT, the IGBT chip area is reduced to one-fourth the first-generation one; the conduction saturation voltage drop is about one-third the first-generation one, the breaking time and power loss are about 30% first-generation one and the power module can be produced. It is developing towards

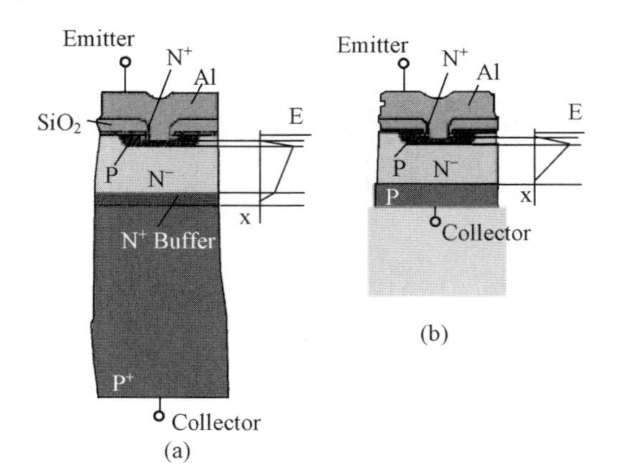

Figure 7.16 Comparisons of PT IGBT and NPT IGBT, PT IGBT (a), NPT IGBT (b).

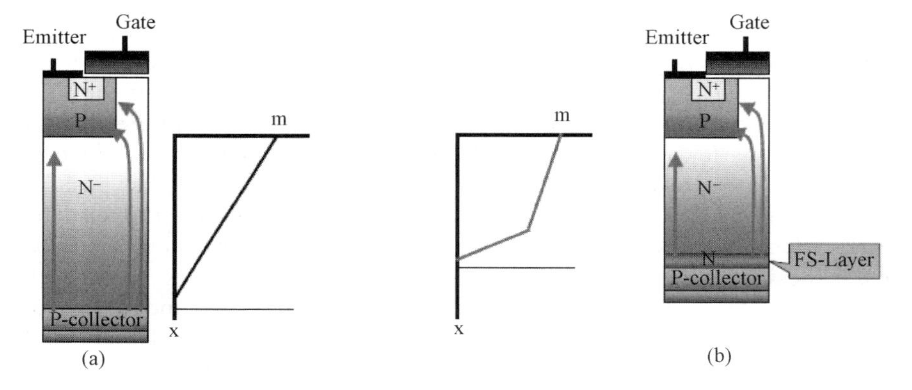

Figure 7.17 Structural comparisons on FS IGBT and NPT IGBT.

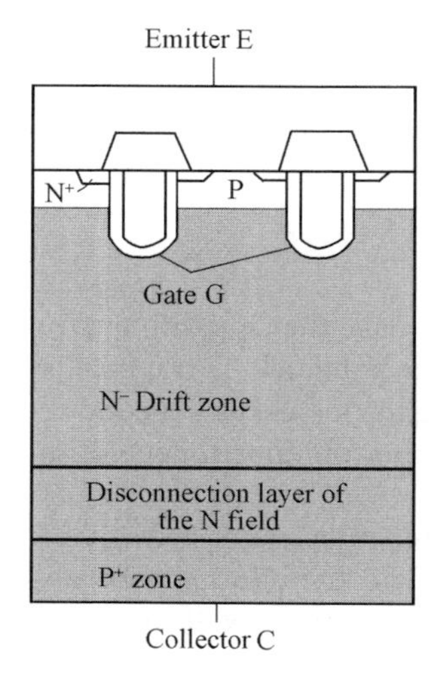

Figure 7.18 Structure of IGBT with trench gate and FS structure (sixth-generation IGBT).

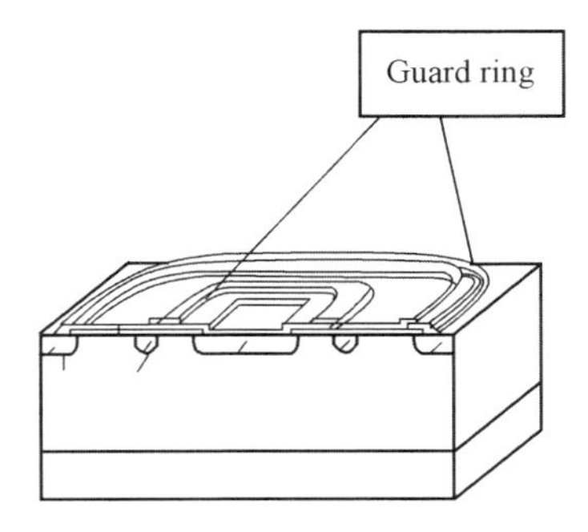

Figure 7.19 IGBT with guard ring.

high voltage and big current. The development of grid-connected inverters for renewable energy of high efficiency and low consumption plays an important role.

The silicon carbide is the most promising power electronic device in the twenty-first century. Its voltage withstanding capacity is 10 times the silicon, and its heat conductivity is three times the silicon with the working junction temperature above 200 °. The switching frequency of the silicon carbide power device can be greatly improved, and the conduction losses and switching losses can be reduced to one-tenth the silicon power device. Thanks to improvement of heat conductivity and junction temperature, the size of the radiator can become smaller, and thus it is especially suited to the field-controlled power device.

The intelligent power module (IPM), emerging in 1990s, consists of many single power devices in shunt, which is integrated with the drive, protection, electrically isolating circuits to one silicon wafer or substrate, forming the concept of power electronic integration. Since the surface of the outer layer has the maximum field intensity, it is easy to be broken down. As a result, the guard ring (see Figure 7.19) is added to disconnect the path of the leak current, and reduce the surface field intensity of the outer layer, further improving the reverse voltage withstanding capacity.

7.4 Inverters

7.4.1 Role of the Inverter in the PV System

Conversion of DC power to AC power is called as inversion. The inverter is the core of the PV power system. The inverter is used to convert the DC power generated by the solar cell array to AC power and then transmitted to the AC load equipment connected. For the grid-connected inverter, it shall flow back the surplus power to the grid to make full use of the MPPT of the solar module (array) and protect the grid in case of abnormal conditions (or failures).

After connected with the grid, the PV grid-connected inverter will regulate the phase difference between the grid-connected inverter output voltage and the grid voltage. When the voltage phase of the grid-connected inverter is leading than the grid, the power will be transmitted to the grid; otherwise, when it is lagged behind the grid, the power will be transmitted to the grid-connected inverter. When the DC side is provided with energy storage devices, the grid side can charge the energy storage devices, via the grid-connected inverter. Figure 7.20 shows, the DC power generated by the PV array is

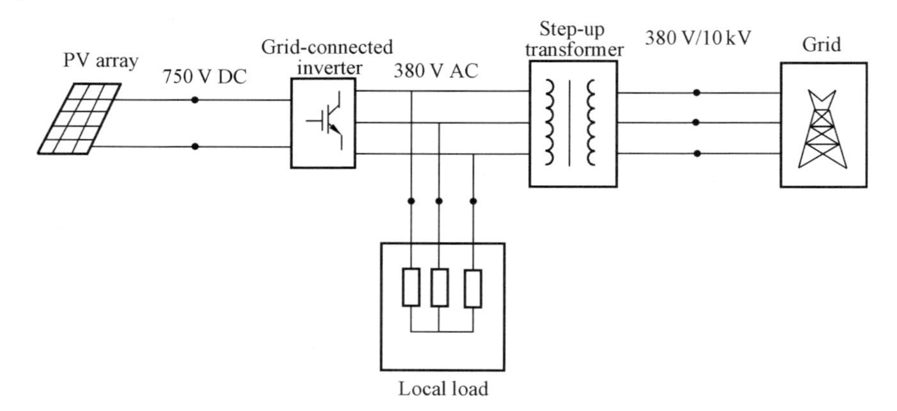

Figure 7.20 Example of the inverter in the PV system.

first converted to AC power by the inverter and then supplied to the system 380 V load and the surplus power can be transmitted to the 10 kV utility grid after stepping up by the step-up transformer.

For the large grid-connected inverter, several inverters shall run in shunt. When the real-time PV power generation is small, the number of inverters shall be reduced properly to make the inverters run efficiently, improve the inverter generation efficiency and reduce the harmonic content. In addition, the large inverter shall be also designed with such functions as reactive power compensation, grid frequency and voltage control, over/under-voltage (frequency) and short-circuit protection, automatic disconnection from the system in case of abnormal system conditions, as well as automatic recovery etc.

7.4.2 Working Principles of the Inverter

The inverter is a conversion device to convert the DC power to AC power via making/breaking operation of the switchgear. The output waveform of the inverter can be divided into rectangular wave and sinusoidal wave with sinusoidal wave output more popular. The inverter can change its output pulse sequence width to adjust the output sinusoidal wave characteristics, which is the so-called pulse width modulation (PWM) technology. The PV system grid-connected technology has two related technologies: One is the sinusoidal wave pulse width modulation (SPWM), that is, the making/breaking time and switching sequence of the control power switch can be controlled to make the inverter voltage track the given sinusoidal reference waveform and achieve the sinusoidal wave of the inverter output voltage; and the other is selective harmonic elimination (SHE-PWM) technology. In the PWM technology, the making/breaking of the inverter switch is controlled to make the inverter output the electrical pulse sequence and then control the width of each pulse in the pulse sequence to make the inverter voltage track any given reference waveform and realise control over the output power, enabling it good dynamic performance.

Below will introduce the PWM technology for the single-phase inverter. Figure 7.21 shows a single-phase half-bridge voltage inverter circuit. The switches on the two arms of the inverter (S_+ and S_-) can be switched on/off to regulate the output voltage V_0.

Figure 7.21 PWM technologies with single-phase inverter.

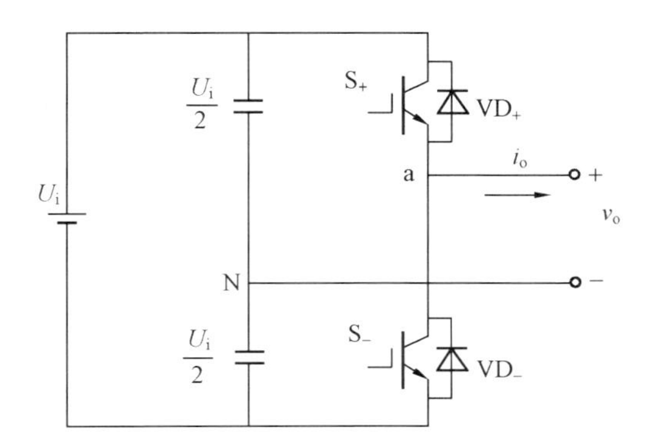

The making/breaking status of the switch tube on each arm of the voltage inverter is determined by the comparison result of the modulation signal $V_m(t)$ (i.e., the given reference waveform signal output by the desired inverter) and the triangle carrier wave $V_c(t)$. When $V_m(t) > V_c(t)$, the switching tube S_+: making; the switching tube S_-: breaking; and vice versa. Figure 7.22 shows the process where the comparison between the modulation signal $V_m(t)$ and the triangle carrier signal $V_c(t)$ is carried out to convert the DC input to sinusoidal wave AC output.

Here the following definition is introduced: amplitude modulation index $m_a = V_m / V_c$, where V_m is the amplitude of the modulation signal $V_m(t)$, and V_c is the amplitude of

Figure 7.22 Process where the comparison between $V_m(t)$ and $V_c(t)$ is carried out to convert the DC input to sinusoidal wave AC output.

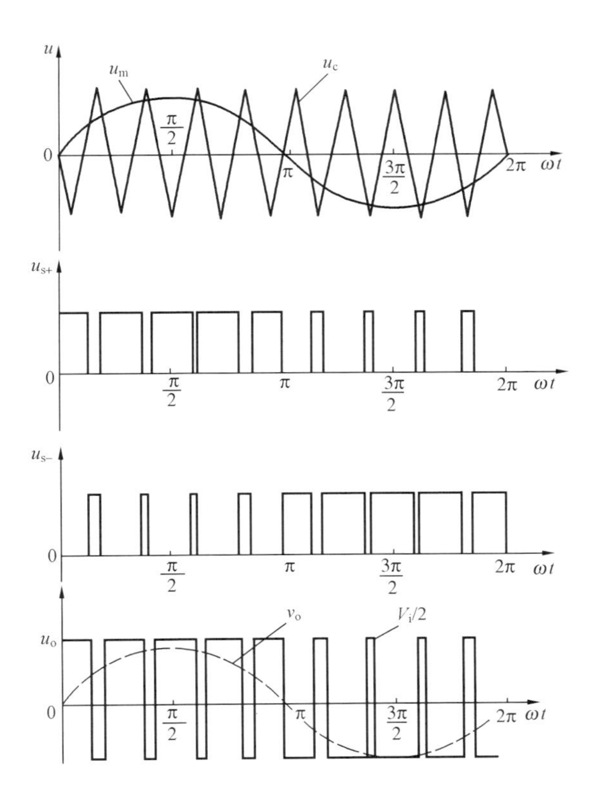

$V_c(t)$ triangle carrier signal (generally, the triangle carrier signal amplitude is constant while the modulation signal amplitude is adjustable in the control design). Then, another definition is introduced: frequency modulation index $m_f = f_c / f_m$, where f_c is the frequency of the triangle carrier signal and f_m is the modulation signal frequency (generally, the carrier signal frequency is much larger than the modulation signal). Figure 7.22 shows the inverter output voltage is the pulse sequence with pulse width variation, and the pulse width varies with change of transient value of the sinusoidal wave modulation signal. The SPWM control technology can make the inverter output power become one mostly with sinusoidal wave fundamental component. The output signal, however, also has resonant wave component in addition to the sinusoidal wave component where the frequency of the sinusoidal wave component is that of the modulation signal. The resonant wave in the output voltage V_o occurs around the frequency of the carrier and its times, that is, $H = m_{i \pm k}$, where i = 1,2,3,...

Further, the SHE-PWM technology can be used, that is, the making/breaking moment of the switchgear can be worked out in an accurate manner by the mathematical method, to eliminate the resonant wave of the specific order so as to get a near-sinusoidal AC output.

Figure 7.23 shows the structural diagramme of three-phase full-bridge voltage inverters and the sinusoidal wave output after PWM.

It shall be pointed out, in the three-phase DC-AC inverter control, the three vectors of the three phases can be expressed by space vector. As a result, in the three-phase voltage inverter, the space vector modulation technology has become the PWM control technology with widest applications. It is also realised by selection of the appropriate switching status and calculation of the switching time of the power switching device.

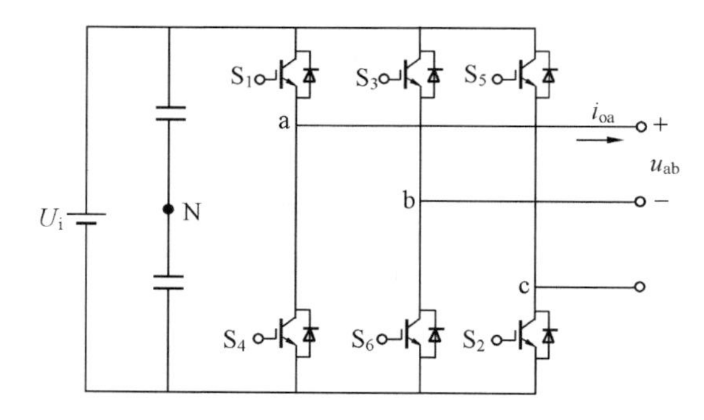

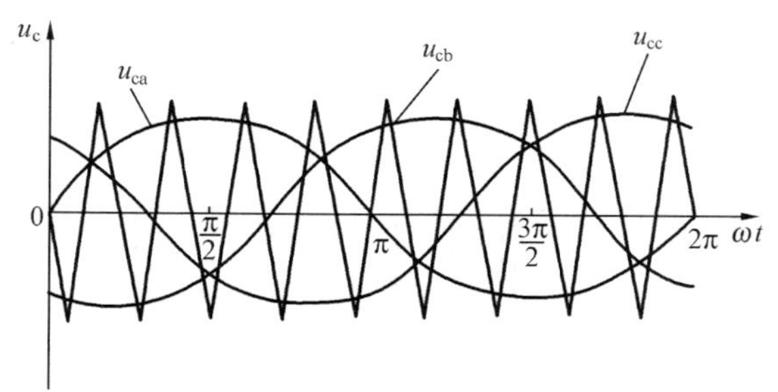

Figure 7.23 Schematic diagramme of three-phase full-bridge voltage inverters and output waveform of three-phase SPWM sinusoidal wave.

7.4.3 Control of the Inverter

The control objectives of the inverter are to improve the stable and dynamic performance of the inverter output voltage. The stable performance mainly refers to the stable accuracy of the output voltage and the capacity with imbalanced loads, and the dynamic performance refers to the total harmonic distortion (THD) of the output voltage and the dynamic response time in case of sudden load changes.

The early inverters are mostly provided with multiple actions to improve the output waveform due to limitations of low frequency of the power switching device. As the power device performance is improved and the PWM technology is widely used, the inverter output voltage has seen better and better quality. In recent years, as the DSP with high-speed computation capacity comes into existence, the advanced modern control theories and methods have been applied to the inverter, making fully digital inverters available.

The PV station is very demanding for the inverter as described below:

① The inverter shall output the sinusoidal wave current. The power fed by the PV station to the utility grid must conform to the indexes specified by the grid, for example, the output current of the inverter cannot contain any DC component, and in the inverter output current, the high-order harmonics must be minimised and mustn't bring harmonic pollution to the grid.
② The inverter shall run efficiently in case of big variation of loads and solar irradiation.
③ The inverter shall keep the solar cell array working at the maximum power point (MPP). Because the output power of the solar cell has relations to the solar rays, temperature, load variations, that is, its output characteristics are nonlinear, the inverter shall be designed with maximum power tracking function so that the array can run at the optimum working point via the automatic regulation of the inverter no matter what changes happen to the solar rays and temperature.
④ The inverter shall be characterised by small size and high reliability. For the PV station for residential purposes, the inverter is usually installed indoors or on the wall so that its size and weight are limited.
⑤ The inverter shall be able to supply power independently in case the solar rays are available when the mains power supply fails.

7.4.4 Inverter Circuit and Inverter Types

In addition to the power electronic switchgear, the inverter circuit is also a key component of the inverter. The core of the inverter circuit is the inverter switching circuit, which carries out the inverter function by making and breaking of the semiconductor switchgear. For a complete inverter circuit, it shall also be designed with the control circuit, the input circuit, the output circuit, the auxiliary circuit and the protection circuit. The basic structure is shown in Figure 7.24.

The main functions of each circuit are described below:

(1) Input circuit: The input circuit shall provide the main inverter circuit with DC voltage to ensure its normal service.
(2) Output circuit: The output circuit shall correct, compensate and regulate the quality of the AC power output by the main inverter circuit (including waveform, frequency,

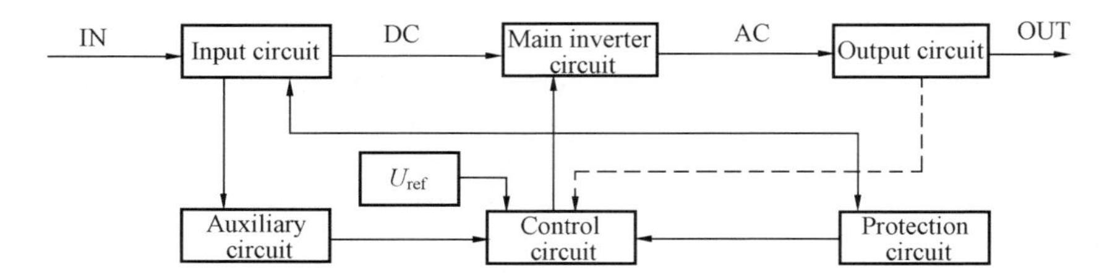

Figure 7.24 Basic Schematic diagramme of inverter circuit.

voltage, current amplitude and phase position, etc.) to make it compliant with the user's demand.

(3) Control circuit: The control circuit shall offer a series of control inverter pulses for the main inverter circuit to control the making and breaking of the inverter switch, and cooperate with the main inverter circuit to fulfil the inversion function. In the inverter circuit, the control circuit plays an important role like the main inverter circuit.
(4) Auxiliary circuit: The auxiliary circuit shall convert the input voltage to the DC voltage applicable to the circuit. It includes several detection circuits.
(5) Protection circuit: The protection circuit consists of the input over/under-voltage protection, output over/under-voltage protection, overload protection, overcurrent protection, short-circuit protection and overheating protection and the like.
(6) Main inverter circuit: The main inverter circuit is a conversion circuit composed of the semiconductor switchgear. It consists of two types: isolated and non-isolated. Both of them are hybrid of two circuits—the step-up circuit Boost and the step-down circuit Buck. These circuits can form the single-phase inverter and the 3-phase inverter.

The inverter can be classified by the following criteria. It can be divided into grid-connected and off-grid inverters by destination of the inverter output energy; the single-phase inverter, the three-phase inverter and the several-phase inverter by the number of phases; single-ended (including forward and fly-back inverters), pull-push, half bridge and full bridge inverters by the type of the main circuit; voltage and current inverters by stable output reference; sinusoidal wave and non-sinusoidal wave inverters (rectangular wave, step wave, quasi-rectangular wave, etc.) by output AC waveform; and power frequency modulation (PFM) and pulse width modulation (PWM) inverters by control methods.

7.4.5 Selection and Requirements of Inverters for PV Applications

The following requirements shall be followed during selection of PV inverters:

1. The disturbances to the grid caused by the inverter, including the current harmonics, voltage fluctuation, voltage flicker, reactive power, grid impedance, superimposed disturbance and so on, must conform to the criteria.
2. The disturbances to other electric equipment caused by the inverter must conform to the relevant criteria, including conduction disturbance, space radiation disturbance and so on, especially the disturbances to other electric equipment caused by the LV grid-connecting system.

3. As for the disturbances caused by the grid to the inverter, for example, voltage flicker, electrical noise, surge, high-frequency component and the like, the inverter shall be designed with certain bearing capacity.
4. The efficiency evaluation indexes of the inverter (e.g., the maximum efficiency, and the maximum power point tracking (MPPT) efficiency) and the reliability indexes (e.g., MTBT and MTBR) shall conform to the requirements.
5. The inverter for PV power purposes shall conform to the following codes and standards: 'Special inverters for grid-connected PV power systems (GB/T20321-2006)', IEC62109 or UL1741. As for safety concern, it shall meet the following codes and standards: IEC62109-1 'Safety requirements on PV inverters', IEC62116 'Test procedure of islanding prevention measures for utility-interconnected PV inverters' and GB/T19939 'Grid-connecting specifications on PV systems'.
6. The inverter circuit shall be connected in accordance with 'Technical specification on connection of PV stations to grids' and the relevant Chinese certifications (e.g., Golden Sun Certificate).
7. The PV system shall be designed with AC switchgear, which is used to switch the standby inverter, ensure normal power supply to the system and measure the line energy.

7.5 Controllers: Module Power Optimisation and Intelligent Monitoring

7.5.1 Functions of the PV System Controllers

The controllers of the PV system shall be designed with detection, control, regulation and protection functions. In recent years, the module power optimisation and intelligent monitoring have been added as the intelligent grid develops.

1. The grid-connected PV station shall be designed with the following detection functions:
   ① AC/DC electrical and generation capacity test: Including the voltage, current and power of the PV array; the output voltage, current and power of the inverter; cumulative power, efficiency and losses.
   ② Power quality test for the station: Including working voltage deviation and fluctuation, frequency deviation, phase position deviation, harmonic and waveform distortion, voltage imbalance, DC component and flickering and so on.
   ③ Safety test: Including over/under-voltage protection, over/under frequency, islanding protection test, lightning protection and earthing, insulation performance, and current leakage and so on.
   ④ Meteorological data test technology study: Including solar radiation, ambient temperature, array temperature, wind velocity/direction and so on.
   ⑤ Data computation storage and monitoring.
2. Control of PV station output power quality
   For the PV system, its output power fluctuation and voltage deviation shall be controlled within $\pm 5\%$, and its THD shall be controlled within $\pm 3\%$, and the reactive power compensator shall be provided in the principle 'local reactive power balance'. The PV system shall be connected to the utility grid in three phases if possible where

the allowed three-phase voltage imbalance mustn't exceed 1.3%. The DC component generated by the inverter system mustn't be larger than 1% of the inverter output current. For the small-sized PV station, it shall immediately stop supplying power to the gate lines to ensure synchronous operation with the grid during PV station grid connecting when the grid-connecting point frequency is larger than 49.5–50.2 Hz.

3. Regulation of PV station output power and voltage
   The large-sized PV station shall be able to control its active power output according to the demand of the grid dispatcher, limit the output power variation rate and handle the output power reduction caused by fast reduction of solar radiation. In addition, the large/medium-sized PV stations shall be designed with the capacity to enable the reactive power compensator and regulate the ratio of the PV station step-up transformer.
4. Low voltage ride through (LVRT) capacity
   When the utility grid is in abnormal conditions, the large/medium-sized PV stations shall be designed with some response capacity for abnormal voltage. The so-called 'low voltage ride through (LVRT)' means that the PV station shall be still able to keep grid-connected operation and even supply some reactive power to the utility grid to support its recovery until the grid recovers normal operation when the grid-connecting point has voltage drop. This is a specific operation functional requirement for the grid-connected PV station when utility grid has voltage drop. Figure 7.25 shows the LVRT requirements on the PV system. When the voltage at the grid-connecting point falls in the zone in and above the voltage contour, the PV station must keep uninterrupted grid-connected operation; and the PV station is allowed to stop supplying power to the gate lines only when it falls below the voltage contour.
5. Prevention of overcurrent and islanding effect
   Many relay protection functions shall be provided to ensure equipment and human safety in case the PV station or the grid is in abnormal conditions or fails. The protection functions shall meet the requirements on reliability, selectivity, sensitivity and

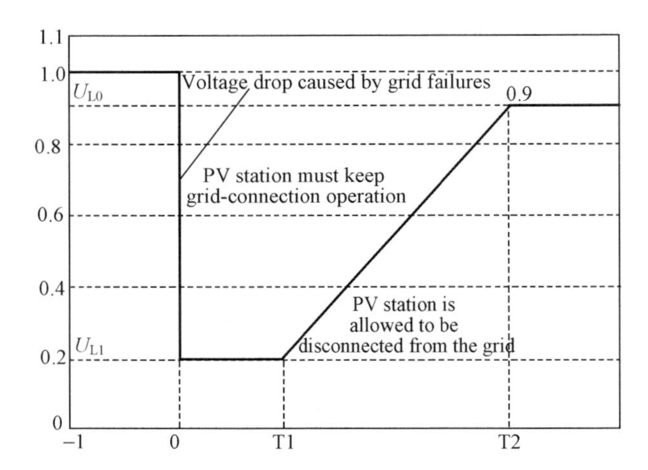

Figure 7.25 LVRT requirements on PV systems.

fast action so as to keep safe operation of the grid and the PV system. The PV station shall be designed with overcurrent capacity, for example, the PV station shall be able to work for minimal 1 min when it is below 1.2 times rated current.

The islanding effect refers to the effect that the grid-connected PV system can still supply power to some lines of the utility grid when the utility grid has sudden voltage loss. For instance, when some lines of the utility grid are blackout due to failure or maintenance, they can be powered by the grid-connected PV system, forming a self-powered ring with the surrounding loads. In this case, it may put the safety of the maintenance staff and the grid quality at risk. And it may also damage the equipment when it resumes power supply to the islanding grid after maintenance. As a result, the PV station must be designed with the capacity to rapidly detect the islanding effect and disconnect with the grid immediately. The active islanding protection of the PV station is mainly based on frequency shift, phase position jump and harmonic monitoring, active power change, reactive power change, and impedance change caused by current pulse injection and so on.

6. Dispatching automation and communications

   In accordance with 'Technical specification on connection of PV stations to the grid, State Grid Corporation of China', the large-/medium-sized PV stations must be designed with the capacity to carry out data communications with the grid dispatcher, laying foundation to the future intelligent grid.

7.5.2 Maximum Power Point Tracking Technology (MPPT) of Solar Cell Controllers

The conversion efficiency of the solar cell varies with changes of the solar intensity and ambient temperature. To make the PV system output the maximum power, the MPPT has become the key subject for PV utilisation. It is known that in the traditional linear circuit, the appropriate load matching can be selected to make the load have maximum power. Theoretically, the load can achieve the maximum power when the load resistance is equal to the inner resistance of the power supply system. In the PV system, however, it is not so simple since the inner resistance of the solar cell is under the influence of the solar intensity, ambient temperature and loads. At present, a DC/DC converter is usually added between the solar cell array and the load, and the conductivity of the power switch tube in the DC/DC converter can be changed to regulate, control and ensure the solar cell array to work at the maximum power point, realising maximum power point tracking and control. Figure 7.26 shows the output power characteristic curve of the solar cell array. It indicates, with the maximum power point as the boundary, when the array working voltage is larger than the maximum power voltage U_{PMAX}, the array output power will rise with reduction of the output voltage of the solar cell; and when the array working voltage is less than the maximum power voltage U_{PMAX}, the array output power will rise with growth of the output voltage of the solar cell. As a result, the end voltage of the solar cell array can be controlled to make the array intelligently output the maximum power in various solar rays and temperature conditions. The MPPT is an automatic optimisation course. For various PV systems, it can be realised by associated topology circuit of the DC/DC converter circuit, for example, the control algorithm method, active power disturbance observation method, the incremental conductance algorithm, hysteresis comparison

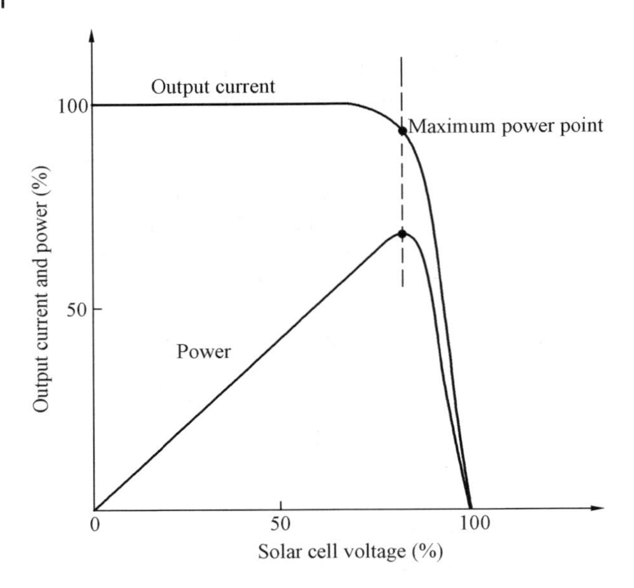

Figure 7.26 Power-voltage curve of solar cell arrays.

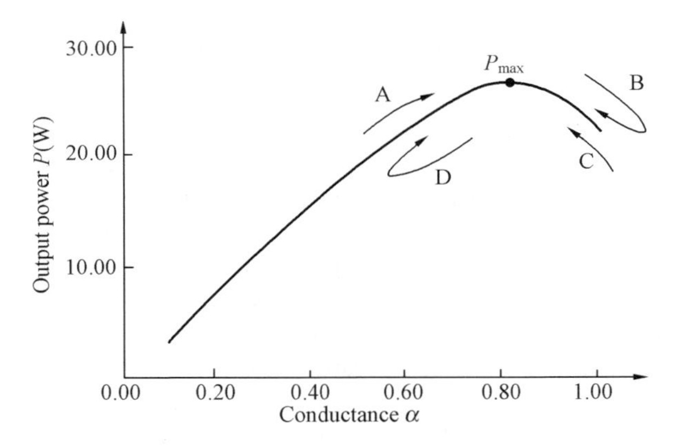

Figure 7.27 Identification of maximum output working point of solar cell array by power disturbance observation method.

and fuzzy control method. The power disturbance observation method is based on the principle: First a disturbance output signal (U_{PV}+ΔU) is given, and measure its power change, and compare it with the power before disturbance presence; the power rise shows the disturbance direction is correct, and if the power after disturbance presence is less than before disturbance presence, it shall be disturbed in the opposite direction. Figure 7.27 shows the diagramme where the maximum output working point of the solar cell array is identified by micro adjustment (Δα) of the conductance of the DC/DC converter.

As for the maximum power point control technology, the following two proposals are available: In the first proposal, each module is installed with one inverter (a micro inverter), and then under MPPT management and DC/AC conversion and finally connected to the grid in series. It has a disadvantage that if some module has small output current due to shading or performance deterioration, the output current of the modules in series will be determined by the module with poor performance (the so-called 'buckets effect'). In the second proposal, each module is added with DC/DC converter (a power optimiser) for MPPT management and then connected in shunt to output to one terminal inverter for DC/AC conversion and finally to the grid. Since the power optimiser is a DC/DC converter, it can optimise the maximum power of each module independently, and then the maximum power collected by the power optimisers can be collected. If a module has lower power than other modules due to shading or performance deterioration, its associated optimiser can reduce the output voltage to improve the output current and thus it will not affect the power output of other modules in series, improving the system total power generation. The power optimiser is based on the scheme that the maximum power is collected from the modules in series and then output to the inverter in a coordinated manner. In this way, it can settle the problem that some modules are shaded. It is reported that the overall efficiency of the PV system can be improved by 5–20%. Moreover, the quantities of the modules in series are unnecessary to be equal for the system with multiple modules, facilitating the design of PV stations.

The power optimiser, which is not an inverter, is a DC-DC buck/boost converter. Or in other words, it is a MPPT at single module grade. The power optimiser shall conduct maximum power optimisation for a single module, and then transmit it to the terminal inverter for DC-AC processing and finally supplied to the civil users or connected to the grid. Usually, the terminal inverter can be a pure inverter without MPPT or an inverter with secondary MPPT. At present, the dominating power optimisers consist of two types in the market: the series type and the shunt type, and the control topologies used are also different.

GNE, China, has integrated the data wireless transmission function to its module power optimiser, which can conduct remote real-time data acquisition and monitoring at the module level, and if the big data analysis and cloud computation functions are provided, it can realise intelligent management. In this way, the inverter can serve as the DC/AC converter of the PV cell and the terminal for data acquisition of the energy Internet (see Figure 7.28).

At present, Solargiga Energy, China's largest inverter manufacturer, is building the 'iSolarCloud' platform. Huawei, a well-known giant in the communication sector of China, is integrating digital information, wireless broadband, Internet and multi-media to the PV technology to improve the power generated by the PV stations and the operation and management efficiency.

7.5.3 Installation Angle and Position Regulation of Solar Cell Arrays by the Controller

The solar cell is a typical nonlinear semiconductor active component, and its output is a set of curves with solar illuminance as the reference parametric variable. For the solar PV system installed in a fixed position, the solar angle varies at every moment in each day from spring to winter. If the irradiating surface of the solar cell module can be always vertical to the solar incidence rays, the generation efficiency can reach the optimum

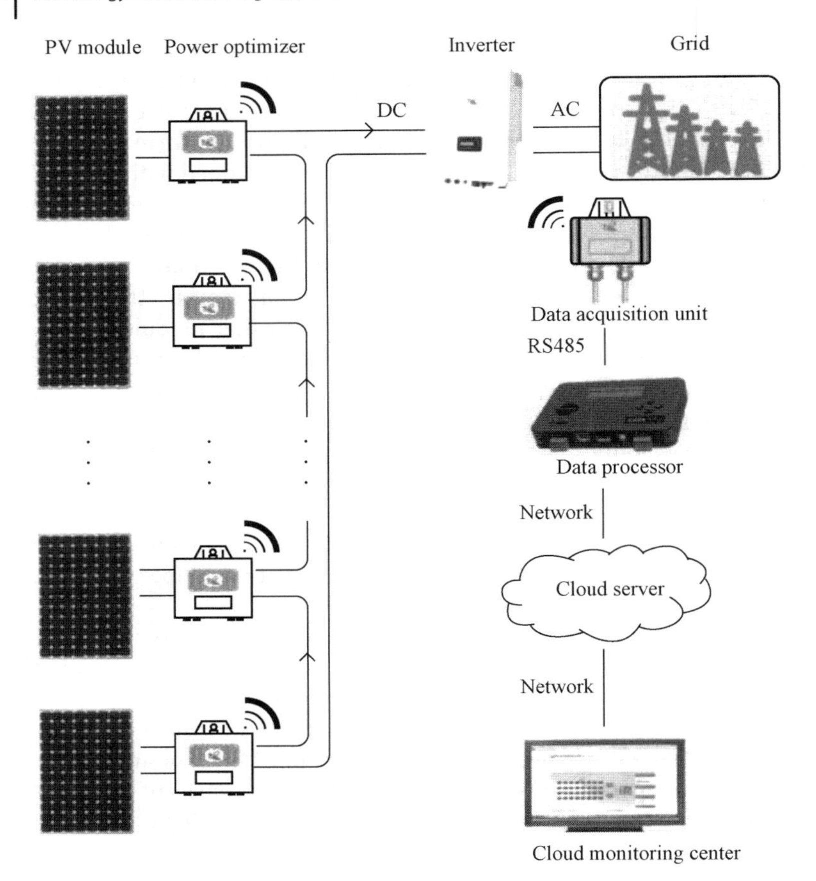

Figure 7.28 Power optimisation and module intelligent monitoring management.

state. As a result, the installation position and angle of the PV module (array) have great impact on illumination of solar radiation, and they will exert direct influence over the power generation of the PV system. No matter the PV module (array) is designed with fixed installation or automatic tracking method, it is a common concern to select the optimum tilt angle. These technologies and devices are also a part of the solar cell MPPT technologies, which are of great importance for concentrator solar cells.

For installation position of the PV module (array), there are two relevant angle references: the tilt angle and the azimuth. The tilt angle refers to the angle between the PV module (array) plane and the horizontal ground, and the azimuth to the angle between the vertical plane of the module (array) and the south in the northern hemisphere. If the azimuth is constant, the solar radiation energy received by the PV module (array) will be significantly different in the months. If the tilt angle is only based on the local latitude, it will often result in surplus power generation of solar cells in summer and insufficient voltage of the energy storage devices in winter, and thus it is not the best choice. If the tilt angle is constant, the solar radiation energy received will be of great difference at the moments each day. As a result, it is a key link of the PV system control system to control and regulate the tilt angle and the azimuth of the PV module (array).

The PV module automatic tracking system can improve the received solar radiation via tracking the solar motion curve, and it can continuously regulate the module tilt angle and azimuth to achieve the maximum solar radiation. In most applications, the single-axis tracking systems are designed with 0° as the azimuth, and the PV module fixed on the axis tracks the sun at 15°/hour. The dual-axis tracking system is based on the active tracking concept, that is, two stepper motors are used to simultaneously track the tilt angle and the azimuth so as to make sure the irradiating surface of the module (array) always keeps vertical to the solar rays. One axis rotates in the west-east direction, called as the azimuth tracking; and the other regulates the tilt angle, called as the tilt angle tracking. They can be also called as the horizontal-axis tracking system and the equatorial-axis tracking system. Many solar tracking control systems can calculate the angles of the sun at each moment of a day throughout a year according to the longitude and latitude of the module (array) installation position, and store them to PLCs, SCMs or computers. That is to say tracking is realised by calculation of solar position. The dual-axis tracking system generally adopts light intensity as the sensor (or GPS or satellite ephemeric calculation) for control to make the module irradiating surface always vertical to the solar direct rays and achieve maximum output of the module. The concentrator solar cell module can only adopt the dual-axis tracking system (refer to Section 4, Chapter 3).

The statistics show, the dual-axis tracking system can increase the PV power generation by 30–40%. For the stand-alone PV system, it shall consider the continuity, evenness and maximality of the solar radiation on the PV module (array), and for the large-/medium-sized PV system, the maximum solar radiation throughout a year shall be obtained, and most of them are designed with the single-axis tracking system since the array is huge and the cost is high and it is difficult to carry out outdoor maintenance.

7.5.4 Other Functions of the Controller

The control technology of the PV system shall also include the followings: DC-DC and DC-AC conversion, computer control and remote monitoring technologies.

The DC-DC conversion technology can be applied to charging/discharging control of the energy storage system. For instance, the controller shall specify and control the charging/discharging conditions of the lead-acid battery and control the power output of the solar cell module and the lead-acid battery to the load as demanded. It has been described in Section 2, Energy Storage Battery, this chapter. The DC-AC conversion technology (i.e., inverter technology) is mainly applied to the off-grid or grid-connected power system, which has been introduced in the above Section 4.

The computer control and remote monitoring technologies are new high-tech based on computers, auto-control principles, auto-detection, remote communications and electronic technologies. Based on these control and equipment technologies, the PV module, energy storage devices, the loads and the grid can be connected to form a complete PV station system.

In addition, the controller of the PV system must be provided with such detection, protection and alarm functions as lightning protection, overcurrent, overheating, short-circuit, inverse connections, abnormal DC voltage and abnormal grid voltage and the like.

Finally, it shall be pointed out, the controller will have increasingly stronger functions as the PV industry develops. And it is expected to integrate the traditional controls, the inverter and the monitoring system.

7.6 Applications of PV Systems

7.6.1 Classifications of PV Systems

The PV system can be divided into two types by power supply method. In the first type, the solar cells and the energy storage device form a stand-alone power supply system to supply power to the load. When the solar cell output cannot meet the load demand, the energy storage battery can supply power; and when the solar cell output power is larger than the load demand, the surplus power can be stored in the energy storage battery. In the second type, the solar cell control system is connected in shunt with the utility grid. When the solar cell output cannot meet the load demand, the grid can supply power; and when the solar cell output power is larger than the load demand, the surplus power can be transmitted to the utility grid. The former is mainly applied to the region without electricity but the cost of the energy storage device shall be added; and the latter is only suited to the region with electricity. For the PV systems installed in the region with electricity available, some are also installed with energy storage batteries to ensure uninterrupted power supply for the load and peak levelling for the grid.

From the viewpoint of PV market, two types are available: One is the private PV system, which is characterised by self-generation and self-consumption, off-grid, energy storage device, and small size, and mainly applied to the regions such as field works, border posts, communication base stations and the areas without electricity that the grid cannot cover. The other is investment PV system, which is characterised by benefit by power selling, grid-connected, without energy storage system, large size and mainly used to some BIPV and PV ground stations. The private PV system shall focus on cost and availability and the investment PV system on economic return. Most of the distributed PV stations are the private ones, and the large-sized PV ground stations are mostly investment ones. The PV application is rapidly expanding and becoming more and more mature.

7.6.2 Application Type, Size and Load Types of the PV System

The PV systems can be divided into the following six types by their application type, scale and load types: The first three are those for the small-power users; the fourth is the small-/medium-sized distributed grid-connected PV stations, the fifth is the large-sized grid-connected PV stations and the sixth is the hybrid power system.

7.6.2.1 Small-Power DC PV Systems

The system is characterised by pure DC loads and there is no special requirement on the operation time. The loads are mainly for the day time. The typical ones are the charger of the mobile phones and laptops as well as the small-sized DC water pump system. The system does not need any energy storage device or controller (see Figure 7.29). Those for power supply to the DC compact fluorescent lamps and TV sets in the remote

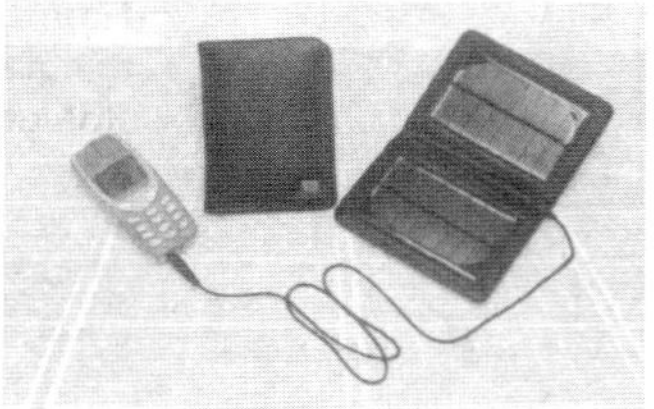

Figure 7.29 Applications of small-power DC PV systems.

regions where no electricity is available, however, are often provided with energy storage devices.

7.6.2.2 DC Power Supply Systems Required of Controllers

The PV system is also only suited to the DC load system, for example, the power supply of communications, tele-measuring and monitoring equipment, beacon lighthouse, street lamps and rural central power supply station. The system, although not large in scale, is very important. Accordingly, the solar cell module (array) shall be provided with controllers and energy storage devices. Figure 7.30 shows the PV power supply system of a GSM communication relay station and the power supply system of a solar lighthouse.

7.6.2.3 AC/DC Power Supply Systems Required of Inverters

The PV system is usually used for the PV stations with AC/DC loads such as the residential buildings in the region without access to electricity and the border posts. Generally, they shall supply power simultaneously to the DC/AC loads, and the load has large power consumption and the scale is also big. It shall be provided with the off-grid inverter and the energy storage device, but the grid-connected inverter is unnecessary since it does not transmit power to or receive power from the utility grid. Figure 7.31 shows the PV power supply system of the border posts that needs inverters and energy storage devices.

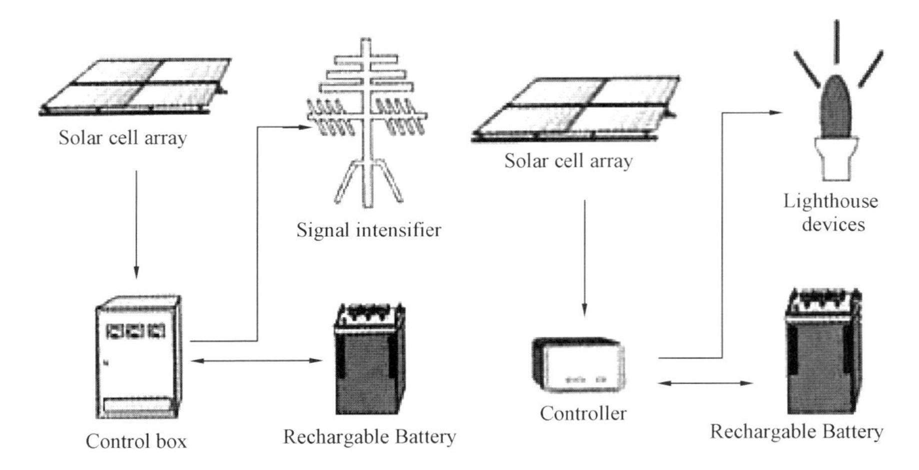

Figure 7.30 Communication relay station and solar lighthouse required of controllers and batteries.

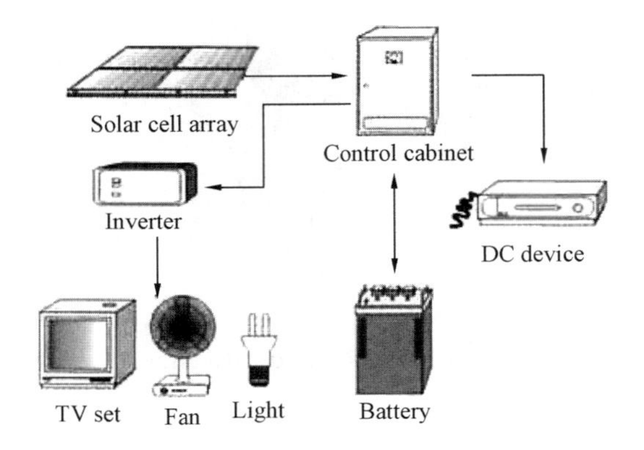

Figure 7.31 Applications of PV systems that need inverters, controllers and energy storage devices.

7.6.2.4 Small-/Medium-Sized Distributed PV Systems with the Grid-Connected Inverter

The system is provided with the grid-connected inverter and the meters for electricity selling metering, for example, the solar parking lot. In the system, the surplus power generated in the day time will be sold to the utility grid. And when the load power is insufficient, it can acquire power from the utility grid, for example, the residential community. In addition to the grid-connected inverter and the meters for electricity selling metering, the system shall be also provided with the meters for electricity purchase metering. Figure 7.32 shows the PV system with the meters for electricity selling metering and the PV system with the meters for electricity selling and purchase metering.

BIPV will be discussed in Section 6.

7.6.2.5 Large-Sized Centralised Grid-Connected PV Stations

The system can convert the DC power generated by the solar PV module by the grid-connected inverter to the AC power compliant with the mains grid requirements, and connect it to the utility grid. Figure 7.33 shows the sketch of the system and the actual system is very complicated. Please refer to Section 7.7 for details.

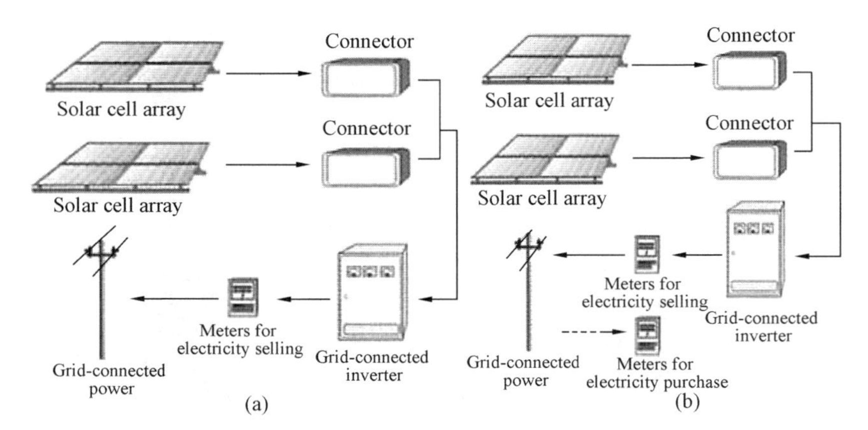

Figure 7.32 Applications of the PV system with the grid-connected inverter and meters for electricity selling/purchase.

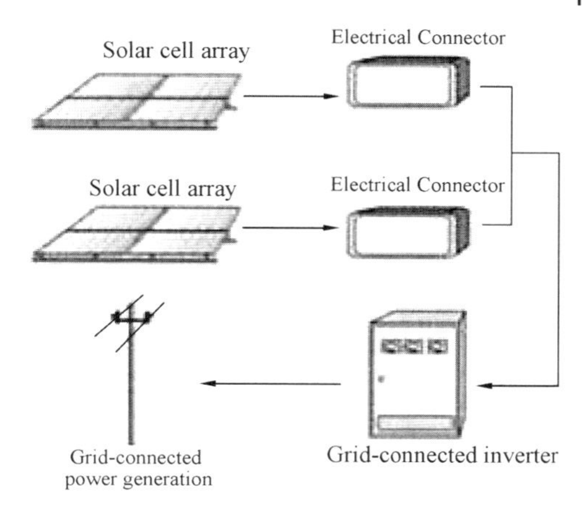

Figure 7.33 Sketch of large-sized grid-connected PV station.

7.6.2.6 Hybrid Power System

In addition to the solar PV module, the system is also designed with other power supplies such as the diesel generator, wind generator, and the grid power supply and so on. The hybrid power supply system can reduce its dependence on the weather or other factors. It can combine the characteristics of several power generation technologies to avoid their shortcomings. The hybrid power supply system of the wind and the PV power is very promising because the wind power and the solar power are mutually complimented for each other as for time and seasons so that the natural resources can be fully utilised.

The hybrid system, however, also has some disadvantages: first, the control system is very complicated. Since several energies are used, the system shall monitor the operation of each energy resource to settle the mutual influence between the sub-systems and coordinate the operation of the whole system. Obviously, the control system is more complicated than the stand-alone system. Second, the primary investment is large because the design and installation works of the hybrid system are larger than the stand-alone one. At present, many communication power supplies in the remote regions and the power supplies of civil aviation equipment adopt the hybrid power supply system (see Figure 7.34).

Figure 7.34 Wind power and PV hybrid power system.

7.6.3 Energy Storage Device Charging/Discharging by Small-/Medium-Sized PV Systems

The small-/medium-sized PV system consists of the solar cell array, the blocking diode, the controller, the inverter, the energy storage devices and the supports as well as the transmission and distribution systems. In the stand-alone PV system, the solar cell array shall charge the energy storage device under the control of the charging/discharging controller with blocking diode to protect the energy storage device from overcharging or over-discharging, and disconnect the load and the charger circuit according to system operation. Figure 7.35 (a) and (b) show the diagramme of energy storage device charging/discharging lines when the stand-alone PV system is provided with DC/AC loads.

In Figure 7.35, the blocking diode is also known as 'the anti-counter-charging diode' or 'the isolation diode'. It is designed to protect the energy storage device by its one-way conductance. Its maximum output current must be larger than the maximum output current of the solar cell array, and its reverse withstanding voltage must be larger than the maximum voltage of the energy storage devices.

When necessary, the stand-alone PV system can be provided with additional diesel generator, wind generator or other backup power supplies. They can also charge the energy storage device, as shown in Figure 7.36.

For the small-power PV system, those only provided with DC loads are usually designed with the voltage corresponding to the energy storage device nominal voltage or its integral multiple. For the PV system provided with AC loads or AC/DC loads, its

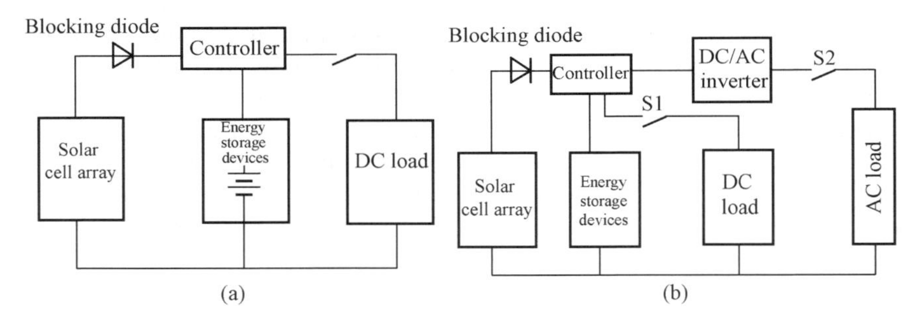

Figure 7.35 Stand-alone (off-grid) PV system charging/discharging diagram. (a) DC load only; (b) DC/AC loads.

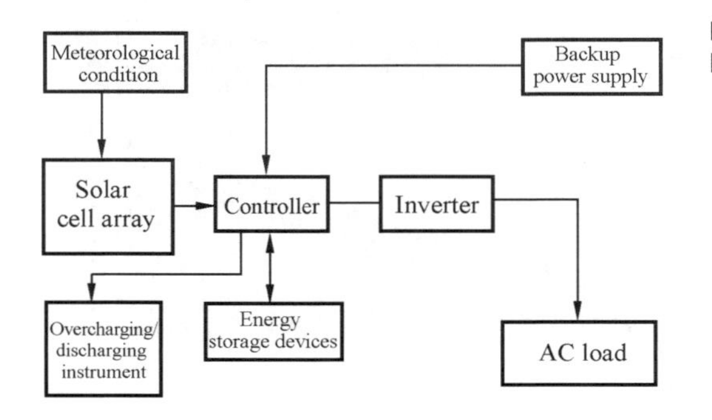

Figure 7.36 Stand-alone PV system with backup energies (e.g., wind power).

voltage often corresponds to the voltage of the appliance, for example, 220 V, 110 V, 48 V, 36 V, 24 V, 12 V.

7.7 BIPV and Distributed PV Stations

7.7.1 BIPV

BIPV refers to the building integrated with the PV system (or building integrated photovoltaic) where the PV system shall be viewed as an indispensable part of the outer structure of the building and it can generate power. In practise, BIPV, however, accounts for only a small portion, and in most cases, the PV system is only installed on the existing building, which is called as BAPV (building attached photovoltaic) where the PV system is not designed with the functions of building structure and materials. As a result, BIPV and BAPV shall belong to BMPV (building mounted photovoltaic), that is, the PV system mounted on the building. The BIPV is only a form of BMPV. Figure 7.37 shows the diagramme of a grid-connected PV system with BAPV.

The building energy consumption accounts for one-third of the total energy consumption in the world where the air conditioner and heat supply account for a big proportion. Since integration of PV and buildings can reduce effectively the energy consumption in the building, it will become the biggest market in the future and an important subject in the distributed PV system. It develops a wide sector for PV application and indicates that the PV power has ushered to the massive application stage. Generally, the solar cell module is installed on the roof of the house or the building, and the outlet terminal is connected with the utility grid via the controller and the inverter. In this way, the PV array and the grid can supply power to the user, and a grid-connected PV system is established. It can reduce the cost of energy storage devices and PV power. Besides, it is peak-levelling and environmentally friendly. Further, the glass curtain wall of the PV system can be used to replace the ordinary curtain wall glass. In this way, the curtain wall, being the building material, can be also used to generate power, which will become beautiful landscape in the city. BIPV has the following virtues: it can save land, improve power efficiency, reduce investment and losses of transmission lines, and replace or partially replace the building materials.

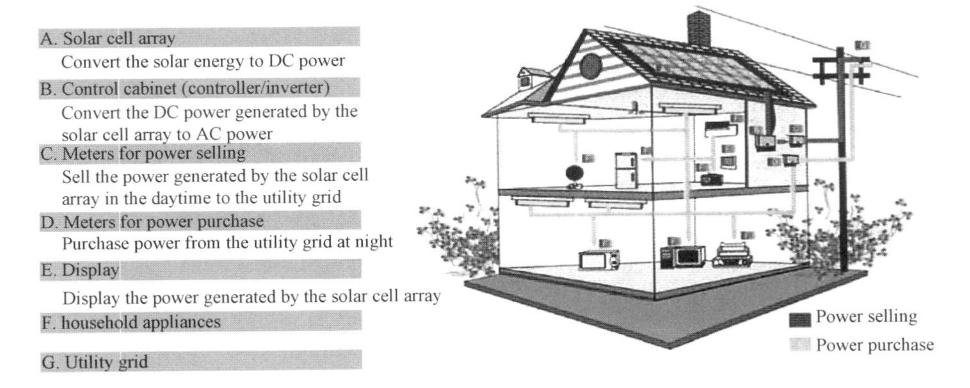

Figure 7.37 Diagram of grid-connected PV system with BAPV.

A BMPV system is often composed of the PV module, wall or roof, cooling air passage and supports. For a complete BMPV system, it shall be provided with energy storage devices, the inverter, and the system controller.

BIPV has the following virtues:

① It can make effective use of building roof and curtain wall instead of the precious land resource, which plays a crucial role for the city with costly land;
② It can effectively reduce the building energy consumption and achieve the goal of energy conservation;
③ It generates and consumes power in the same place, which can save the investment on the grid in a certain distance. The power generated by the grid-connected PV system in BIPV can be supplied to the building loads or transmitted to the grid. In this way, the PV system and the utility grid can jointly supply power to the load, improving power supply reliability;
④ The PV system of BIPV is generally installed on the roof or the wall to absorb the solar energy. Accordingly, the system can supply power while play a role in insulation for the building;
⑤ Since the PV cells are modular, it is easy to install the PV array and select the power capacity;
⑥ The grid-connected PV system is free from noise and pollutant emission. It does not consume any fuel, and it can reduce effectively light reflection on the glass and avoid surface light pollution like the glass buildings. It is based on green and environmentally friendly concept, and thus it can improve the overall quality of the building.

Figure 7.38 (a) and (b) show two applications of roof-mounted solar cells where (a) is the crystalline silicon PV building project, Shanghai World Expo, 2010; (b) the roof provided with thin-film solar cells.

BMPV shall be integrated to building design, which can play a role in building appearance and insulation, improve its value and reduce the cost. Two types of BMPV modules are most promising: One is PV tile and thin-film roof and the other glass PV thin-film curtain wall. It shall pay attention to the special requirements and maintenance when the building is provided with the PV system. If the installation and structure are improper or poor ventilation may result in hot spot and result in early attenuation and failure of the module. To promote BMPV, the PV module and system manufacturers shall develop close cooperation with the building company.

Figure 7.38 Two applications of PV roof. (a) PV building project, Shanghai World Expo, 2010; (b) Roof c/w thin-film solar cells.

7.7.2 Design Principles of BIPV Grid-Connected Power Systems

It shall take into account the surroundings of the building during design to avoid or keep distance from the shades. With full consideration to the building, the orientation and the tilt angle shall be identified for the solar cell module, and good ventilation shall be available for the solar cell module.

The BIPV module shall be designed in accordance with the following requirements: pressure-withstanding, rain-proof, sound-proof, heat-insulation, good ventilation, easy installation, lightning protection, leakage-proof, no hot spot and the like. The power performance shall conform to utility grid access. As a building element, it shall not affect the functions and appearance of the building. For the PV glass curtain wall, it shall also consider the colour and transmittance. During BMPV module installation, it shall pay attention to the mechanical strength, conductor connections (not loose) and diode heat dissipation in the junction box. For the PV curtain walls, it shall also take into account the allowed maximum wind pressure and impulse capacity.

7.7.3 National Policies and Certification of BIPV in China

To encourage development of BMPV, in 2009, the Ministry of Finance of PRC and the Ministry of Housing and Urban-Rural Development of PRC jointly issued 'Temporary management rule on financial subsidies of BIPV' and 'Interpretation on policies on accelerating BIPV applications (Financial Building [2009] #128)', and the Ministry of Finance of PRC published 'Advice on implementation of accelerating BIPV application' and 'Guideline to application of BIPV application pilot projects' to offer subsidies to the BIPV pilot projects.

For codes and certification of PV system and BMPV, it shall be carried out in four stages in accordance with China Quality Certification Centre (CQC): the first stage is to carry out safety certification on the PV module quality of BIPV; the second to conduct safety certification on the component quality of BIPV, including the inverter, the controller, the connector, the cables and lines, glass and back film and so on; the third to conduct evaluation on the off-grid power system of BPIV; and the fourth to conduct evaluation on the grid-connected power system of BIPV, including the wind/solar hybrid power PV system.

7.7.4 Encouragement of Distributed PV Stations by the Chinese Government

The eastern part of China is characterised by dense population, developed industries and huge power demand. The northwestern part, especially Xinjiang and Tibet, is endowed with rich resources and abundant solar rays, which is suited to build large-sized PV stations, and relatively few population and industries. As a result, the surplus power shall be transmitted to the east of China via the HV grid stretching several thousand kilometres. Accordingly, the Chinese government encourages distributed PV stations.

The distributed power system (distributed generation) contains the following meanings:

① Since the power generated is mainly used by the user or locally used, it is near the user, distributed and not transmitted out;
② The installed capacity is small with power ranged from several KW to dozens of MW;

Figure 7.39 PV stations in the agricultural greenhouse and the fish pond.

③ The grid-connecting voltage is 110 KW and below;

It has the following virtues:

1) Good peak-levelling performance, simple operation, fast startup/shutdown, flexible dispatching; and especially suited to adopt the local solar energy;
2) Low cost and high efficiency since there is no need to transmit the power for a long distance; suited to rural countryside, pastoral area and mountainous regions far away from the large grid and the roof power generation in the urban area;
3) High safety and reliability.

In addition to BIPV, another distributed PV station—the solar PV greenhouse has also won full support of the government. In July 2014, the Chinese government issued 'Notice on further implementation of the policies about distributed PV power', which says 'when the PV projects (project capacity ≤ 20 MW) built on such land resources as the deserted land, waste hills and slopes, agricultural greenhouse, mud flat, fishponds, lakes and the like are connected to the grid at the voltage below 35 kV and the power generated is consumed at the grid-connecting point, the financial subsidy is available according to the local benchmark power price, and the grid companies shall offer high efficient and rapid service for the distributed power supply via green channels' (see Figure 7.39 for the PV agricultural greenhouse). The top of the greenhouse is thin-film solar cells (especially the flexible ones). The PV greenhouse will not cover additional cultivated land and it can raise the output of the produce. It is reported that 2 GW PV agricultural projects are under construction or proposed to be built with signed contracts in the eastern part of China in the recent two years (Source: Solarzoom Network, 2014).

In recent years, the large-sized PV stations have won rapid development in China. To encourage the distributed PV system, National Energy Administration of China released in 2014 that the new installed capacity of PV power was 14 GW, which includes 8 GW distributed grid-connected power system, and 6 GW ground power stations. At present, the distributed power supply in China just starts and it has a long way to go.

7.8 Grid-Connected PV Systems and Intelligent Grids

7.8.1 Grid-Connected PV Systems

The grid-connected PV system consists of the PV cell array and the grid-connected inverter, and it can directly input the power to the utility grid via the grid-connected inverter without any energy storage device. Compared with the off-grid PV system, the grid-connected PV system omits the process of energy storage and release, and thus reduces energy consumption, saves land overage and reduces the configuration cost.

There are two types of grid-connected PV systems: centralised grid-connecting and distributed grid-connecting ones.

Centralised grid-connected PV systems: The power generated will be directly transmitted to the grid and dispatched by the grid to the user. The power exchange is one-way. The centralised grid-connected large-sized PV stations are generally the national-level ones, which have big investment, long construction period, and large land coverage and are suited to the desert and the unpopulated regions, which are usually far away from the load point.

Distributed grid-connected PV systems: The power generated will be directly transmitted to the load, and the surplus or deficit power shall be regulated by the connected grid. The power exchange may be double-way. It is suited to the small-/medium-sized PV systems. The distributed small-/medium-sized grid-connected PV system, especially the BIPV system, which have such virtues as lower investment, faster construction, smaller land coverage and more governmental support policies, are the ones with top priority in China's PV power industry.

The numerous small-/medium-sized distributed grid-connected systems can be connected to the large national grid in the form of micro-grid and work together with the grid. They will run in the micro-grid and be connected to the interconnected ultra/super-high-voltage grid via the LV/MV distribution grids, which is deemed as the key characteristic of China's grid-connected PV systems and also the key technical subject in improvement of PV power scale and quality. It can expand the range and flexibility of PV applications.

7.8.2 Technical Specifications of the Grid on the Grid-Connected PV System

The basic requirement for the grid-connected PV system is that the frequency and phase position of the sinosoidal wave current output by the inverter must be identical to that of the utility grid voltage. In addition, either the centralised large-sized PV station or the distributed small-/medium-sized PV station may exert adverse influence over the grid when connecting to the utility grid due to PV power intermittence and instability. As a result, the PV station shall conform to the following requirements:

(1) System voltage deviation
    Since the solar power varies with solar ray, weather, season and temperature changes, the output power is unstable. Accordingly, the associated actions shall be taken when the PV station is connected to the utility grid to ensure the voltage fluctuation of the connected utility grid compliant with GB/T 12325-2008 'Power quality: supply power voltage deviation'. Verification shall be conducted when the allowed voltage deviation at the connecting point between the PV system and the grid is $> \pm 5\%$.
(2) Harmonic problem
    The grid-connected inverter shall first convert the DC power generated by the PV module to the sinusoidal wave current with frequency and phase position identical to that of the grid and then be connected to the grid. Generally, a lot of harmonics will be generated when the DC power is converted from DC power to AC

power. In accordance with GB/T14549-93 'Power quality: utility grid harmonics', the limits of harmonic voltage and harmonic current of the utility grid shall have strict specifications; otherwise, the associated filter shall be provided to prevent pollution on the power quality of the utility grid. After the PV power is inverted by the grid-connected inverter, the total harmonic distortion (THD) shall fall within 3.0%. Actual measurements shall be conducted before the PV station is connected to the grid, and the limits of resonant wave shall be allotted by the actual access capacity.

(3) The voltage fluctuation and flickering shall conform to GB/T12327-2008 'Power quality: Voltage fluctuation and flickering'.
(4) The imbalanced voltage shall conform to GB/T15543-2008 'Power quality: 3-phase voltage imbalance'.
(5) The DC component shall conform to the relevant publications of IEEC-1547.
(6) Reactive power balance: The power factor of the PV station is very high, above 0.98, and it is basically pure active power output. To meet the principle of local balance by layers and regions for reactive power compensation, the PV station shall be provided with appropriate reactive power compensators to meet the requirements of the grid on reactive power, improve the voltage quality and reduce line losses. For the small-sized PV station, the index for power factor shall be given, and the large-/medium-sized PV stations shall be designed with active power control and reactive power regulation capacity.

The PV array is usually 110 V or 220 V. For the grid-connected PV station, several arrays are usually connected in series or shunt and then boosted to the voltage level identical to the grid voltage. Based on the voltage of the grid connected with the PV system, the PV system can be divided into LV grid-connected PV system and HV grid-connected PV system. For the small-sized PV station with installed capacity no larger than 200 KW, it is often connected to the 400 V LV grid via the small-sized inverter; for the medium-sized PV station, it shall be connected to the 10–35 kV grid; and for the large-sized PV station, the DC power generated by the module shall be first converted by the inverter and then connected to the 66 kV HV grid via the step-up transformer. Since the grid-connected PV system is power generating equipment, similar to the thermal power plant and the hydroelectric station, it must follow the relevant laws and regulations on the power industry.

For the small-sized PV station, the utility grid shall be viewed as the load, and the grid has no requirements on the active power, but it shall minimise the reactive power to the grid. When the power generated is insufficient due to cloudy, gloomy or rainy days or the system failure, it shall be automatically switched to the other side and the grid will supply power to the load. In case the grid frequency and voltage are abnormal, it shall be disconnected immediately. For the large-/medium-sized PV station, the utility grid shall be viewed as the power supply, and it shall be designed with the capacity to withstand abnormal grid frequency and voltage and able to offer support for grid stability. When the grid failure results in voltage drop, the PV station must keep grid-connected operation; and it is not allowed to stop supplying power to the grid until the grid voltage drops to the given degree. In this case, the PV system shall be automatically disconnected with the grid, becoming the status of stand-alone PV system, and the power supply shall be connected for the emergency load or the ordinary load. Some PV systems with automatic switching function are provided with energy storage device. When the grid is in blackout, power-limitation or failure, these PV systems can operate stand-alone and supply power to the load. They can work as the power supply system of key or emergency loads such as the emergency communication power supply, the medical equipment, the gas service station, indication and lighting of emergency shelters.

When the PV station or the grid is in abnormal conditions or fails, the associated relay protections shall be provided to ensure equipment and human safety, safe operation of the grid and PV equipment, and the safety of maintenance staff and the public. As for overcurrent and short-circuit protection, the PV station shall be designed with certain overload (overcurrent) capacity, which shall refer to GB/T 19939-2005 'Technical specification on the PV system'. As for anti-islanding protection, it must be closely cooperated with the grid relay protections or refer to GB/T 19939-2005 'Technical specification on grid-connections of the PV system'. The codes, however, only propose the requirements on power quality and basic safety of the small-sized PV station, and the codes on the large-sized PV stations shall be further improved.

7.8.3 Significance of the Intelligent Grid on PV Power and Other New Energy Utilisation

The intelligent grid is based on the integrated, high-speed and double-way communication network. Thanks to such advanced technologies as the advanced sensor and measurement technology, the advanced equipment technology, the advanced control methods and decision-making system, it can achieve the objectives on reliability, safety, economic and high efficient operation as well as environmental protection and operation safety. Its main characteristics include self-healing and excitation, connection of various power generation types, available to supply the power compliant with the users' demand in the twenty-first century and start the optimal and high efficient operation of the power market and assets. The intelligent grid includes the followings: intelligent substation, intelligent distribution grid, intelligent energy meter, intelligent interactive terminal, intelligent dispatching, intelligent appliance, intelligent building, intelligent urban power network, intelligent power generation system, new energy storage system and so on. It is a new grid integrating the advanced sensor and measurement technology, communication technology, information technology, computer technology and control technology with the physical grid. The robust intelligent grid is still at the early stage, and thus no accurate definition is accepted.

The intelligent grid, based on the physical grid, enhances the elasticity, reliability and sensitivity of the user side by means of intelligent metering device, sensor and measurement means, double-way digital communication means, control technologies and computer technology with minimal artificial interference. First, the intelligent grid shall be stable and reliable and self-healing. It shall be safe and has the capacity against external disturbance. It shall be able to integrate the information from various sources. In addition, it shall be able to interact with the power user and regulate for balance in real time. The intelligent grid is no longer a simple transmission system and it becomes an information system. The intelligent grid employs a lot of new technologies in such links as power generation, transmission, distribution and consumption to achieve optimal configuration and energy saving as well as emission reduction.

In the twenty-first century, the new energy sources such as the PV power and wind power will see wide application. The characteristic of the new energy resources is intermittent. The power generation and quality will vary with solar intensity and wind force changes, and the grid connection and dispatching will become a major issue as the PV and wind power projects are built. And the conflict with the traditional utility grid becomes more and more serious. When the PV station and the wind farm are connected to the utility grid, it may exert adverse influence over the utility grid. In the robust intelligent grid, the new energy power stations such as the PV station etc. are the real coupling points. And the study on connection, consumption and regulation capacity of these new energies will significantly push the development of clean energies and finally settle the problems existing in the connection of PV power and wind power.

For the new clean energies such as PV power, the intelligent grid shall be designed with the following functions:

① The advanced control and energy storage technologies shall be employed to improve the technical standard of clean energy power grid-connection and the acceptance capacity of clean energies;
② The grid structure of the large-sized clean energy base shall be reasonably planned and the ultra/super high voltage and flexible transmission technologies shall be employed to transmit the power generated by clean energies;
③ The intelligent distribution and consumption equipment shall be used to accept, coordinate and control the distributed energies and conduct friendly interaction with the user;
④ The Internet of Things can be used to understand the demand on the user's side, regulate the power generation in time and improve the economic performance of energy production;
⑤ The sensor and communication system can be used to identify the accurate position of the power failure and ensure reliable operation of the PV system.

In addition, the energy storage device can play a role in peaking level and valley filling, frequency and voltage stability control as well as power quality regulation in the intelligent grid, help flexible connections of the intermittent new energies and load regulation and emergency handling. The energy storage technology shall become one of the core technologies of the intelligent grid. All in all, the intelligent grid can facilitate connection of renewable energies.

The state grid of China is developing towards the intelligent grid. The intelligent grid in China with the ultra high voltage (UHV) grid as the backbone is based on the coordinated development of the grids at various voltage levels. It can receive various power resources, especially, it can make the PV and wind power and other intermittent power the quality power. The intelligent grid is the key building direction of the state grid. The intelligent grid is planned in three stages in China with expected total investment above RMB 4000 billion yuan. The first stage (2009–2010) is the pilot stage with investment of RMB 550 billion yuan; the second stage (2011–2015) is the comprehensive construction stage with investment of RMB 2000 billion yuan where the UHV grid accounts for RMB300 billion yuan; the third stage (2017–2020) is the upgrade stage with investment of RMB 1700 billion yuan where the EHV grid accounts for RMB 250 billion yuan. In 2020, the state intelligent grid will be completely built.

The intelligent grid can realise optimal configuration of power consumption and more importantly, it can adapt to connection of various renewable energies, including the PV power, and maximum power supply. The western grid in China is weak but the wind power, solar power and hydropower are abundant. The grid connection and dispatching problem is very serious. As the intelligent grid is improved, the renewable energies, including the PV power and the wind power and so on, will witness wide application.

7.8.4 Development of China's PV Industry in the Past 10 Years and Its Outlook

In the twenty-first century, China faces more serious challenges concerning energies: The demand on CO_2 emission reduction and environmental pollution improvement requires China has to change its energy structure (to reduce the proportion of coal); its hydraulic resources are limited by space; its oil and gas resources are insufficient, and its dependence on import rises; most of China's uranium mines are depleted ones and the enriched uranium is dependent on import; the per capita energy consumption rises and the industrialisation is accelerated, which widens the gap between energy demand and supply; the regional distribution of energy is inconsistent with consumption distribution, raising the energy losses and cost; it has not built the nationwide power distributed grid structure, the power emergency response capacity is weak, and the renewable energies account for a small proportion, and it is still not suited to multiple energy power generation and integrated utilisation.

China's PV industry goes through 5 stages over the past 10 years (2004–2014):

1) Rapid development stage (2004–2008): As many countries such as Germany offered subsidies to the PV industry, China made use of the foreign market, technology and capital, and developed rapidly. In 2007, China, surpassing Japan, became the largest producer of PV cells and modules in the world. At this stage, the price of the core raw material of PV power—the polycrystalline silicon was larger than US$400/kg.
2) Great setback stage (2008–2009): As the global financial crisis broke out, China's PV industry went through a hard time with product price falling rapidly where the price of the polycrystalline silicon even dropped below US$40/kg.
3) Rebounding stage (after 2009): China's PV industry was positioned as the strategic rising industry and the market began to get warm again. Having learned from the foreign governments on new energy subsidies, China issued 'Golden Sun Pilot Projects' in 2009, and carried out bidding on the large-sized ground PV stations and gave appropriate subsidies to the PV power.
4) International trade friction stage (2011–2013): The PV products saw sharp reduction on prices due to periodic surplus and technical advance. And the protectionism on international trading rose. In November 2011, Solar World, U.S. branch, officially proposed to carry out the trade tariff protection of 'anti-subsidy and anti-dumping' on China's PV products (implemented in 2012 and 2013). China's PV manufacture industry experienced another setback, and it was rapidly restructured and adjusted.
5) Rational development stage (2013-now): The planning on China's PV industry is adjusted, and the domestic market is steadily developed, and the government issues a series of supporting policies on the PV industry. The core technologies achieve great breakthroughs, and the overall competence of the whole PV industry is improved. The PV markets in other countries such as Japan, India and the South East Asian countries become active. China's PV industry returns to a rational development period. In 2014, the new installed and connected capacity in China was 11 GW, ranking the first in the globe (one-fourth the global new installed capacity). Since

the output of China's solar cell modules is larger than 33 GW, accounting for 75% of the global output, it still needs a lot of imported ones.

The above analysis shows, the PV industry has been always dependent on the governmental subsidies, which affects the periodic change of the industry. As the PV industry develops rapidly over the past 10 years and the technical level on it witnesses great improvement, the PV efficiency has been continuously improved with continuous reduction of module price. Compared with other power generation means, the cost of the PV power, however, has to be further reduced.

In the latter half of 2013, the State Council of China issued 'Several opinions on promotion of healthy development of the PV industry'. Following it, the associated actions have been taken by the relevant departments with the supporting policies in place, including 'Notice on the issues concerning the subsidy policy for the distributed PV power by kWh, the Ministry of Finance', 'Notice on the issues concerning additional criteria regulation for renewable energy power price and the environmentally friendly power price, NDRC' and 'Notice on promotion of healthy development of the PV industry by means of the price leverage, NDRC'. These supporting policies have the following core concepts: During the '12th-five-year plan', the installed capacity target of the PV units was raised from 21 GW to 35 GW; for the large-sized grid-connected PV station, the price of the grid-connected power falls in a range of RMB 0.9 yuan/kWh–1.0 yuan/kWh; for the distributed PV power, the subsidy of 0.42 yuan/kWh is available to all the power generated, and the part for private purposes is suited to the policies on the benchmark grid-connected power of the local coal-fired units; and the time for subsidies is 20 years.

These supporting policies promoted the development of the PV industry in China. It helps to settle the trouble that China's PV industry faces—both the suppliers and the consumers are abroad. China's PV cells and modules ranked the first across the globe in 2007. China's installed capacity of PV units, however, has seen slow growth. In the end of 2009, China's cumulative installed capacity of PV units was only 0.3 GW. Only five years later, in 2014, China's polycrystalline silicon output was 130000 t, accounting for 45% of the global output and ranking the first in the world; China's solar cell output exceeded 33 GW, accounting for more than 75% of the global output, but the price of the PV modules, however, was reducing dramatically. In the top 10 manufacturers of solar cell modules across the world, 7 are China-based. In 2014, China saw 10.95 GW (44 GW in the world) of new grid-connected PV power, accounting for 25% of the global new capacity and also ranking the first. In 2014, China saw about 25 billion kWh of annual PV power, which only accounted for 0.4% of the national total power generated (5550 billion kWh). Obviously, the PV industry has a bright future in China and it has a long way to go.

The Chinese government has issued policies to encourage application of PV power and also restrain and regulate the PV manufacture industry, and it has obviously raised the access threshold. In 2015, National Energy Administration (NEA), Ministry of Industry and Information, and Certification and Accreditation Administration of PRC jointly issued 'Advice on promotion of advanced PV technical product application and upgrade', and implemented the 'Pacemaker' programme of PV products to raise the market access standard of PV product and push progress of PV technology and industrial upgrade. Each year, NEA arranges some projects with certain market scale to adopt the advanced technical products and organises the pilot projects of PV power. It specifies to adopt the 'Pacemaker' products as the first priority. In 2015, the market access standards for PV products are: The PV conversion efficiency of polycrystalline and the monocrystalline silicon solar cells shall be no less than 15.5% and 16%, respectively; the PV conversion efficiency of silicon-based, CIGS, CdTe and other thin-film solar cells shall be no less than 12%, 13%, 13%, 12%, respectively. Concerning the attenuation rate, the strict regulations are also available.

We shall also notice that the PV power technology is still at a period with rapidly technical change when the industry faces dramatic ups and downs. In this case, new technologies and processes may emerge at any moment, for example, the renovation of polycrystalline silicon preparation process, or the technical breakthrough on thin-film materials, or the emergence of high efficient solar cells and promising new concept solar cells and so on. Accordingly, the outdated technology may face the risk to be replaced at any moment.

As for investment on PV power stations, it has the following virtues: one investment and long-term return, low follow-up operation and maintenance cost, and stable benefit. Especially, the latest subsidy policy was issued in China, which clarifies the subsidy period of 20 years and further solidifies the stability of investment return.

As the PV power is rapidly promoted, it is estimated that the grid-connected power generation will be the main market of the PV power in the upcoming period, which will last for a rather long time, and its market share will continuously rise. The gradual growth of grid-connected power generation application symbolises that the human being is launching a new energy system

As the rapid promotion of PV power, the grid-connected power is expected to serve as the main market in the PV industry for a long term, and its market share will rise continuously. The application proportion of grid-connected power will grow, which indicates the human society is developing the new energy system with sustainable development in the twenty-first century, playing a significant role in the human energy change.

7.9 Codes and Test Verifications of the PV System

7.9.1 Necessity and Main Contents of PV Product Certification

The purpose of the users to purchase the PV product or build the PV station is the power generated, instead of the conversion efficiency or power of the solar cell. What they are concerned is: 1) How much power can be generated each year on every square kilometre? 2) What is the cost of each kWh? 3) How many years can the product work in a stable and reliable manner? 4) What is the cost of annual maintenance? 5) What is the return period of energy? 6) Initial system investment and grid-connected power price. Or, the PV system must keep stable in the acceptable time range and conform to the performance requirements in the given environment and time periods.

During the service life of the PV station (20–30 years), it shall take into account the solar UV radiation and temperature change in seasons as well as corrosion of seawater, acid rain, sand dust, ice and snow, hail, mold, bird droppings, NO_X, SO_X and so on. And the system is likely to have power attenuation due to the following factors: 1)

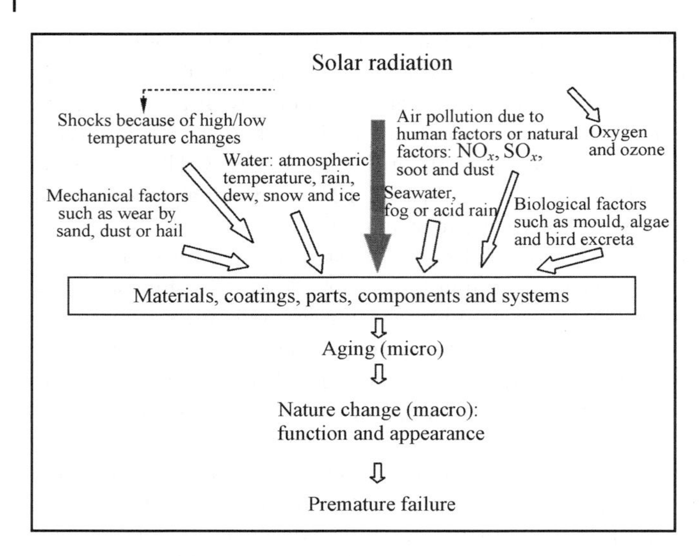

Figure 7.40 Field environment of PV system.

Packaging material performance attenuation (e.g., EVA yellowing); 2) Cell and module interconnecting performance deterioration; 3) Moist invasion; 4) Light-induced degradation of the PV component; 5) Hot spot present in the module in the cold/hot alternative environment and so on (see Figure 7.40 for the working environment of the PV system). As a result, the material, coating, component, module or system of the solar cell must be monitored and certified by the professional agencies in an environment with the above environmental factors considered.

The certification of PV products consists of safety certification and performance certification. The following PV products shall be tested: the controller, the inverter, integrated controller and inverter; the grid-connected inverter; the energy storage device for the PV system; the small-sized stand-alone PV system; testing and evaluation of stand-alone stations; testing and evaluation of the grid-connected PV system; testing of components and raw materials.

The following shall be conducted in the lab: (1) temperature circulation; (2) hot impulse test; (3) freezing/melting circulation; (4) temperature test; (5) temperature/humidity circulation; (6) hail test; (7) sand dust test; (8) vibration and etc. The following methods shall be used: (1) acceleration lifetime test (ALT); (2) environmental stress scan (ESS) and so on.

The outdoor system shall be subject to the following in-situ tests and certifications: evaluation of rated power, system output and reliability. For the DC array measurement, the array shall be removed or disconnected from the inverter, and the portable curve tracer shall be connected with the array positive/negative terminals to measure the I-V curve in the testing period, and the portable Infrared imaging cameras shall be adopted. The energy storage device, module, inverter, controller and the auxiliaries shall be subjected to safety test to determine the weak links in the system and offer reference for further improvement. Finally, the overall evaluation shall be made for the PV system, including the effective power generation time and power quantity.

As for the certification agencies on the PV products and systems across the globe, TUV is mainly oriented to the European market and UL for the USA (Canada) market. China Quality Certification Centre (CQC), the certification agency for PV products in China, has been authorised to develop the certification work on solar PV products. It can control the product quality and help the company to improve the PV product quality, enhancing the market competence.

As for the difference between testing and certification, in the former, the sample is sent by the manufacturer to the testing lab for relevant tests according to the associated standards; in the latter, after the sample is tested, the agency legally authorised will inspect the factory and then issue the certificate and after that, it will carry out aperiodic supervision and inspection to the factory and market spot check to make sure the products produced by the manufacturer compliant with the sample. Before the final PV product is identified, it must be first demonstrated by the relevant authority.

7.9.2 TUV Certification Oriented to the European Market

The PV testing lab of TUB, Germany, can test and certificate the PV products (including the PV module, component and system) for the companies both at home and abroad. The certification covers lab test and field test. Especially, it can evaluate the production-related quality assurance method in the initial and the follow-up tests.

Take testing of PV modules as an example. It is based on the IEC 61215/IEC 61646 publications and safety level II (IEC 61730). The performance measurement of the PV module is carried out in the standard testing conditions (STC). The power generation of the PV module is measured in the medium climatic conditions of Germany, or in Europe or any place in the world. As for characteristic test of the solar cell, it can determine the dark characteristics of the solar cell in addition to the solar ray characteristics, and record the hot spot present with the remote infrared imaging technique. In the production site, it can use the tunable spectrum of the calibration module and the solar simulator to measure the performance.

As for the PV component and system, they must conform to the operation environment and electrical conditions of the PV module/system and work safely in the service life of the system. The testing contents include the followings:

- Mechanical performance test (drop test, plugging force test, etc.);
- High/low temperature impulse test: −40−+80°C, 200 cycles;
- High-temperature high-humidity test: 85°C, 85% of humidity, 1000 hr;
- Anti-UV rays test: as per ISO4892-2, 500 hr;
- Anti-corrosion test (5 days);
- Bypass diode test.

RWE, Germany, has set up the safety lab of RWE (Guangdong), which has been accepted by China National Accreditation Board for Laboratories (CNAL) on complete testing capacity. The lab testing items of the PV component can be conducted and certificated locally.

7.9.3 UL Certification Oriented to the U.S. and Canadian Market

UL is the first third testing agency engaged in PV product demonstration and research in the world. UL1703, published in 1986, has become the basis for PV module safety demonstration of the United States, which is also accepted by the Canadian government. When the client carries out the demonstration in accordance with the same set of standards, they can simultaneously obtain the market access of the United States and Canada. UL offers two sets of dominant PV module performance standards: 'Crystalline silicon terrestrial photovoltaic (PV) modules—Design qualification and type approval' (IEC61215); and 'Thin-film terrestrial photovoltaic (PV) modules—Design qualification and type approval' (IEC61646). UL1703, IEC61215, and IEC61646 have some differences on evaluation and testing methods of PV products. The safety demonstration based on UL1703 puts the focus on the potential risks of the relevant staff and surroundings during normal installation, usage and maintenance of the PV product, for example, electric shock, fire and so on. IEC61215 and IEC61646, however, mainly evaluate the performance stability and reliability of the module during long-term operation in outdoor applications.

In UL1703, the four tests series—temperature cycle test, humid/hot humid/cold cycle test, water spray test and pressure test are independent from each other. After the associated environmental conditions are tested, the performance of the product on voltage withstanding, leakage current, wet insulation and earthing will be evaluated. In this way, it can identify the potential safety problems that may harm the staff when the module operates in a long time in outdoor applications. In IEC61215 and IEC61646, a set of correlated test series are carried out to test the electrical performance and other performance of the module in the site after each key link, and then the data shall be compared with the initial testing data of the sample to judge whether the variation of module stability and reliability falls in an acceptable range.

The hot spot withstanding test and temperature test, which has low qualification rate in UL1703 tests, will be described below:

(1) Hot spot withstanding test: It is aimed to evaluate the capacity of the module on heat generated by the cell in case of the hot spot effect. It is known that some cell may have output current less than other cells due to some reasons (e.g., shaded, damaged), and in this case, the cell in the hot spot will become a load in the whole module and consume the energy generated by other cells and generate a lot of heat, resulting in broke cell and cover glass or bubbling, lamination and combustion of glue coating of the module laminations, and then electric shock or fire and so on. In UL1703, the hot spot withstanding test will test the energy withstood by single cell in the module in case of the hot spot effect and hold on for 1 hr, and 100 times will be conducted intermittently. In IEC61215 and IEC61646, the hot spot withstanding test is targeted to the whole module where some cells will be shaded for the hot spot effect, and hold on for 5 hr.

(2) Temperature test: It is aimed to evaluate whether the selected raw material and component can conform to the temperature limits in various operation environments. Since many plastic materials are used in the PV module, for example, the back board material, the junction boxes and the connectors, and they will become rapid ageing and lose the insulation and mechanical protection functions when exceeding their normal working temperature range. The excessively high temperature may result in combustion of the plastic materials, resulting in fire. UL1703 requires, additional 20 ° shall be added to the working temperature limits of the back board material and other plastic. In IEC61215 and IEC61646, no similar requirements on the raw material selected are presented.

IEC has accepted some contents of UL1703 and published IEC61730-1 and IEC61730-2. The PV product is expected to obtain the safety and performance evaluation simultaneously, shortening the market access time.

7.9.4 Certification of PV Products in China

As the demotic PV market develops, China Quality Certification Centre begins to test and certificate the PV product in accordance with the national or international standards see Figure 7.41).

The demonstration accreditation and management system and execution system of China are shown in Figure 7.42 where CNCA is short for Certification and accreditation Administration of PRC, CNAB for China National Accreditation Service for Conformity Assessment, and CNAL for China National Accreditation Board for Laboratories.

The following labs have signed with CQC to carry out PV product and system testing: Tianjin Power Supply Research Institute (Institute 18), Shanghai Institute of Space Power Sources (Institute 811), Solar Power System and Wind Power System Quality Testing Centre of CAS, Shenzhen Electronic Product Testing Centre, National Centre of Supervision and Inspection on Solar Photovoltaic Product Quality (Wuxi) and Key Lab for National-Level PV Product Testing(Yangzhou), which can test and certificate the PV system, the PV module, the PV component, the PV application product and raw/auxiliary materials.

China Quality Certification Centre adopts three modes for certification of PV products: 'Type test+Factory inspection+Supervision and inspection'. (1) Type test: The samples that fail in the type test are allowed to be improved until they are accepted. (2) Factory inspection: Inspection in factory is generally conducted after the type test is passed with aim to check the production capacity of the factory. (3) Supervision and inspection: After the certificate is issued, aperiodic inspection will be carried out to the products produced by the factory.

Many PV manufacturers in China have passed the certification of China Quality Certification Centre and they are allowed to attach the 'Golden Sun' mark to their product, indicating that China's PV industry is gradually integrated to the world. It is expected to improve the PV product standards, normalise the domestic market and promote product export. To connect with the international industry at a faster pace, China is enhancing the input on the field and improving the hardware and software so as to set up an internationally accepted testing and certification agency as soon as possible.

As for the concrete items on safety and performance tests, they are similar in China and the United States. The Chinese codes and standards on PV systems and the relevant IEC publications are listed in Table 7.2 through Table 7.6.

At present, China's PV products, especially the PV system, are still gradually improved. Only the national standard for the user's PV system is available but that on the stand-alone PV station is not. The requirements on the power quality and basic safety performance are proposed for the grid-connected PV utility interfaces but no testing methods are presented. BIPV is the main market of the PV power, but no relevant IEC publications are available. Under the leadership of China Electricity Council, a standard work team, composed of China Electric Power Research Institute, State Grid Corporation of China, and the associated institutes, is set up to prepare the technical standards (more than 40) on PV grid connection. The draft for comment

SOLAR CELL PRODUCT CERTIFICATE

Certificate No.: xxxxxxxxxxxxxxxx

Name and address of applicant
xxxxxxxxxxxxxxxxxxxxxx
xxxxxxxxxxxxxxxxxxxxxx

Name and address of manufacturer
xxxxxxxxxxxxxxxxxxxxxx
xxxxxxxxxxxxxxxxxxxxxx

Name and address of production plant
xxxxxxxxxxxxxxxxxxxxxx
xxxxxxxxxxxxxxxxxxxxxx

Product name and series, specification and model
xxxxxxxxxxxxxxxxxxxxxx
xxxxxxxxxxxxxxxxxxxxxx

Product standards and technical requirements
xxxxxx

Certification model

Product type test + initial factory inspection + supervision after certification

The above products comply with the requirements of CQCXXXX - 20XX certification rules, and specially issued this certificate

Certification date: xxxxxx Term of validity : xxxxxx

On the basis of regular supervision, validity of the certificate is valid based certificate issuing authority.

Director:

CHINA QUALITY CERTIFICATION CENTRE

Section 9, No.188, Nansihuan Xilu, Beijing 100070 P.R.China
http://www.cqc.com.cn

Figure 7.41 Certificate issued by CQC for PV products.

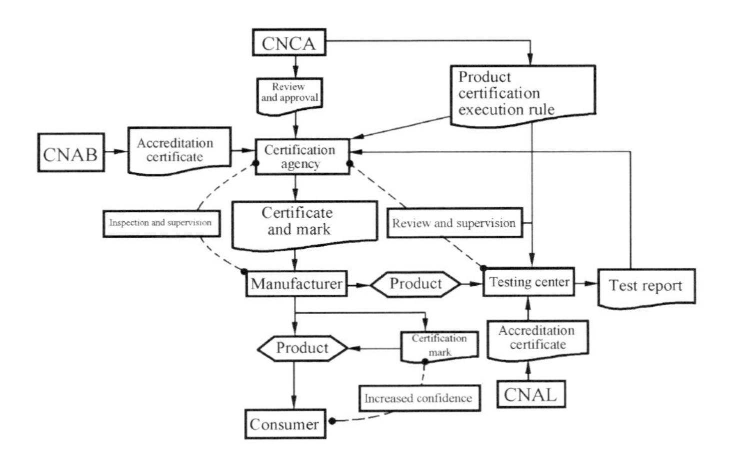

Figure 7.42 Certification and accreditation management system and execution system of China.

Table 7.2 Chinese national standards on the off-grid PV system.

1	GB/T18479-2001, equivalent to IEC 61277	Ground PV systems—introduction and guideline
2	GB/T19064-2003	Technical specification and testing method on solar PV power supply systems for domestic purposes
3	GB/T20321.1-2006	Technical specification on the inverter for off-grid wind/solar power systems
4	GB/T20321.2-2006	Testing method for the inverter for off-grid wind/solar power systems
5	GB/T19115.1-2003	Off-grid type wind/solar hybrid power system for household purposes
6	GB/T19115.2-2003	Testing method for off-grid type wind/solar hybrid power system for household purposes
7	GB/T19393-2003 Eqv.IEC61702(1995)	Rating of direct coupled photovoltaic (PV) pumping systems
8	GB/T20513-2006 Eqv.IEC61724(1998)	Photovoltaic system performance monitoring—Guidelines for measurement, data exchange and analysis
9	GB/T20514-2006 Eqv.IEC61683(1999)	Photovoltaic systems—Power conditioners-Procedure for measuring efficiency
10	SJ/T11127-1997 Eqv.IEC61173(1992)	Overvoltage protection for (PV) power generating systems—Guide

Table 7.3 Technical codes on grid-connected PV power in China.

1	GB/T19939-2005	Technical specification on grid connections of PV systems
2	GB\Z19964-2005	Technical specification on PV station connecting with the power system
3	GB/T20047-2006 Eqv.IEC61727(1995)	Photovoltaic (PV) systems—Characteristics of the utility interface

Table 7.4 Relevant Chinese national technical codes under approval.

1	GB/TXXXX–2004	Technical specification on stand-alone PV systems (Yunnan Normal University)
2	GB/TXXXX-2006 Eqv.IEC62124-2004, IDT	Stand-alone PV system—Design verification (IEC62124-2004, IDT)
3	CGC/GF004:2007(PVGPAP recommended standard)	Technical specification on lead-acid battery certification for solar PV energy systems
4	DB11/TXXXX-2008	Technical specification on solar PV outdoor lighting fixtures (Beijing)
5	GB/TXXXX-2006	Technical specification and testing method on the inverter specially for grid-connecting PV power generation
6	GB/TXXXX-2006	Technical specification on solar PV system applied to civil buildings
7	Local standard, Shanghai	Technical code on solar power applied to civil buildings (PV power generation)
8	GB/TXXXX-2006	Testing method for PV system grid-connecting performance

Table 7.5 IEC publications related to the off-grid PV system.

1	IEC61837-2007	Solar photovoltaic energy system—terms, definitions and symbols
2	IEC61194-1992	Characteristic parameters of stand-alone photovoltaic systems
3	IEC61277-1995	Terrestrial photovoltaic Generating Systems—General and Guide
4	CEI/IEC 62124:2004	Photovoltaic stand—alone systems—Design verification
5	CEI/IEC 1173	Over-voltage protection for photovoltaic power generating systems—Guide
6	IEC61702-1995	Rating of Direct Coupled photovoltaic—Pumping Systems
7	IEC62253 Ed.1.0	Direct Coupled photovoltaic—Pumping Systems—Design Qualification and Type Approval
8	IEC61724(1998)	Photovoltaic system performance monitoring—Guidelines for measurement, data exchange and analysis
9	IEC62257-1 (2003)	Recommendations for small renewable energy and hybrid systems for rural electrifications Part 1: General Introduction to rural electrification
10	IEC 62257-2 Ed.1.0	Recommendations for small renewable energy and hybrid systems for rural electrifications Part 2: from requirements to range of electrification systems
11	IEC62257-3 TS Ed.1.0	Recommendations for small renewable energy and hybrid systems for rural electrifications Part 3: Project development and management
12	IEC61427(2005)	Secondary cells and batteries for photovoltaic energy systems (PVES)—General requirement and methods of test
13	IEC62108 Ed.1.0	Concentrator Photovoltaic Modules and Assemblies—Design Qualification and Type Approval

Table 7.6 IEC publications related to the grid-connected PV system.

1	IEC62109-1 (2005)	Safety of power converters for use in Photovoltaic power systems—Part 1: General requirement
2	IEC62109-2 (2005)	Safety of power converters for use in Photovoltaic power —Part 2: Particular requirement for inverters
3	IEC61683(1999)	Photovoltaic systems—Power conditioners—Procedure for measuring efficiency
4	IEC 60364-7-712 (2002)	Electrical installations of buildings—Part 7–712: Requirements for special installations or locations—Solar photovoltaic power supply systems
5	IEC 62109 Ed.1.0	Electrical safety of static inverters and charge controllers for use in photovoltaic power systems
6	IEC61727 Ed.2.0(2004)	Characteristics of the utility interface—Photovoltaic systems
7	IEC62116 Ed.1.0	Testing Procedure—Islanding Prevention Measures for Power Conditioners used in grid connected photovoltaic generation systems
8	IEC62234 Ed.1.0	Safety Guidelines for Grid Connected Photovoltaic systems mounted on Buildings

is prepared for more than 10 standards, including 'Code on design of PV stations connected to the power system', 'Code on design of PV power connected to the power distribution system', 'Technical specification on reactive power compensation for PV stations'. And the draft for the rest (nearly 30), including the project building code, the code on power station operation, technical requirement and specification on power stations, is available and ready for publication.

It is urgent to prepare and improve the national standards on the PV power. When the national standards are available, it can help normalise the market, ensure product quality and assure the user to remove the doubt. Otherwise, it is difficult to be accepted by the building departments and the grid corporations. It is believed that the IEC publications can help internationalisation and export of the PV products and systems.

Bibliography

1 EEPW. (2014) Development of China's PV industry over the past decade. http://www.eepw.com.cn/article/246134.htm (last accessed October 4, 2017).
2 CCID. (2013–2014) Annual report on China PV Market Research. http://www.ccidreport.com/report/content/3654/201402/637330.html (last accessed October 16, 2017).
3 Tangkun, Zhoul, Yanfang, *et al.* (2013) MWT Solar Cell. *Solar Power (the Chinese Journal)*, **3**, 31–33.
4 SEMI. (2014) *International Technology Roadmap for Photovoltaic, Fifth Edition.*
5 King RR, Edmondson KM, *et al.* (2007) 40% efficient metamorphic GaInP/GaInAs/Ge multijunction solar cell. *Appl. Phys. Lett.*, **90**, 183516.
6 Schneider K. (2009) World record: 41.1% efficiency reached for multi-junction solar cells. *Fraunhofer ISE online.* https://wenku.baidu.com/view/25164d680722192e4536f654.html (last accessed October 16, 2017).
7 Romeo A, *et al.* (2006) High efficiency flexible CdTe solar cell on polymer substrate. *Science Direct Solar Energy & Solar Cells*, **90**, 3407–3415.
8 Solar China Publishing House. (2012) *Solar Energy International (the Chinese magazine)*, **3**, 17–18.
9 Stolt L, *et al.* (1993) ZnO/CdS/$CuLnSe_{vc}$ thin film solar cells with improved performance. *Appl. Phys. Lett.*, **62**, 579–599.
10 Bhattacharyaa RN, *et al.* (2000) 15.4% CuIn1-xGaxSe2-based photovoltaic cell from solution-based precursor films., *Thin Solid Films.* **361,362**, 396–399.
11 Bremaud D, Rudmannl D, Bilger G, *et al.* (2005) Towards the hing house development of flexible CIGS solar cells on polymer films with efficiency exceeding 15%. *Proc. Of 31th PVSC*, **2005**, 223–226.
12 Huangwei, Baoxiu, Mi, Zhiquiang, Gao. (2011) *Organic Electronics (in Chinese).* Beijing: China Science Press, pp. 224–238.
13 Shuyang, Wang, Yineng, Xiao, Liuji, Mingze, Sun. (2013) *Principles and Applications of Organic Solar Cells (in Chinese).* https://max.book118.com/html/2011/0624/326509.shtm (last accessed October 16, 2017).

Technology, Manufacturing and Grid Connection of Photovoltaic Solar Cells, First Edition. Guangyu Wang.

14 Maathew S, Yella A, Gao P, *et al.* (2014) Dye-sensitized solar cells with 13% efficiency. *Nature Chemistry*, **6**(8), 242–247.
15 Green MA, Ho-Baillie A, Snaith HJ. (2014) The emergence of perovskite solar cells. *Nature Photonics*, **8**, 506–514.
16 Solar Be. (2013) Power optimizer: The ideal solution for future distributed home solar systems. http://magazine.solarbe.com/201311/05/1149.html (last accessed October 16, 2017).

Index

a

AC/DC power supply systems 295–296
adhesives and modified EVA film 232–234
air mass (AM) 6
1 air mass (AM1) 6, 7–8
anti-reflection coating
  ellipsometer 122–123
  LPCVD 113
  objectives and principles for 112–113
  PECVD
    chain-type indirect 116
    plate-type direct PECVD system 117–118
    plate-type indirect system 118–119
    silicon nitride coating 113–115
    tubular direct PECVD system 115–117
  PVD 119–122
a-Si:H/μC-Si cells, stacked TFSCs 146–147
a-Si/μC-Si thin film
  HWCVD method 75–76
  LPCVDE technique 76–77
  PECVD method 74–75
  silane
    ammonia 74
    chlorosilane disproportionation 72–73
    fluoride method 73
    magnesium silicide 73
    TFSSC industry 74
  silicon film growth methods 78–79
a-Si thin-film
  advantages 68–69
  characteristics 67, 68
  energy band diagramme of 67
  Fermi level pinning 69–70
  physical vapour deposition 68
  polycrystalline silicon film
    linear laser technique 78
    metal-induced method 77
    RTP technique 77–78
    SPC technique 77
  silicon film growth methods 78–79
Astro-SiN50 122
ATON sputtering system 122
auxiliary circuit 286

b

BACCINI test and sorting system 131, 132
back contact (BC) sputtering system of Leybold Optical 163–164
back finger electrodes 192–193
back surface field (BSF) 27–29, 85, 149
band tail structure 67, 68
bath-type PSG-removal system 111
boron diffusion N-type crystalline silicon solar cells 192–193
Bridgman method 50
BTU chain-type diffusion furnace 103–105
building integrated photovoltaic (BIPV) system 143
  applications of PV roof 300
  design principles 301
  diagram of grid-connected PV system 299
  encouragement by the Chinese government 301–302
  energy storage devices 300

building integrated photovoltaic (BIPV) system (*contd.*)
  national policies and certification in China 301
  solar cell modules 225–230
Bururi gripper 133

c
cadmium telluride (CdTe) solar cell 201–204
  TFSCs
    CIGS and CIS 204
    non-vacuum method 207–208
    roll to roll method 209–210
    solar absorption coefficient 204
    structure of 205
    vacuum co-evaporation and vacuum-sputtering methods 205–207
casting polycrystalline silicon
  crystal boundaries and dislocations in 57–58
  disadvantages 49
  fabrication of 50
  furnace 59–60
  metal impurities and gettering 56–57
  non-metal impurities in 54–56
  preparation process of 50–54
  principle of 49
  pseudo-single crystal 58–59
CdTe polycrystalline thin film 80
cell conversion efficiency 85, 86
cell wafer sorting 237
cell welders 240–241
centralised grid-connected PV systems 303
chain-type PSG-removal production system 112
China's coal-fired power price 4
chlorosilane disproportionation 72–73
Code on design of PV power connected to the power distribution system 317
Code on design of PV stations connected to the power system 317
compound semiconductor solar cells
  CdTe and CdS thin film materials 80
  CdTe TFSCs 201–204
  CIGS TFSCs 204–210
  $CuInS_2$ 80–81
  $CuInSe_2$ 80
  fabrication methods of
    LPE technique 198
    MBE 198–200
    MOCVD technique 198
    vacuum evaporation technique 197–198
  GaAs solar cell 79–80
  III–V compound materials 199–201
concentrator solar cells 217–219
continuous grain silicon (CGS) 77
control circuit 286
controllers
  functions 287–289, 293–294
  installation angle and position regulation of solar cell arrays 291–293
  maximum power point tracking technology 289–291
conventional fossil energy, China 1, 2
crystal boundary impurity gettering 58
crystalline silicon PV building project 300
crystalline silicon solar cells (CSSCs) 22
  anti-reflection coating preparation
    LPCVD 113
    PECVD (*see* plasma-enhanced chemical vapour deposition (PECVD))
    refractive index 113
  back finger electrodes 192–193
  boron diffusion N-type crystalline silicon solar cells 192–193
  cascading/chain-type production lines 133–136
  conversion efficiency of 187–188
  cost analysis for 142
  diffusion junction preparation
    chain-type diffusion furnace 102, 103–105
    dilute phosphoric acid 102
    phosphorus impurity gettering 104–106
    $POCl_3$ gaseous diffusion, tubular furnace 100–102
    principles 99–100
    thickness of 103–104

laser edging isolation 109–110
MWT 193–195
passivation emitter silicon solar cells 190–192
plasma edging isolation 106–109
production flow of 87–88
PSG removal (*see* phosphorus silicon glass (PSG) removal)
PV conversion efficiency 137, 138
raw silicon wafer
conduction type 88–89
high-speed multi-purpose silicon wafer testers 92–93
minority carrier lifetime 90–92
resistivity and thin layer square resistance 89–90
thickness and homogeneity 92
silicon wafer by robots 133, 136
silicon wafer surface cleaning and texturing
chemical corrosion texturing 94–96
laser texturing 97–98
principles of 93–94
RIE techniques 97–99
solar cell production line (*see* solar cell production line)
solar cell testing and sorting
equipment 130–132
objectives of 130
solar simulator 136
spectral response 139
structure of 85–87
top/bottom electrodes preparation
BACCINI 125–126
dry and sintering procedures 125
electrode slurry 128–130
fast sintering furnace system 126–128
technical requirements 124
V-I characteristics 136–139
crystalline silicon solar module production 237–238
crystalline solar cells 4, 5
CSSCs. *see* crystalline silicon solar cells (CSSCs)
$CuInSe_2$ (CIS) thin film materials 80–81
CZ monocrystalline silicon
forced convection 63
heat transfer of grain growth 60–62
impurity segregation 64–66
natural convection 62–63
temperature distribution 63–64

d

deep-discharging closed valve-regulated lead-acid (VRLA) battery 266
deep-discharging electrolyte-absorption lead-acid battery 266
direct PECVD system 115–116
distributed grid-connected PV systems 303
doped silicon 11
double-crucible CZ method 65
double-sided glass solar cell modules 225–226
dye-sensitised solar cells 213–215

e

electrode slurry 128–130
electronic-grade silicon (EGS) 35
ellipsometer 122–123
energy sources 1–4
energy storage batteries
application to PV system 265–267
capacity design of battery packs 274–275
fuel cells 272–274
lead-acid batteries 266, 268
liquid flow energy storage batteries 269–271
lithium ion batteries 268–269
super capacitors 271–272
epoxy 232
'Exploration on $SiCl_4$ hydrogenation reaction,' 48

f

fast sintering furnace system 126–128
Fermi level pinning 69–70
field-effect transistor (FET) 78
filed stop (FS) technology 279–281
filling factor 26
flat plate solar cell modules 224–225
fluidised bed reactor (FBR) 37, 45

fluoride method 73–74
forbbiden band 9
FORTRIX silicon wafer tester 92, 93
fossil energy sources 1
fossil fuels 1
Fourier transform infrared spectroscopy (FTIR) 84
fuel cells 272–274

g
generation-transfer-recombination process 13
Glow Discharge Mass Spectrometry (GDMS) analysis 81–82
grid-connected power system 260–261
grid-connected PV system
  centralised 303
  development of China's PV industry 307–309
  distributed 303
  intelligent grid 305–307
  technical specifications 303–305

h
heat exchange method (HEM) 50
heterojunctions
  composition of 32
  construction and working principle 32–34
hollow, double-sided glass PV modules 228
hot wire chemical vapour deposition (HWCVD) 68, 75–76
hybrid power system 297
hydrogenation 150

i
indirect PECVD system 116
inexhaustible energy 1
input circuit 285
insulated gate bipolar transistor (IGBT) 275
  filed stop technology 279–281
  non-punch-through technique 278–279
  structure and working principles 275–277
  trench gate technology 278–279
intelligent power module (IPM) 281
interconnection of cell wafers 237–238
interdigitated back-contact (IBC) cells 192–193
intrinsic absorption semiconductor material 15–16
intrinsic thin-layer (HIT) high-efficiency solar cells
  a-Si/monocrystalline silicon heterojunctions 196
  characteristics of 197
  double-surface cells 196
inverters
  control 285
  inverter circuit and inverter types 285–286
  role in PV system 281–282
  selection and requirements 286–287
  working principles 282–284

l
lamination sealing 238
large-sized centralised grid-connected PV stations 296–297
laser grooved buried contact (LGBC) cells 189
laser scribers 240
laser texturing 97–98
lead-acid batteries 266, 268
light absorption area 188, 189
liquid flow energy storage batteries
  sodium-sulphur battery 269–270
  super capacitors 271–272
  vanadium redox batteries 270
  zinc-bromine flow batteries 270–271
liquid phase epitaxy (LPE) technique 78, 198
lithium ion batteries 268–269
low voltage ride through (LVRT) capacity 288
LPCVD technique 76–77

m
magnetically controlled (MC) sputtering 181–182, 184
magnetic-field-applied CZ method (MCZ) 65

main inverter circuit 286
maximum power point tracking technology (MPPT) 289–291
μC-Si thin film (μc-Si)
  LPCVD technique 76–77
  nature of 70–72
metal gate electrode contact area 188, 189
metallisation wrap-through (MWT) silicon solar cell 193–195
metal organic chemical vapour deposition (MOCVD) technique 198
metal oxide semiconductor field effect transistor (MOSFET) 275
modified EVA film 233
module packaging
  materials 229
    adhesives and modified EVA film 232–234
    back plate and localisation 233–236
    frameworks and junction boxes 235–236
    inspection and sorting of cell wafers 230–231
    upper cover glass 231–232
  system
    cell welders 240–241
    laser scribers 240
    main equipment 239
    solar cell module laminators 242
    solar simulators, turnover trolleys and frame machines 242–243
  techniques 237–239
modules, solar cell. *see also* module packaging
  attenuation 253
  BIPV 225–230
  classification 223
  common specifications 252
  development 253–254
  indoor tests of PV cells 245
  indoor tests of PV modules 245
  module packaging and PV system reliability 242–244
  outdoor tests of PV modules 245–246
  packaging structure of flat plate 224–225
  packaging techniques 224
  power and efficiency 250–252
  testing methods and verification standards 246–247
  testing principles of electroluminescence 249–250
  tests of macro defects 247–249
  tests of PV performance 247
molecular bean epitaxy (MBE) 198–200
monocrystalline silicon solar cells 4
monocrystalline silicon wafer 58
Mott-CFO model 67
multiple quantum well (MQW) solar cells 219, 221
multiple stepped evaporation method 205

n
next-generation solar cells
  concentrator solar cells 217–219
  dye-sensitised solar cells 213–215
  MQW solar cells 219, 221
  organic solar cells 210–213
  perovskite solar cells 215–217
nonequilibrium carrier 13
non-intrinsic absorption semiconductor material 16
nonopaque solar cell modules 226–227
non-punch-through (NPT) technique 278–279

o
optimum operating voltage and current 25–26
ordinary EVA film 232
organic silica gel 232
organic solar cells 210–213
output circuit 285–286

p
perovskite solar cells 215–217
phosphorus silicon glass (PSG) layer 104
phosphorus silicon glass (PSG) removal
  bath-type 111
  chain-type 111–112
  principles and processes of 110–111
photo chemical vapour deposition (photo-CVD) 68
photoconduction attenuation method 91

photo-generated current 22–23
photovoltaic (PV) system
  AC/DC power supply systems 295–296
  BIPV 299–302
  characteristics 259
  classification 259–260
  codes and test verifications 309–317
  composition and simple working principles 263–264
  controllers 287–294
  DC power supply systems 295
  energy storage batteries 265–275
  energy storage device charging/discharging 298–299
  grid-connected power system 260–261
  grid-connected PV system 302–309
  hybrid power system 297
  inverters 281–287
  large-sized centralised grid-connected PV stations 296–297
  power-switching devices 275–281
  shortcomings 259
  small-/medium-sized distributed 296
  small-power DC 294–295
  stand-alone power system 260
  working principle 262
physical vapour deposition (PVD)
  Al(Ag) back electrode 154
  ATON sputtering System 122
  vs. PECVD 121–122
  principles of 119–121
plasma-enhanced chemical vapour deposition (PECVD) 68, 69, 70
  a-Si/μC-Si thin film 74–75
  chain-type indirect 116
  plate-type direct PECVD system 117–118
  plate-type indirect system 118–119
  vs. PVD 121–122
  silicon nitride coating 113–115
  TFSSCs
    a-SiC:H P-type thin film 152
    a-Si thin film grown mechanism 177–179
    depositing system, linear series 160, 162
    deposition chamber, Leybold Optical 160, 163
    deposition temperature 168
    electrode and the substrate 168
    frequency excited by plasma 169
    gas pressure 167–168
    glow discharge and plasma generation 175–177
    horizontal clustering 160, 161
    hydrogen dilution 167
    intrinsic zone (a-Si:H) 152–154
    linear parallel, Leybold Optical 160, 162
    linear series and linear parallel types 160
    N-type (a-Si:H) thin film 154
    power excited by plasma 169
    P-type (a-SiC:H) film deposited 152
    vertical folding 160, 162
  tubular direct PECVD system 115–117
plate-type direct PECVD system 118
P-N junction energy band structure
  bending of 18, 19–20
  external voltage effect 20
  photo-generated current 22–23
  solar radiation effect 20–22
polycrystalline silicon film
  linear laser technique 78
  metal-induced method 77
  RTP technique 77–78
  SPC technique 77
power conversion efficiency 26
prevention of overcurrent and islanding effect 288–289
protection circuit 286
pseudo-single crystal 58–59
pulse width modulation (PWM) technology 282
PV effect 20–22

q

quantum devices 4
quartz crucible 50, 51–52

r

radio frequency high pressure depletion (RF-HPD) 74

rapid thermal process (RTP) 77–78
reactive ion etching (RIE) techniques 97–99
renewable energy source 1–4
RF sputtering system 183, 184
roof c/w thin-film solar cells 300
Roth-Rau PECVD system 118

s

SCHMID cell-sorting machine 131
SCHMID phosphorus coating machine 102
secondary ion mass spectrometry (SIMS) analysis 82–84
selective harmonic elimination (SHE) 282
semiconductor materials
- compound 79–81
- direct and indirect transitions 16–18
- impurities in
  - carbon/oxygen content in silicon wafer infrared spectroscopy 83, 84
  - GDMS 81–82
  - SIMS analysis 82–84
- intrinsic absorption 15–16
- light absorption of 14–15
- non-intrinsic absorption 16

semiconductors
- communisation motion of electrons 9, 10
- directional movement of electrons and holes 12–13
- energy bands image 9
- energy band structure 10–11
- energy levels
- Fermi Level 11–12
- generation and recombination of carriers 13–14

short-circuit current, solar cell 24–25
$SiCl_4$ zinc reduction method 45–47
Siemens method 37, 46, 48
$SiHCl_3$ hydrogen reduction method 46
silicon nitride coating
- PECVD 113–115
- PVD 119–122

silicon nitride physical sputtering 119, 120
silicon solar cells 29
simple energy storage stand-alone PV system 262
single wafer welding 237
sinusoidal wave pulse width modulation (SPWM) 282
SiNx deposition 115, 116
small-/medium-sized distributed PV systems 295–296
small-power DC PV systems 294–295
sodium-sulphur battery 269–270
solar cell arrays
- design 254–255
- electrical connections and hot spot effect 255–256
- installation and measurement 256–257

solar cell industry 3, 4
solar cell module laminators 242
solar cell modules. *see also* module packaging
- attenuation 253
- BIPV 225–230
- classification 223
- common specifications 252
- development 253–254
- indoor tests of PV cells 245
- indoor tests of PV modules 245
- module packaging and PV system reliability 242–244
- outdoor tests of PV modules 245–246
- packaging structure of flat plate 224–225
- packaging techniques 224
- power and efficiency 250–252
- testing methods and verification standards 246–247
- testing principles of electroluminescence 249–250
- tests of macro defects 247–249
- tests of PV performance 247

solar cell production line
- on-line inspection 140
- quality control
  - cleaning and texturing process 140–141
  - diffusion process 141
  - in PECVD procedure 141

solar cell production line (*contd.*)
in print and sintering procedures 141–142
PSG removal procedures 141
in wafer edging isolation 141
working environment 140
solar cells
BSF 27–29
casting polycrystalline silicon (*see* casting polycrystalline silicon)
CZ monocrystalline silicon
forced convection 63
heat transfer of grain growth 60–62
impurity segregation 64–66
natural convection 62–63
temperature distribution 63–64
development stages 4, 5
filling factor 26
heterojunctions
composition of 32
construction and working principle 32–34
I-V characteristic curve 24
open-circuit voltage 24
optimum operating point 25
optimum operating voltage and current 25–26
P-N junction crystalline silicon solar cells and analysis 29–31
P-N junction energy band structure 18–23
power conversion efficiency 26
short-circuit current 24–25
solar-grade polycrystalline silicon (*see* solar-grade polycrystalline silicon)
temperature characteristics 27
solar generator 223
solar-grade polycrystalline silicon
FBR method 45
PV industry 35–39
semiconductor P-N junction 38–39
$SiCl_4$ zinc reduction method 45–47
Siemens method 48–49
solar cell conversion efficiency 40
UMG silicon 41–45
VLD method 47–48
solar-grade silicon (SOG silicon) 38
solar PV greenhouse 302
solar radiation
in China 6, 7
photoelectric conversion 6
wavelength of 7–9, 8
solid phase crystallisation (SPC) 77
sputtering 120–121
Staebler-Wronski effect (S-W effect) 69–70
stand-alone power system 260
super capacitors 271–272

t

TDR-105 silicon monocrystalline furnace 65, 66
Technical specification on reactive power compensation for PV stations 317
Temporary rule on PV power operation and management 263
TFSCs
CdTe 201–204
CIGS 204–210
thin film silicon solar cells (TFSSCs) 72
advantages of 143–144
Al(Ag) back electrode, PVD 154
a-Si:H and μC-Si
light absorption of 150–151
structures of 146–147
a-Si:H/a-SiGe:H stacked solar cells 150–151
back metal preparation system 179–182
BC sputtering system, Leybold Optical 163–164
deposit silicon film
HWCVD method 170–171
VHF-PECVD Method 169–170
energy band structure adjustment for
a-SiC carbon material to widen the gap 172
a-Si Ge material to narrow down the gap 172
a-SiGe TFSCs 172–174
a-Si hydrogen content and deposition temperature 172
boron, phosphorous and hydrogen 175
significance of 171–172

glass cleaning and surface texturing equipment 158
history and prospects of 144–146
laser scriber 164–165
module of 154–157
PECVD method
a-Si thin film grown mechanism 177–179
depositing system, linear series 160, 162
deposition chamber, Leybold Optical 160, 163
deposition temperature 168
electrode and the substrate 168
frequency excited by plasma 169
gas pressure 167–168
glow discharge and plasma generation 175–177
horizontal clustering 160, 161
hydrogen dilution 167
intrinsic zone (a-Si:H) 152–154
linear parallel, Leybold Optical 160, 162
linear series and linear parallel types 160
N-type (a-Si:H) thin film 154
power excited by plasma 169
P-type (a-SiC:H) film deposited 152
vertical folding 160, 162
power generation principle of 147–150
production system of 155, 157
TCO
glass substrate 151–152
parallel metal DC diode sputtering system 182–183
performance, preparation and testing of 166–167
physical sputtering 179–182
RF and MC Sputtering Systems 183–185
sputtering equipment 158–160
testing equipment 165, 166
ZAO target 158–160
thin-film solar cells 4, 5
thin-film transistor liquid crystal display (TFT-LCD) 72
three-phase full-bridge voltage inverters 284
transparent collector 278
transparent conductor oxide (TCO) 147
glass substrate 151–152
parallel metal DC diode sputtering system 182–183
performance, preparation and testing of 166–167
physical sputtering 179–182
RF and MC Sputtering Systems 183–185
sputtering equipment 158–160
trench gate technology 278–279

u

UL1703 tests 312
upgraded metallurgical grade (UMG) silicon 37, 41–45
UV solidification packaging technique 224

v

vanadium redox batteries (VRB) 270
vapour to liquid deposition (VLD) method 47–48

w

world energy consumption 1–3

z

zero air mass (AM0) 6
zinc-bromine flow batteries 270–271
zinc oxide doped with aluminum (ZAO) 151